STRONG LIGHT-MATTER COUPLING

From Atoms to Solid-State Systems

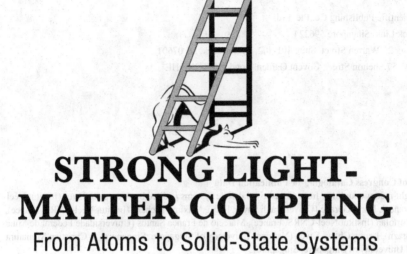

STRONG LIGHT-MATTER COUPLING
From Atoms to Solid-State Systems

Editors

Alexia Auffèves
Institut Néel-CNRS, France

Dario Gerace
Università di Pavia, Italy

Maxime Richard
Institut Néel-CNRS, France

Stefano Portolan
Institut Néel-CNRS, France

Marcelo França Santos
Universidade Federal de Minas Gerais, Brazil

Leong Chuan Kwek
National University of Singapore and Nanyang Technological University, Singapore

Christian Miniatura
INLN-CNRS, University of Nice Sophia, France
CQT, National University of Singapore, Singapore

World Scientific

NEW JERSEY • LONDON • SINGAPORE • BEIJING • SHANGHAI • HONG KONG • TAIPEI • CHENNAI

Published by

World Scientific Publishing Co. Pte. Ltd.

5 Toh Tuck Link, Singapore 596224

USA office: 27 Warren Street, Suite 401-402, Hackensack, NJ 07601

UK office: 57 Shelton Street, Covent Garden, London WC2H 9HE

Library of Congress Cataloging-in-Publication Data
Strong light-matter coupling : from atoms to solid-state systems / edited by Alexia Auffèves (Institut Néel-CNRS, France), Dario Gerace (Università di Pavia, Italy), Maxime Richard (Institut Néel-CNRS, France), Stefano Portolan (Institut Néel-CNRS, France), Marcelo de França Santos (Universidade Federal de Minas Gerais, Brazil), Leong Chuan Kwek (National University of Singapore, Singapore) & Christian Miniatura (National University of Singapore, Singapore).
    pages cm
  Includes bibliographical references and index.
  ISBN 978-9814460347 (alk. paper) -- ISBN 9814460346 (alk. paper)
  1. Quantum optics. 2. Quantum electrodynamics. I. Auffèves, Alexia, editor of compilation. II. Gerace, Dario, editor of compilation. III. Richard, Maxime, 1977– editor of compilation. IV. Portolan, Stefano, editor of compilation. V. Santos, Marcelo de França, editor of compilation. VI. Kwek, Leong Chuan, editor of compilation. VII. Miniatura, C. (Christian), editor of compilation.
  QC446.2.S773 2014
  535'.15--dc23

                                                                    2013042521

British Library Cataloguing-in-Publication Data
A catalogue record for this book is available from the British Library.

ISBN 978-981-4460-34-7

Typeset by Stallion Press
Email: enquiries@stallionpress.com

Printed in Singapore

Preface

It all started in Singapore. Well, unless it was in France. No actually, if you think about it properly, it's rather in Brazil that the story began. Or else maybe not....

Ok, let's put it differently. During winter 2008, in Grenoble, Maxime and Alexia wanted to compare the signatures of strong light-matter coupling in the classical and in the quantum regime, and spent hours drinking coffees to discuss how to describe this quantum-classical boundary — they even thought that a school could be great, but they forgot.

But then Alexia went to Brazil to visit Marcelo, Marcelo went to France to visit Alexia and he met Maxime, and they spent hours drinking coffees, bringing together theory of quantum optics and solid state physics. In the meantime, Alexia met Dario in Italy, and Dario who already knew Maxime came to Grenoble too. They spent hours drinking coffees, putting together laser physics, nanophotonics and quantum optics. And one day, in May 2010, a few months after Stefano had joined Alexia's group, Marcelo and Dario arrived in Grenoble at the same period and they met each other. They all spent hours drinking coffees, trying to have every strongly coupled things make sense at the same time, and they thought that a school would help. But they forgot.

At that time, Marcelo was taking a sabbatical in Singapore in Kwek's group which led to Marcelo inviting Alexia, Dario and Stefano to CQT. It was August 2010, and they spent hours drinking Spinelli's coffees. During this stay, Alexia met Christian, who told her about the Singaporean miracle for fundamental research. Alexia told Christian about the school idea. Something about strong light-matter coupling, talking about quantum and classical physics, bringing together different communities, different concepts and tools, dealing with atoms and solid-state systems, from semiconducting systems to superconducting devices....

Christian listened, nodded and took everybody to Kwek's office. It is maybe there, during enthusiastic conversations dotted by Kwek's good, inimitable laugh, that the school started to become real, turning hours of scientific conversations between friends working on different but connected subjects into an actual international school dealing with the physics of the strong coupling regime.

This little story reflects a recent and fruitful trend in the scientific world: blending topics and scientific communities. In that respect, the notion of strong-light matter coupling constitutes a textbook example. Depending on who you are talking to this notion refers to a genuinely quantum effect or to a purely classical one, the term "Rabi splitting" thus having at least two different meanings. Indeed, this notion has been employed in very diverse scientific communities in the last three decades.

Since the early eighties, a few atoms, down to a single one, have been coupled to optical and microwave field in cavities, leading to pioneering demonstrations of cavity quantum electrodynamics, actual realization of textbook Gedanken experiments, and building blocks for quantum information processing. With the emergence of semiconductor nanotechnology in the early nineties, the strong coupling regime in its classical sense could be demonstrated, opening the door to the physics of Bose gases in solid state environment. Later, the quantum version of the strong coupling regime could also be achieved in semiconductor nanostructures, offering prospects to exploit light-matter interaction at the single photon level in scalable architectures. More recently, impressive developments in the so-called *circuit QED*, involving superconducting quantum bits coupled to microwave cavities, have allowed to apply the strong coupling concepts to the fields of quantum communication and information processing. It is nice and enlightening to notice that solid-state devices, initially designed to develop quantum information technologies in a potentially integrable and scalable framework, now appear as marvellous tools to investigate quantum optics in its most fundamental aspects, and open up brand new fields of research that atomic physicists had not explored, once again closing the gap between fundamentals and applications.

The right place to organize the school emerged naturally from this blending spirit. Research in the old Temasek city has been increasingly successful over the past decade, gaining more and more visibility thanks to the audacious and inspired support from the Singapore government. Center for Quantum Technologies is one of those success stories. It is in this stimulating context that we were given the chance as well as a substantial financial support to launch the school project. It rapidly became an event exceeding by far our best expectations: we had the means to invite top-level lecturers from all over the world, one of them even being awarded the Nobel Prize a few months later. We also had the means to organize a three-week school, allowing us to set up a program covering as many different approaches and interpretations of the strong light-matter coupling physics, including forefront research topics. Considering the enthusiasm expressed both by the lecturers and the attendees, it was decided that this event would be the first one among many others to come, as regular sessions of the Singapore School of Science.

This international school on "strong light matter coupling" has been organized in the following way: Four fundamental courses were delivered by five distinguished speakers on atomic cavity QED (Serge Haroche), cavity QED with superconducting circuits (S. Girvin), solid state cavity QED with semiconductor nanostructures (J.-M. Gérard and L. C. Andreani), and theoretical methods for open quantum systems (H. J. Carmichael), respectively. These lecturers were carefully chosen among leading world experts in these fields, and for their renowned abilities at delivering clear and pedagogical lectures at a graduate level. These fundamental courses were then complemented by ten advanced courses dealing with the most recent developments of strong light-matter coupling in various and disparate fields. In particular, we have selected a non-comprehensive list of topics of current research interest, such as: quantum plasmonics (D. Chang), atoms in strongly focused laser beams (C. Kurtsiefer), recent experiments in circuit QED (P. Bertet), cavity QED with quantum dots and photonic crystal cavities (A. Badolato), strong phonon-photon coupling in opto-mechanical nanostructures (S. Groeblacher), polariton condensation (J. Bloch), quantum polaritonics (S. Savasta), cold atoms in cavities (I. Carusotto), the ultra-strong coupling regime of light-matter interaction (C. Ciuti), and recent developments in nonlinear photonic circuits (A. De Rossi).

The present book is mostly inspired from the lectures delivered at the school, but partly detaches from them, thus constituting a complementary manual introducing some of the topics covered. The volume is organized as follows: the first 4 chapters present fundamental topics showing the generality of strong light-matter interaction in different contexts, while the following chapters are devoted to specific experiments or theoretical developments exploiting cavity QED as a common playground. The first chapter from S. Haroche and J.-M. Raimond is meant to give an introduction to the basic physics and the most recent experimental developments in atomic cavity quantum electrodynamics, such as the essential description of atom-cavity coupling through the Jaynes-Cummings model, and the direct observation of quantum non-demolition measurements of photon number states in a cavity, which are among the most exciting demonstrations of quantum jumps dynamics in the real world. The second chapter from L. C. Andreani, is giving an overview of light-matter coupling in the classical sense in solid-state cavity QED, particularly emphasizing on the effects of dimensionality on the coupling constants, oscillator strengths and field confinement lengths. The rich physics connected with exciton-polaritons in bulk, two-dimensional and zero-dimensional semiconductors/insulators is presented under a unified theoretical formalism. The latter system describes the solid-state analogue of the ultimate cavity QED system, where artificial atoms (quantum dots) are coupled to photonic nanocavities in the near-infrared range. In the third chapter, P. Bertet gives a comprehensive overview of the emerging field of circuit QED, namely the transposition of Jaynes-Cummings-related physics to the world of microwave circuits, where atoms are made of Josephson qubits, and cavity photons are stored in superconducting resonators. The fourth chapter is a broad overview by H. J. Carmichael into theoretical methods for quantum open systems, which is common to analyzing an incredibly broad class of physical systems, ranging from atomic cavity QED to solid state and circuit QED presented before. Especially relevant is the description of quantum trajectories, which directly connects to the experiments highlighted in the first chapter.

In the following chapters we target a few specific topics where enhanced light-matter coupling is likely to play a crucial role in the future, such as quantum information, quantum plasmonics, quantum polaritonics, nonlinear and quantum optics. The fifth chapter is an introductory and very pedagogical presentation by S. Girvin, a leading expert in the field of quantum information, where direct application of the cavity QED systems finds its potential usefulness. The next chapter by A. Amo and J. Bloch is devoted to a presentation of the latest exciting experiments exploiting the quantum fluid nature of exciton-polaritons in low-dimensional nanostructures.

In chapter seven, D. Chang gives an introduction to the emerging field of quantum plasmonics, where enhanced light-matter coupling is induced by strong field confinement at dielectric/metallic interfaces. The eighth chapter is an in-depth introduction to the wide possibilities offered by polaritonic systems to generate quantum states of radiation, and it is contributed by S. Portolan (a co-organizer of the school), O. Di Stefano, and S. Savasta (school lecturer). Finally, the last chapter is devoted to the state-of-art in photonic crystal circuits and enhanced nonlinear effects owing to diffraction-limited light confinement, as explained by A. De Rossi.

Acknowledgments

This school has been made possible by the financial support of Singaporean and French institutions, whose contribution is gratefully acknowledged:

— the Nanyang Technological University (NTU)
— the National University of Singapore (NUS)
— the Center for Quantum Technologies (CQT)
— the French Embassy in Singapore
— the joint French-Singaporean Merlion program
— the Centre National pour la Recherche Scientifique (CNRS), and especially the DERCI and its representatives Minh-Hà Pham-Delègue and Luc Le Calvez
— the Fondation Nanoscience of Grenoble
— the joint French-Singaporean international research unit CINTRA
— the Centre pour l'Energie Atomique, CEA-INAC of Grenoble
— the Institute Néel — CNRS of Grenoble
— the French Groupe de Recherche (GDR) of Quantum Information IQFA

The event took place at NTU and was locally and more than efficiently organized by IAS, in particular Chris Ong Lay Hiong, Lim Yoke Meng Louis, Alex Wu Zhiwei, Zhongzhi Hong are all gratefully acknowledged, as well as Toh-Miang Ng for setting-up and maintaining the School website. We thank CQT, especially its director Artur Ekert and its administrative executive, Evon Tan, for hosting the first lecture and dinner of the school, and for co-organizing the public lecture of Serge Haroche together with representatives of the French Embassy, namely Aurélie Martin, Charlie Berthoty and Walid Benzarti.

The Editors

Contents

Contents

Chapter 1

Cavity QED in Atomic Physics

Serge Haroche[*,†] and Jean-Michel Raimond[*,‡]

*Laboratoire Kastler-Brossel, ENS, UPMC-Paris 6, CNRS, 24 rue Lhomond
75005 Paris, France

†Collège de France, 11 place Marcelin Berthelot, 75005 Paris, France
†haroche@lkb.ens.fr, ‡jmr@lkb.ens.fr

1. Foreword

Cavity Quantum Electrodynamics (CQED) deals with the interaction of atoms and electromagnetic fields in a confined space.[1] Originally devoted to the investigation of spontaneous emission of atoms between mirrors, it has evolved into a general study of light-matter interaction in the strong coupling regime, where the coherent coupling of atoms with the field dominates all relaxation processes. In quantum information,[2] CQED has become an important research domain, dealing with the coupling of qubits (realized with two-level atoms) with quantum harmonic oscillators (realized by a field mode in a cavity). The domain has recently expanded into new areas, in which the atomic qubits are replaced by artificial atoms such as quantum dots and Josephson circuits. In the latter case, CQED has led to the new field of Circuit-QED which is reviewed in another series of lectures in these proceedings.

Manipulating states of simple quantum systems has become an important field in quantum optics and in mesoscopic physics, in the context of quantum information science. Various methods for state preparation, reconstruction and control have been recently demonstrated or proposed. Two-level systems (qubits) and quantum harmonic oscillators play an important role in this physics. The qubits are information carriers and the oscillators act as memories or quantum bus linking the qubits together. In microwave CQED, the qubits are Rydberg atoms and the oscillator is a mode of a high-Q cavity. After a general presentation of this physics, the lectures given by one of us (S. H.) at the Singapore School of Physics on "Strong Light-Matter Coupling" have described various ways to synthesize non-classical states of qubits or quantum oscillators, to reconstruct these states and to protect them against decoherence. Experiments demonstrating these procedures with Rydberg atoms were described. These lectures have been an opportunity to review basic concepts of measurement theory in quantum physics and their links with classical estimation theory.

A similar set of lectures entitled "Exploring the quantum world with photons trapped in cavities and Rydberg atoms" was given by the other author of this chapter (J.-M. R.) in the Les Houches Summer School of July 2011 on "Quantum machines: measurement and control of engineered quantum systems".[3] Since these lecture notes covered in detail, albeit in a different order, the same topics as the ones discussed in the Singapore school, it has seemed to us appropriate to reproduce them in this book, with the permission of M. Devoret, B. Huard and R. Schoelkopf the directors of this Les Houches School session and of Oxford University Press, the publisher of the les Houches lecture notes. A PDF version of the powerpoint presentation of S. H. at the Singapore School can be found the site of our group www.cqed.org.

2.   Introduction

Cavity Quantum Electrodynamics (CQED) experiments implement the simplest matter-field system, a single two-level emitter, a 'spin-1/2', coupled to a single quantum field mode represented by a unidimensional harmonic oscillator, a 'spring'. The history of CQED originates in a seminal remark by Purcell.[4] He noted that the spontaneous emission of a spin can be modified by changing the boundary conditions imposed to the vacuum field. The first CQED experiments in the 80s observed modifications of the emission rate, either an enhancement[5] or an inhibition.[6,7]

Since these early studies, CQED has thrived. It reached the strong coupling with the first micromaser experiments, using transitions between Rydberg levels coupled to millimetre-wave superconducting cavities.[8] It has been extended to the optical domain, with atomic transition coupled to optical Fabry–Perot cavities,[9] then with solid-state emitters (quantum dots for instance) coupled to monolithic cavities using Bragg mirrors[10] or photonic bandgap structures.[11] It even led to industrial applications. The widespread VCSEL lasers, for instance, rely on cavity QED effects in the weak coupling regime. A recent and fascinating avatar of CQED is circuit QED, with superconducting qubit circuits coupled to high-quality stripline cavities (see other Chapters in this Book).

This Chapter will focus on the microwave CQED experiments realized in our laboratory, using circular Rydberg atoms coupled to superconducting millimetre-wave Fabry-Perot cavities. Both the spin and the spring systems have in this context an extremely long lifetime, of the order of a second for the spring. The excellent controllability of both systems, the slow pace of the experiments, with a typical time step in the few tens of microseconds range, allow us to realize complex sequences.

We first present (Section 3) an overview of the experimental techniques and of the theoretical tools describing the atom-cavity interaction and cavity relaxation. We then show in Section 4 that this system is ideal for the illustration of the quantum measurement postulates. It can be used for an ideal, projective, Quantum Non Demolition (QND) measurement of the number of photons in the cavity. Atoms probing repeatedly the cavity count the photons without absorbing them! By performing repeated measurements on the cavity field, we witness for the first time the quantum jumps of light, as photons are lost one by one due to cavity damping.

With the addition of controlled displacements of the field, this QND measurement can be turned into a complete determination of the cavity quantum state, providing an unprecedented insight into its quantum nature (Section 5). Reconstructing the state of 'Schrödinger cats', quantum superpositions of mesoscopic components with different classical properties, we observe their decoherence. They are rapidly transformed into mere statistical mixtures by their coupling to the cavity's environment.

This experiment sheds light on the quantum-to-classical transition, one of the main open questions in contemporary quantum physics. Why is it that we observe a tiny fraction of all possible quantum states at our scale? Why is it that we never see a cat[12] in a superposition of being dead and alive at the same time? Why is it that our macroscopic meters, while measuring quantum systems in state superpositions, never evolve into absurd superpositions of different indications? Why is it that the classical world, ruled by quantum physics, is classical at all? Environment-induced decoherence[13] provides partial answers to these important questions.

In order to harness the quantum for useful purposes, we have to face the decoherence obstacle, which must be combated efficiently. We discuss in Section 6 a quantum feedback experiment. Transposing to the quantum realm the concept of feedback, it prepares on demand non-classical states with a well-defined photon number (Fock states). It also protects them against decoherence, by reverting rapidly the quantum jumps due to relaxation.

Another strategy against decoherence uses reservoir engineering. We propose in Section 7 to prepare and protect non-classical field states, by coupling strongly the cavity to a carefully designed environment. We then conclude (Section 8) by discussing a few perspectives for these microwave CQED experiments.

3.   A spin and a spring

3.1.   *Experimental set-up*

The scheme of our microwave CQED set-up is presented in Fig. 1 (for a detailed description see[1,14]). Circular Rydberg atom samples are prepared in box B at precisely defined time intervals, with a selected velocity v ($v = 250\,\text{m/s}$ in most experiments described here), out of a thermal Rubidium beam by laser and microwave excitation. They propagate through the experiment, cooled down to $0.8\,\text{K}$ by an ^3He refrigerator. They interact with the superconducting Fabry Perot cavity C and are finally detected by field ionization in D.

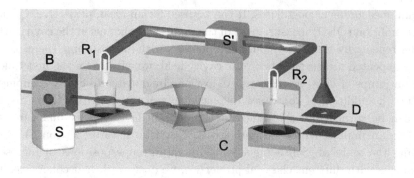

Fig. 1.   Scheme of a microwave cavity QED set-up.

Circular Rydberg states have a large principal quantum number and maximum orbital and magnetic quantum numbers. Their wavefunction is located around Bohr's circular orbit. As a consequence of the correspondence principle, their properties can be computed classically. However, they are extremely well suited for experiments unveiling the quantum.

We mainly use two adjacent states, $|e\rangle$ and $|g\rangle$, connected by a dipole transition at the frequency $\omega_a/2\pi = 51.1\,\text{GHz}$. These states have no residual degeneracy, making them a nearly perfect implementation of a two-level system (qubit) whose quantum state is pictorially described as a unit vector evolving on the Bloch sphere (with level $|e\rangle$ and $|g\rangle$ at the north and south poles respectively). The atomic Hamiltonian is simply $H_a = \hbar\omega_a\sigma_z/2$, where σ_z is the standard Pauli operator for the z axis of the Bloch sphere.

Circular states have a long lifetime, about 30 ms, when placed in a small homogeneous static electric field defining a preferred axis for their highly anisotropic wavefunction and lifting their degeneracy with the huge hydrogenic manifold. They are very strongly coupled to resonant microwave radiation, due to the large size of the electron orbit ($0.12\,\mu\text{m}$ radius), resulting in a huge transition dipole matrix element.

Highly efficient field-ionization detection provides a measurement of the atomic state in the $\{|e\rangle, |g\rangle\}$ basis. Finally, the atomic transition frequency can be tuned, with a nanosecond time resolution, by the Stark effect in a static electric field. The selective preparation of these levels is complex, using laser, radio-frequency and static fields.[1] The number of atoms in each sample is random and obeys a Poisson statistics, whose average value is kept low, minimizing the effect of samples containing more than one atom.

In order to get information on the atom-cavity interaction, we use a Ramsey atomic interferometry technique. Before and after their interaction with C, the atoms, initially prepared in $|g\rangle$, undergo two classical $\pi/2$ resonant pulses in the low-Q cavities R_1 and R_2 driven by the source S'. Using a proper phase reference, the first pulse realizes the transformation $|g\rangle \rightarrow (|g\rangle + |e\rangle)/\sqrt{2}$ and prepares a state represented by a vector along the Ox axis in the equatorial plane of the Bloch sphere. In a proper interaction representation, the second pulse realizes the transformations $|g\rangle \rightarrow (|g\rangle + \exp(i\phi_r)|e\rangle)/\sqrt{2}$ and $|e\rangle \rightarrow (|e\rangle - \exp(-i\phi_r)|g\rangle)/\sqrt{2}$, where the phase ϕ_r of the Ramsey interferometer can be tuned by adjusting the relative phase of the pulses in R_1 and R_2 or by applying a transient Stark shift on the atomic transition between R_1 and R_2. The final probability $\pi_e(\phi_r)$ for observing the atom in $|e\rangle$:

$$\pi_e(\phi_r) = [1 + \cos(\phi_r)]/2, \qquad (1)$$

exhibits 'Ramsey fringes', oscillating between 0 and 1 as a function of ϕ_r.

The small static electric field required for the stabilization of the circular states makes them incompatible with closed metallic cavities. We use thus a Fabry–Perot design, with two mirrors facing each other. A field across the mirrors stabilizes the atoms and makes it possible to tune their transition via the Stark effect while they interact with C.

The photon (or energy) storage time T_c is the most critical parameter in these experiments. A long T_c requires a very high mirror conductivity, which can only be provided by superconducting metals at cryogenic temperature. It also requires an excellent surface state to minimize losses induced by diffraction on defects.

The quest for high-quality cavities has been a long process, since these two requirements are somewhat incompatible. We finally developed a fabrication technique based on diamond-machined copper substrates, with an extremely smooth surface (10 nm roughness), covered with a thin (12 μm) layer of high-purity Niobium deposited by cathode sputtering.[15]

The mirrors have a 50 mm radius and are 27 mm apart. They have a toroidal shape in order to lift the polarization degeneracy. They sustain a non-degenerate TEM$_{900}$ Gaussian mode with a linear polarization orthogonal to the cavity axis and a waist $w = 6$ mm. The frequency of the mode is adjusted by piezoelectric elements changing the cavity length. The damping times of the cavities used in recent experiments range from 65 ms to 0.13 s, a macroscopic time interval. The latter corresponds to a quality factor $Q = 4.5\,10^{10}$ and to a finesse $\mathcal{F} = 4.9\,10^9$, a thousand times better than that of the best optical cavities.

The mode is a quantum harmonic oscillator, with the Hamiltonian $H_c = \hbar\omega_c(N + 1/2)$, where ω_c is the field's angular frequency and $N = a^\dagger a$ is the photon number operator (a is the photon annihilation operator). The eigenstates of H_c are the non-classical Fock states $|n\rangle$, with a well-defined number n of photons. The vacuum, $|0\rangle$, is the ground state.

A classical source S, weakly coupled to the mode via the diffraction loss channels (Fig. 1) can be used to inject in C a coherent semi-classical state $|\alpha\rangle$ defined by the complex amplitude α. This injection is represented by the unitary displacement operator $D(\alpha) = \exp(\alpha a^\dagger - \alpha^* a)$, with $|\alpha\rangle = D(\alpha)|0\rangle$. The coherent state is an eigenstate of the annihilation operator a (with eigenvalue α). It can be expanded on the Fock states basis as $|\alpha\rangle = \sum_n c_n |n\rangle$ with $c_n = \exp(-|\alpha|^2/2)\alpha^n/\sqrt{n!}$. Its photon number distribution, $P(n) = |c_n|^2$, is Poissonian with an average $\bar{n} = |\alpha|^2$. Note that the displacement operator describes a global translation in phase space. In particular, it acts on an initial coherent state $|\beta\rangle$ according to $D(\alpha)|\beta\rangle = \exp(\alpha\beta^* - \beta\alpha^*)|\beta + \alpha\rangle$, a quantum version of the addition of classical fields in the Fresnel (phase) plane.

3.2. Atom-field interaction

The atom-field interaction is described by the Jaynes and Cummings[16] model. The complete Hamiltonian is $H = H_a + H_c + H_{ac}$. The interaction term H_{ac} is:

$$H_{ac} = -i\hbar\frac{\Omega_0}{2}f(vt)[a\sigma_+ - a^\dagger\sigma_-], \qquad (2)$$

where $\sigma_+ = |e\rangle\langle g|$ and $\sigma_- = |g\rangle\langle e|$ are the atomic raising and lowering operators.

The 'vacuum Rabi frequency', Ω_0, measures the strength of the atom-field coupling when the atom is located at the cavity centre, where the electric field amplitude is maximal. It is proportional to the dipole matrix element of the atomic transition and to the amplitude of a single photon field stored in the cavity. In our experiments, $\Omega_0/2\pi = 50\,\text{kHz}$. The function $f(vt)$ reflects the variation of the atom-field coupling with time while the atom crosses the Gaussian mode at right angle with the cavity axis. Taking the origin of time when the atom reaches the axis, f simply writes $f(vt) = \exp(-v^2t^2/w^2)$.

The eigenstates of H, the atom-field 'dressed states' can be straightforwardly expressed in the basis of the uncoupled atom-cavity states $\{|e,n\rangle\}$ and $\{|g,n\rangle\}$, eigenstates of $H_a + H_c$.[1] We will only consider here either exact atom-cavity resonance $\delta = \omega_a - \omega_c = 0$ or the non-resonant dispersive case $|\delta| \gg \Omega_0$.

At resonance ($\delta = 0$), the uncoupled states $\{|e,n\rangle\}$ and $\{|g,n+1\rangle\}$ have the same energy $[(n + 3/2)\hbar\omega_c]$. Note that $|g,0\rangle$ is apart and is not affected by the atom-field coupling. The eigenstates of $H_a + H_c$ form a ladder of twofold degenerate multiplicities. The coupling H_{ac} lifts this degeneracy. The dressed states with $n+1$ excitations are $|\pm,n\rangle = (|e,n\rangle \pm i|g,n+1\rangle)/\sqrt{2}$, with energies separated at cavity centre ($f = 1$) by $\hbar\Omega_n$ where $\Omega_n = \Omega_0\sqrt{n+1}$.

An atom initially prepared in $|e\rangle$ with an n-photon Fock state in the cavity corresponds to an initial quantum superposition of the two non-degenerate dressed states $|\pm,n\rangle$. The later evolution is thus a quantum Rabi oscillation between $|e,n\rangle$ and $|g,n+1\rangle$ at frequency Ω_n, the atom periodically emitting and reabsorbing a photon in C.

Note that, at most times during this evolution the atom-cavity system is in an entangled state. Quantum Rabi oscillations have been used in a variety of CQED experiments to create and engineer atom-field entanglement, culminating in the generation of a three-particle entangled state of the GHZ type, with three two-qubit quantum gate actions.[14, 17]

Note that the atomic motion through the mode is simply taken into account in this regime. A complete crossing of the mode at a constant velocity v is equivalent to a constant coupling at cavity centre during an effective interaction time $t_r = \sqrt{\pi} w/v$.

Far from resonance ($|\delta| \gg \Omega_0$), the dressed states nearly coincide with the uncoupled levels. Energy exchange between the atom and the field is prohibited by mere energy conservation. The only effect of the mutual interaction is a slight shift of the joint atom-cavity levels.

A simple second-order calculation shows that the atomic frequency is shifted at cavity centre ($f = 1$) in the field of an n-photon state by $\Delta\omega_a = 2(n+1/2)s_0$, where $s_0 = \Omega_0^2/4\delta$. This energy shift includes a constant part, s_0, which corresponds to the Lamb shift of $|e\rangle$ induced by the vacuum fluctuations in C. The other part, $2ns_0$, proportional to the field intensity, is the few-photons limit of the light shifts usually observed in strong laser fields.

When the atom crosses the mode containing n photons in a state superposition $(|e\rangle + |g\rangle)/\sqrt{2}$ (produced in the Ramsey zone R_1), the transient shift of the atomic frequency results in a phase shift of this superposition, by an amount $\phi_0(n + 1/2)$, where $\phi_0 = 2s_0 t_d$. The effective dispersive interaction time is $t_d = \sqrt{\pi/2}(w/v)$. The Ramsey fringe signal is accordingly shifted. The probabilities $\pi_e(\phi_r|n)$ and $\pi_g(\phi_r|n)$ for detecting the atom in $|e\rangle$ or $|g\rangle$ conditioned to the presence of n photons in the cavity are:

$$\pi_e(\phi_r|n) = 1 - \pi_g(\phi_r|n) = \frac{1}{2}\{1 + \cos[\phi_r + \phi_0(n + 1/2)]\}. \tag{3}$$

The detection of an atom thus carries information on the photon number. Since this atom is unable to absorb the cavity field, this information can be used for a QND measurement of n (Section 4).

As a reciprocal effect, the cavity frequency is shifted when an atom is located in the mode. An atom in $|e\rangle$ at cavity centre shifts the cavity resonance by $\Delta_e\Omega_c = s_0$. An atom in $|g\rangle$ shifts it by an opposite amount $\Delta_g\Omega_c = -s_0$. This is the single-atom limit of an index of refraction effect. The non-resonant atom, unable to absorb or emit light, behaves as a piece of transparent dielectrics that modifies by its index the effective cavity length and thus its resonance frequency. With the strongly coupled circular atoms, this effect is large even though there is a single atom in the mode.

This index effect results in a global phase shift for a coherent state $|\alpha\rangle$ injected in C before the atom crosses it. The dispersive interaction realizes then the transformations

$$|e, \alpha\rangle \rightarrow \exp(-i\phi_0/2)|e, \exp(-i\phi_0/2)\alpha\rangle \tag{4}$$

$$|g, \alpha\rangle \rightarrow |g, \exp(+i\phi_0/2)\alpha\rangle. \tag{5}$$

Interaction with an atom in a state superposition results in an entangled atom-cavity state, where the two atomic energy eigenstates are correlated with mesoscopic coherent fields having the same energy but different classical phases. This situation is quite reminiscent of the famous Schrödinger[12] cat, cast in a superposition of its 'dead' and 'alive' state through its interaction with a single atom. In this context, cavity QED experiments are particularly well-suited for studying the decoherence of mesoscopic quantum states.

3.3.   Field relaxation

The atom-cavity effective interaction time is at most in the 100 μs range, since the atoms are crossing the millimetre-sized cavity mode at thermal velocities. This is much shorter than the lifetime of the atomic levels (30 ms). Atomic relaxation can be safely neglected, and cavity damping is the main source of decoherence in these experiments.

The damping of a cavity mode, of a spring, has been described since the early days of quantum relaxation theory. When C is coupled linearly to a large environment, including many degrees of freedom, whose eigenfrequencies span continuously a large interval around ω_c, the master equation ruling the evolution of the field density operator, ρ, can be cast in the general Lindblad[18] form:

$$\frac{d\rho}{dt} = \frac{1}{i\hbar}[H_c, \rho] + \sum_{i=1}^{2}\left[L_i\rho L_i^\dagger - \frac{1}{2}(L_i^\dagger L_i\rho + \rho L_i^\dagger L_i)\right]. \tag{6}$$

The operators L_i are simply $L_1 = \sqrt{\kappa(1+n_{th})}\,a$ and $L_2 = \sqrt{\kappa n_{th}}\,a^\dagger$, where $\kappa = 1/T_c = \omega_c/Q$ is the cavity energy damping rate and $n_{th} = 1/[\exp(\hbar\omega_c/k_BT)-1]$ the mean number of blackbody photons per mode at the mirrors temperature T, as given by Planck's law (k_B is the Boltzmann constant). Note that L_2 vanishes at zero temperature. In our experiment, $T = 0.8\,\text{K}$ and $n_{th} = 0.05$. The L_1 and L_2 operators, proportional to a and $a^\dagger$ respectively, describe the modifications, the 'quantum jumps' of the cavity state when a photon leaks out from the cavity into the environment or when a thermal photon is created.

Coherent states being eigenstates of the annihilation operator, they do not change when a photon escapes into the environment. They are thus rather immune to relaxation. A direct resolution of Eq. (6) at zero temperature shows that an initial coherent state $|\alpha_0\rangle$ remains coherent, its amplitude being exponentially damped with the time constant $T_c/2$. Thus the average photon number decays as the classical field energy, with the time constant T_c. This feature qualifies the coherent states as (approximate) 'pointer states' for the cavity-environment coupling, according to the definition by Zurek.[13]

For the photon number distribution $P(n) = \langle n|\rho|n\rangle$, the master equation reduces to:

$$\frac{dP(n)}{dt} = \kappa(1+n_{th})(n+1)P(n+1) + \kappa n_{th}nP(n-1)$$
$$- [\kappa(1+n_{th})n + \kappa n_{th}(n+1)]P(n), \tag{7}$$

whose steady-state can be obtained by a detailed balance argument. It coincides obviously with the equilibrium blackbody state. If we consider an initial Fock state

$|n_0\rangle$ and the $T = 0$ K case, $P(n_0)$ is ruled at short times by:

$$\frac{dP(n_0)}{dt} = -\kappa n_0 P(n_0),$$ (8)

showing that the lifetime of the $|n_0\rangle$ state is of the order of T_c/n_0.

The larger the Fock state, the smaller its lifetime. This is a decoherence effect. As the Schrödinger cats, the Fock states are extremely fragile, more and more so when their energy increases. In fact, any Fock state with a large n_0 can be expressed as a superposition of n_0 non-overlapping coherent components with the same average energy. Each coherent component is rather immune to decoherence, but their quantum superposition is reminiscent of the Schrödinger cat. It is fragile and decays rapidly into a statistical mixture, spoiling the Fock state.

4. Ideal QND measurement of the photon number: The quantum jumps of light

Most actual measurements are quite far from the ideal, projective measurement described in elementary quantum physics textbooks. This is particularly true when it comes to counting the number of photons in a light field, for instance in a laser pulse. Modern photodetectors can count photons with a high reliability. However, instead of projecting the field on the Fock state corresponding to the result obtained, they cast it onto the vacuum, since all detected photons are absorbed, their energy being used to create the measurable electronic signal.

This total demolition of the light quantum state is not a requirement of quantum physics which allows Quantum Non Demolition measurement processes.[19] They are simply ideal projective measurements of an observable that is a constant of motion in the evolution ruled by the system's Hamiltonian. Repeated QND measurements thus always give the same result as the first one in the series. A sudden transition between two eigenvalues of the measured observable, a 'quantum jump', can be traced to an extra source of evolution for the system, a relaxation process for instance. Projective measurements of the photon number are obviously QND, since the photon number operator, N, is a constant of motion.

QND measurements of a light beam intensity have been realized with non-linear optical systems (for a review, see[20]). The intensity of the signal beam changes the index of a non-linear material (an atomic vapor for instance). This index of refraction modification is read out by a weak meter beam in an interferometric arrangement. Due to the weak optical non-linearities, these measurements only operate with macroscopic signal and meter beams, whose quantum fluctuations get correlated. They are not adapted for QND measurements at the single photon level, nor of course for the observation of the quantum jumps of the light intensity.

We have realized a QND measurement at the single photon level with our CQED set-up.[21, 22] We probe the cavity field with non-resonant atoms in a state superposition which, as discussed above, provide information on the photon number n in the cavity without absorbing or emitting light.

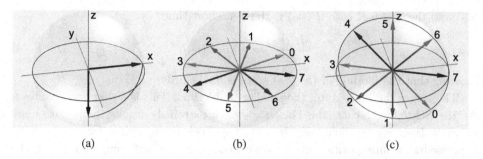

Fig. 2.   Evolution of the atomic Bloch vector in a QND photon-number measurement process. (a) First Ramsey $\pi/2$ pulse in R_1. (b) After the dispersive interaction with C, 8 orientations of the Bloch vector are correlated to the photon numbers from 0 to 7, for $\phi_0 = \pi/4$. (c) After the second Ramsey pulse in R_2.

Figure 2 presents the Bloch vector describing the atomic state at successive stages in this QND process. The atom is originally prepared in $|g\rangle$ (south pole). The first Ramsey pulse in R_1 casts it in a superposition of $|e\rangle$ and $|g\rangle$ represented by a vector along the Ox axis [Fig. 2(a)]. The atom then interacts dispersively with the cavity mode, resulting in a phase shift of the atomic coherence. We have represented in Fig. 2(b) the $\phi_0 = \pi/4$ situation, using an implicit interaction representation such that the atomic state evolution is only due to the interaction with C. After crossing the mode in the Fock state $|n\rangle$, the Bloch vector has rotated by an angle $(n+1/2)\phi_0$.

Eight different orientations of the Bloch vector at the exit of the cavity are entangled with the photon numbers from 0 to 7 (eight photons produce the same rotation as the vacuum, $\phi_0/2$). A single atomic detection is clearly not sufficient to pin-down the photon number. After a second $\pi/2$ rotation in R_2 [Fig. 2(c)], which maps one state of the equatorial plane onto $|e\rangle$, the atom is detected in the $\{|e\rangle, |g\rangle\}$ basis, providing a single bit of information. This does not allow for a complete discrimination of the eight non-orthogonal atomic states at cavity exit. In simple terms, a single bit is not enough to count from 0 to 7!

A single atom detection realizes thus a weak measurement of the cavity field, which nevertheless changes the cavity state. Since the atom cannot emit or absorb photons, the cavity state modification is only due to its entanglement with the atom. The corresponding (completely positive) map for the field density matrix ρ is:

$$\rho \longrightarrow \rho_j = \frac{M_j \rho M_j^\dagger}{\pi_j(\phi_r|\rho)}, \tag{9}$$

where the index $j = \{e, g\}$ indicates the measured atomic state.

The measurement operators M_e and M_g are:

$$M_g = \sin\left[\frac{\phi_r + \phi_0(N + 1/2)}{2}\right], \tag{10}$$

$$M_e = \cos\left[\frac{\phi_r + \phi_0(N + 1/2)}{2}\right], \tag{11}$$

and define a Positive Operator Valued Measurement (POVM) with elements $E_j = M_j^\dagger M_j$, with $E_e + E_g = 1$. The denominator $\pi_j(\phi_r|\rho)$ in Eq. (9) is the probability for detecting the atom in state j conditioned by the field state ρ:

$$\pi_j(\phi_r|\rho) = \text{Tr}(M_j \rho M_j^\dagger). \qquad (12)$$

Since the M_js are diagonal in the Fock state basis, this probability can also be written:

$$\pi_j(\phi_r|\rho) = \sum_n P(n)\pi_j(\phi_r|n), \qquad (13)$$

$P(n) = \langle n|\rho|n \rangle$ being the photon number distribution. The conditional probabilities $\pi_j(\phi_r|n)$ are given by Eq. (3).

The new photon number distribution after a detection in state j for the Ramsey interferometer setting ϕ_r, $P(n|j, \phi_r)$, is thus[22]:

$$P(n|j, \phi_r) = \frac{\pi_j(\phi_r|n)}{\pi_j(\phi_r|\rho)}P(n). \qquad (14)$$

We recover here the usual Bayes law of conditional probabilities. The probability for having n photons, conditioned to an atomic detection in j with the Ramsey interferometer phase set at ϕ_r is, up to a normalization factor (denominator in the r.h.s.), the initial photon number distribution $P(n)$ multiplied by the conditional probability for detecting the atom in j with this Ramsey phase setting when there are n photons in the cavity. Note that if the atom escapes detection, due to the finite efficiency of D, it does not change the photon distribution, since the two density matrices corresponding to a detection in $|e\rangle$ or $|g\rangle$ should then be summed, weighted by the respective detection probabilities.

After each atomic detection, $P(n)$ is thus multiplied by a sinusoidal function of n, proportional to $\pi_j(\phi_r|n)$. Some photon numbers are less likely to be found after the measurement (those which correspond to the highest probability for detecting the atom in the other state). Sending N_a atoms one by one through C iterates this photon number 'decimation' operation. The final $P(n)$ is:

$$P_{N_a}(n) = \frac{P_0(n)}{Z} \prod_{i=1}^{N_a} \pi_{j_i}(\phi_{r,i}|n), \qquad (15)$$

where j_i is the detected state for the atom labelled i in the series, $\phi_{r,i}$ the Ramsey interferometer phase used for this atom and $P_0(n)$ the initial photon number distribution.

Each factor in the product on the r.h.s. decimates some photon numbers. Numerical simulations and mathematical arguments[23] show that, after many atoms have been detected, the final photon number distribution reduces to a single photon number n_m: $P_{N_a}(n) = \delta_{n,n_m}$, provided that the initial distribution $P_0(n)$ spans a photon number range such that no two photon numbers lead to the same detected atomic state (for instance, the $\{0, 7\}$ range when $\phi_0 = \pi/4$). The accumulation of partial information provided by weak measurements ends up in a complete, projective measurement of the photon number.

The final outcome is independent of the initial distribution $P_0(n)$. We can thus choose it freely, provided it does not cancel in the relevant photon number range. A possible choice is a flat distribution, reflecting our lack of knowledge of the photon number distribution in the 0–7 interval.

We realized such a measurement on a coherent field with an average photon number $\bar{n} = 3.82$ (the probability for having 8 photons or more in this field is a few % only). We record the detection of $N_t = 110$ atoms, in a time of the order of 26 ms, with $\phi_0/\pi = 0.233\,\mathrm{rad}$. The Ramsey phase ϕ_r is randomly chosen among four values differing by about $\pi/4$ to speed up convergence. We use a slightly modified decimation law to incorporate the finite contrast of the Ramsey interferometer [see[22] for details].

Figure 3 presents the evolution of the photon number distribution after the detection of N_a atoms ($N_a = 0 \ldots N_t$) for two realizations of the experiment. Let us discuss the one shown on the left. The initial $P_0(n)$ is flat. After one atomic detection, it has a sine shape. After about 20 atomic detections, the decimation process leaves only two possible photon numbers, 5 and 4. For a short time interval, 4 dominates, but, after about 70 atomic detections, only 5 is left, completing the measurement. The other realization (right part of the figure) exhibits a similar behaviour, the final photon number being then 7. We observe here the progressive collapse of the field state due to an accumulation of weak measurements.

The final photon number is selected by the random atomic detection outcomes. Each realization of the experiment leads, after recording N_t atoms, to a different, random photon number according to the basic postulates of quantum physics. Once again, "God is playing dice" in a quantum measurement. The probability for

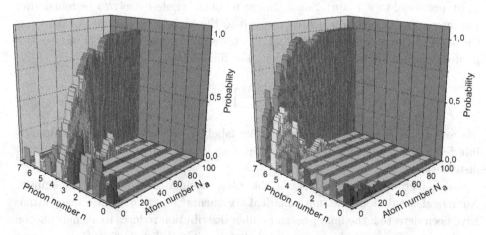

Fig. 3.   Two individual realizations of the QND measurement of a coherent field. The histograms show the evolution of the photon number distribution with the number N_a of detected atoms. Reprinted by permission from Macmillan Publishers Ltd: Nature.[22]

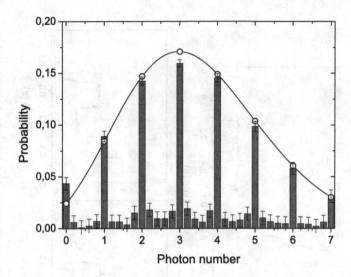

Fig. 4.   Reconstructed photon number distribution in the initial cavity field. Histogram of the average value $\langle n \rangle$ of the final photon distribution $P_{N_t}(n)$. The error bars reflect the statistical uncertainty (2000 realizations of the experiment have been used). The circles (and the continuous line joining them for visual convenience) represent a Poisson law with 3.46 photons on the average. Reprinted by permission from Macmillan Publishers Ltd: Nature.[22]

obtaining the final value n_m must be ideally given by the initial photon number distribution $P_0(n_m)$.

We have recorded 2000 individual realizations of the detection sequence and computed, for each of them, the average photon number $\langle n \rangle$ for the final distribution $P_{N_t}(n)$. The histogram of the measured $\langle n \rangle$ values is shown in Fig. 4. In about 80% of the cases, we do finally obtain an integer photon number, showing that the decimation process has converged. The background in the histogram corresponds to 20% of the experiments for which either N_t atoms have not been sufficient to grant convergence or for which a quantum jump due to cavity relaxation occurred during the measurement process itself. When convergence is obtained, the observed probabilities fit with a Poisson law (open blue circles). Its average photon number, 3.46, coincides nicely with an independent calibration of the average photon number in the cavity at the time when the measurement is complete.

Since the probe atoms do not absorb the cavity field, we can follow the photon number evolution over a long time interval. We keep sending dispersive atoms in C. At each time t, we infer a photon number distribution and its average $\langle n \rangle(t)$ from information provided by the last N_t atomic detections. Figure 5(a) present the time evolution of the inferred average photon number over 0.7 s for two realizations of the experiment whose initial phase (first N_t atoms) is shown in Fig. 3.

The initial phase corresponds to the $\simeq 26$ ms time interval required to perform the state collapse. Then, we observe a plateau with a constant photon number (5 or 7). In both cases, the duration of this plateau is long enough to allow for two

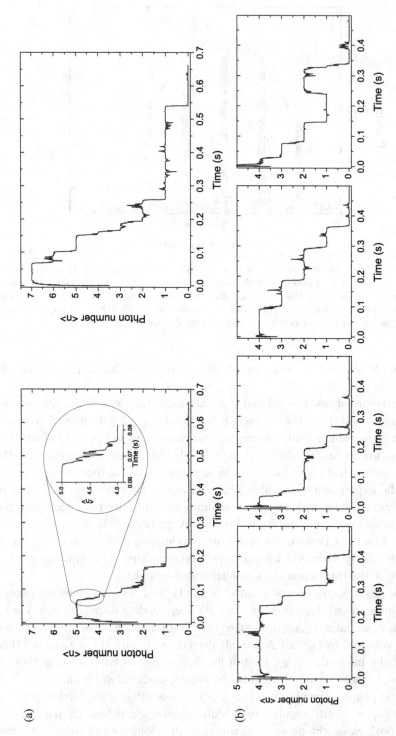

Fig. 5. Repeated measurements. (a) Time evolution of the mean photon number $\langle n \rangle(t)$ for the two sequences whose collapse is shown in Fig. 3. (b) Four other trajectories with an initial collapse in the $n_m = 4$ Fock state. Reprinted by permission from Macmillan Publishers Ltd: Nature.[22]

measurements of the photon number with two independent samples of $N_t = 110$ atoms. That these two measurements lead to the same result is a clear indication of the QND character of this process.

Our photon-counting procedure illustrates thus all the basic postulates for an ideal projective quantum measurement. It provides a quantized result (an integer photon number), whose value is randomly chosen with a statistical distribution given by the initial cavity state. The repeatability of the measurement shows that it does project the field state onto an eigenstate of the measured observable, the $|n_m\rangle$ Fock state, a situation radically different from that of all standard photo-detections.

After the initial plateau, we observe a stepwise relaxation of the photon number towards zero. We monitor here the quantum jumps of light, while the photons escape one by one from the cavity into the loss channels. The inset in the first curve shows a zoom on the quantum jump between the 5 and 4 photon states. Recording it requires the accumulation of information from a few tens of atomic detection and is performed in about 10 ms. Note that the one-photon state in the trajectory on the right has an exceptionally long lifetime (about 300 ms, nearly three average lifetimes T_c).

Figure 5(b) presents four other trajectories, with an initial collapse in the 4-photon state, exhibiting the randomness of the quantum jump occurrences. Note on the rightmost trajectory an upwards quantum jump corresponding to the creation of a thermal excitation in C.

The master equation (6) predicts a smooth relaxation of the average energy, the photon number decaying exponentially with a time constant T_c. This evolution is quite different from the sudden jumps of Fig. 5. This illustrates the difference between a single realization of a quantum experiment and the ensemble average predicted by the density operator. Averaging thousands of quantum jump trajectories, starting from randomly selected photon numbers, all exhibiting quantum jumps at different times, we indeed recover a smooth evolution in excellent agreement with the predictions of the master equation.

We have performed a detailed study of field relaxation.[24] By a careful analysis of all quantum jump trajectories, we reconstruct the damping-induced evolution of the photon number distribution, starting from all Fock states with 0 to 7 photons. We get clear evidence of the fast decay of the high-lying Fock states, whose lifetime is T_c/n. By a fit of the complete data, we extract the coefficients of the most general linear master equation ruling the evolution of $P(n,t)$. The results of this partial quantum process tomography are in excellent agreement with Eq. (7). Similar measurements have been performed simultaneously in the context of circuit QED.[25]

Finally, the QND measurement process leads naturally to the observation of the quantum Zeno effect.[26] Whereas frequently repeated measurements do not affect the dynamics of incoherent damping, they inhibit the coherent evolution of a quantum system. An observed system does not evolve under the action of its Hamiltonian, in clear correspondence with the second Zeno paradox negating motion. Quantum

Zeno effect has been observed on the evolution of a variety of two-level systems, first on a trapped ion.[27]

We have observed it for the coherent classical evolution of the cavity field.[28] We realize a long series of identical coherent injections with a very small amplitude in the initially empty cavity. These amplitudes add up and the field is finally in a mesoscopic coherent state whose average photon number grows quadratically with the number of injections, or with time. When we perform, between injections, QND measurements of n, we project repeatedly, with a high probability, the field back onto $|0\rangle$, since the probability for finding $n = 1$ after a single injection is low. We observe that the field energy effectively remains nearly zero, exhibiting the Zeno effect for the classical runaway process of field injection.

5. Monitoring the decoherence of mesoscopic quantum superpositions

The short lifetime of the Fock states is a first indication of the fragility of non-classical states when they reach the mesoscopic scale. An even more striking example is provided by a Schrödinger cat's decay.

A cat state, quantum superposition of two mesoscopic coherent fields in C with the same average photon number but with different classical phases, can be prepared by a single atom interacting dispersively with a coherent state γ, initially injected in C by the source S (Fig. 1). The atom, initially in $|g\rangle$, is prepared in $(|e\rangle + |g\rangle)/\sqrt{2}$ in R_1. It interacts dispersively with the cavity field, resulting in the atom-cavity entangled state $\exp(-i\phi_0/2)|e\rangle|\gamma\exp(-i\phi_0/2)\rangle + |g\rangle|\gamma\exp(+i\phi_0/2)\rangle$. This is a situation quite reminiscent of the Schrödinger cat metaphor, with a two-level quantum system entangled to a 'macroscopic' degree of freedom (the classical phase of the field).

Setting $\phi_r = -\phi_0/2$, a final atomic detection after R_2 projects (within an irrelevant global phase) the field onto:

$$|\Psi_\pm\rangle = \frac{1}{\mathcal{N}}(|\gamma e^{+i\phi_0/2}\rangle \pm |\gamma e^{-i\phi_0/2}\rangle), \tag{16}$$

where the $+(-)$ sign applies for a detection in e (g), $\mathcal{N}$ being a normalization factor. When $\phi_0 = \pi$, the cat state with the $+(-)$ sign expands only on even (odd) photon number states. In reference to this simple situation, we call even (odd) cat the state with the $+(-)$ sign for all ϕ_0 values.

The cat state is, when $|\gamma|$ is large, a quantum superposition of two quite distinct classical states. It is expected to decay rapidly due to its coupling to the environment via the cavity losses. We had performed in an early experiment[29] a first investigation of this decoherence process. It had shown that the decay rate of the coherence between the superposed states increases rapidly with their separation in phase space, in good agreement with theoretical expectations.

A much more detailed insight of the cat coherence can be obtained through a complete reconstruction of the field density operator as a function of time.

As discussed in the previous Section, the QND measurement of the photon number leads to a partial reconstruction. Many measurements performed on the same quantum state allow us to reconstruct the photon number distribution, i.e. the diagonal of the density operator in the Fock state basis.

We have no access, however, to the non-diagonal elements, which contain information on the field phase distribution. We are unable, for instance, to distinguish a statistical mixture of Fock states with no phase information from a coherent state, with a well-defined phase. Since we use non-resonant probe atoms, they cannot extract directly information about the field phase.

A simple modification of the QND scheme allows us to reconstruct the full density operator.[30] Before sending the QND probe atoms, we first perform a displacement of the cavity field, by letting the source S inject in C a coherent amplitude β. The source S has of course a well defined, controlled phase with respect to the initial cavity field amplitude γ. This 'homodyning' procedure turns the cavity field density operator into $\rho_\beta = D(\beta)\rho D(-\beta)$. We then send many QND probe atoms through C. By repeating the experiment many times, we measure the probability $\pi_e(\phi_r|\rho_\beta)$, conditioned to the translated cavity state. It simply reads $\pi_e(\phi_r|\rho_\beta) = \mathrm{Tr}[G(\beta, \phi_r)\rho]$, where $G(\beta, \phi_r) = D(-\beta)M_e^\dagger M_e D(\beta)$. We thus obtain finally, within statistical noise, the average of the observable $G(\beta, \phi_r)$ in the initial cavity state.

Resuming the experiment for very many (up to 600) different translation amplitudes β carefully chosen in phase space, we obtain the average value of many different observables $G(\beta, \phi_r)$ in the state ρ. We infer from these results all the matrix elements of ρ, using an approach based on the maximum entropy principle.[31] This procedure determines the density matrix that best fits the data, while having a maximum entropy $-\mathrm{Tr}\rho \ln \rho$. We thus make sure that the reconstructed state does not contain more information than that provided by the experimental data.

We represent the reconstructed state by its Wigner function, which is more pictorial than the density operator. It is a quasi-probability distribution $W(\alpha)$ defined over the phase plane by:

$$W(\alpha) = \frac{2}{\pi}\mathrm{Tr}[D(-\alpha)\rho D(\alpha)e^{iN\pi}]. \tag{17}$$

In other words, $W(\alpha)$ is the expectation value of the parity operator $\exp(iN\pi)$ (eigenvalue $+1$ for even Fock states, -1 for odd Fock states) in the cavity field translated by $-\alpha$. The Wigner function contains all information on the field state, since ρ can be directly computed from W.[1] For all ordinary states (coherent, thermal, squeezed), W is a positive Gaussian. Negativities in the Wigner function are a clear indication of the non-classicality of a state. Note that another method for measuring W has been used in trapped ions[32] and circuit QED.[33]

Figure 6(a) presents the measured Wigner function of the even cat (preparation atom detected in $|e\rangle$), for $|\gamma| = \sqrt{3.5}$ and $\phi_0 = 0.74\,\pi$. The Wigner function presents

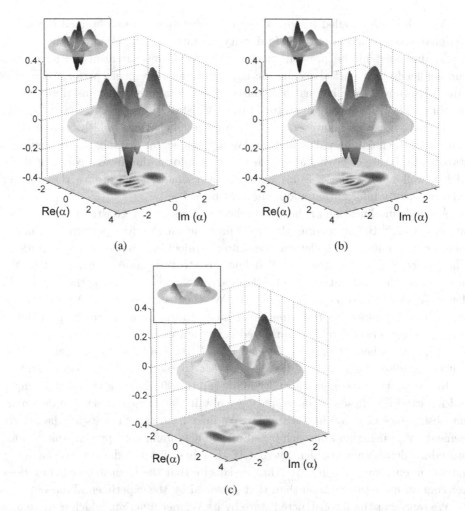

Fig. 6.   Experimental Wigner functions of an even cat (a), odd cat (b) and statistical mixture of two coherent components (c). The insets present the theoretical expectations. The average photon number is 3.5 and the phase shift is $\phi_0 = 0.74\pi$. Reprinted by permission from Macmillan Publishers Ltd: Nature.[30]

two positive bumps centred on the superposed classical amplitudes, $\gamma \exp(\pm i\phi_0/2)$ (the cat's 'ears'). In between these two classical features, we observe a high contrast interference pattern that reveals the quantum nature of the superposition (the cat's 'smile'). The observed Wigner function is quite close to the theoretical expectation (shown in the inset).

The 'size' of this cat is conveniently measured by the square of the distance $\mathcal{D}$ between the two classical components in phase space: $\mathcal{D}^2 = 11.8$ photons, since the decoherence time scale is expected to be $T_d = 2T_c/\mathcal{D}^2$ at zero temperature.[1]

Note that the ears of our cat are not exactly Gaussian (as should be for a superposition of coherent states). This is not due to an imperfection in the reconstruction

procedure, but to the cat preparation stage. Since we are not deeply into the dispersive regime (the atom-cavity detuning is $\delta/2\pi = 51\,\text{kHz}$, not very large compared to
$\Omega_0/2\pi$), the phase shift of a Fock state depends in a non-linear way on the photon
number, leading to a slight deformation of the coherent components.

Figure 6(b) presents the Wigner function of the odd cat. The classical components are the same, but the interference pattern is phase-shifted by π as compared
to that of the even cat. Finally, Fig. 6(c) presents the Wigner function of a statistical
mixture of the two coherent components. It is obtained by mixing data corresponding to different detected states of the preparation atom. The cavity state is then a
statistical mixture of the odd and even cat, or equivalently a statistical mixture of
the two coherent components. The classical ears are still there, but the smile has
gone, as expected.

For the reconstruction of ρ, we can repeat the experiment many times for each
β value. We thus use only a few QND atoms in each realization and still achieve
good statistics. We chose to detect about 20 atoms in a 4 ms time interval. We are
thus able to measure the time evolution of the field state (or of its Wigner function)
with a decent resolution.

Figure 7 presents four snapshots of the even cat Wigner function evolution [the
conditions are those of Fig. 6(a)]. The quantum feature, the cat's smile, decays much

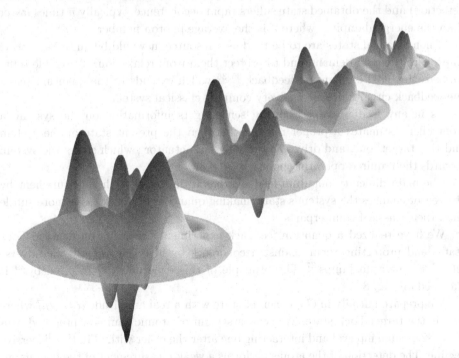

Fig. 7.   Four snapshots of the Wigner function of a decaying Schrödinger cat state. The initial
state is the even cat of Fig. 6(a). The snapshots are taken 1.3 ms, 4.3 ms, 15.8 ms and 22.9 ms (from
front to back) after cat preparation.

faster than the energy (the latter decay corresponds to a slow motion of the ears towards the origin, with a time constant $2T_c$). After about 23 ms, the contrast of the interference pattern is considerably reduced and we are left with a mere statistical mixture. From these data, we deduce the decoherence time scale, T_d, defined as the damping time of the non-diagonal elements of the density matrix in the coherent state basis. We get $T_d = 17 \pm 3\,\text{ms}$, in excellent agreement with the theoretical expectation, 19.5 ms, taking into account the finite mode temperature.

We get in this way a detailed insight into the cat state decoherence. More precise measurements, with quite larger cats, could allow us to realize a full quantum process tomography of the cavity relaxation. Experiments on 'decoherence metrology' are interesting to test alternative interpretations of the conspicuous lack of quantum superpositions at the macroscopic scale.[34]

6.   An experiment on quantum feedback

It is fairly easy in our cavity QED experiments to prepare non-classical states: Schrödinger cats conditioned to the detection of the preparation atom or Fock states, conditioned to the outcome of the QND measurement process. However, this preparation is non-deterministic (being conditioned to an initial random atomic detection) and the obtained state suffers rapid decoherence, typically $\bar{n}$ times faster than the energy damping, where $\bar{n}$ is the average photon number.

If non-classical states are to be used as a resource, it would be quite interesting to prepare them on demand and to protect them from relaxation. A possible route towards this goal is quantum feedback,[35,36] which extends to the quantum realm the feedback circuits present in every complex classical system.

As in any feedback operation, a 'sensor' gets information on the system, a 'controller' estimates a proper distance between the present state of the system and the 'target' one, and drives accordingly an 'actuator', which steers the system towards the required operating point.

The main difficulty quantum feedback has to face is that the measurement by the sensor changes the system's state, making quantum feedback loops more subtle than their classical counterparts.

We have realized a quantum feedback experiment preparing on demand Fock states and protecting them against decoherence by inverting rapidly the adverse effects of quantum jumps.[37] The principle of this experiment,[38] inspired by,[39] is sketched on Fig. 8.

We prepare initially in C a coherent state with a real amplitude $\alpha = \sqrt{n_t}$ where $|n_t\rangle$ is the target Fock state. We send a stream of atomic samples, prepared at a $T_a = 82\,\mu\text{s}$ time interval and interacting one after the other with C in the dispersive regime. The detection of the atoms performs a weak measurement of the field intensity, described by the operators M_e and M_g [see Eq. (11)]. Setting the dispersive phase shift ϕ_0 near $\pi/4$ allows us to distinguish the photon numbers between 0 and 7.

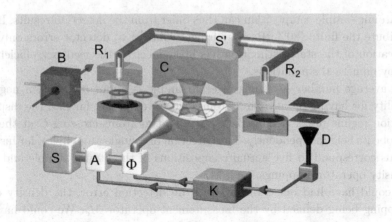

Fig. 8.   Principle of the quantum feedback experiment for the preparation and protection of Fock states. Reprinted by permission from Macmillan Publishers Ltd: Nature.[37]

Instead of piling up information provided by many atoms to perform a complete QND measurement of n, which would prepare a randomly selected Fock state, we act on the cavity field after each atomic detection. Information provided by the atom is used by the controller K (a fast real-time classical computer ADwin-ProII by Jäger Messtechnik) to infer the cavity field density operator ρ. The controller then computes the amplitude α of a displacement minimizing a proper distance, $d(n_t, \rho_\alpha)$, between the translated density operator $\rho_\alpha = D(\alpha)\rho D(-\alpha)$ and the target Fock state. It then sets the amplitude and phase controls A and Φ acting on S to apply this displacement. Numerical simulations show that iterations of this elementary feedback loop drive efficiently the cavity state towards the target.[38]

The atomic samples should efficiently distinguish the Fock states around n_t. A proper choice for the Ramsey interferometer phase is $\phi_r(n_t) = (\pi/2) - \phi_0(n_t + 1/2)$, corresponding to a mid-fringe setting when C contains n_t photons. We have shown by numerical simulations that this choice is well adapted for $n_t = 1, 2$.

For larger target photon numbers, however, quantum jumps due to dissipation are more frequent. We then alternate between two values of the Ramsey phase, $\phi_r(n_t)$ providing a good sensitivity and $\phi_r(n_t - 1)$ providing a fast detection of the most frequent downwards quantum jumps.

The state estimation is a critical part of the feedback loop. It must include all available information on the cavity field. It must take into account our initial knowledge of the field state, all information provided by the atomic detections and all feedback actions so far. It must also include all the known experimental imperfections, carefully calibrated before each experimental run. They include cavity relaxation in the time interval between two atomic detections (the cavity damping time is $T_c = 65$ ms in this experiment) and the finite contrast of the Ramsey interferometer.

The estimation must also include the imperfections of the atomic detector D, which has a 39% detection efficiency and a finite state attribution error rate. The actual atomic sample composition can thus differ from the detection results. In fact, we attribute the finite (80%) Ramsey fringes contrast to detector errors only. This simplification of the state estimation has little effect on the feedback efficiency, as shown by numerical simulations.

The average number of atoms in each sample is about 0.6, with a negligible probability for having more than 2 atoms. Since we operate far into the dispersive interaction regime ($\delta/2\pi = 245\,\text{kHz} \gg \Omega_0/2\pi$), two atoms crossing C at the same time probe its field independently. The detection of a single atom in $|e\rangle$, for instance, may thus correspond to five actual compositions of the atomic sample and hence five density operator mappings.

We could have had one atom $|e\rangle$ with no detection error, the density operator mapping being defined by the measurement operator M_e. We could have had one atom in $|g\rangle$ (measurement operator M_g) with an erroneous detection, or two atoms in $|e\rangle$ (M_e^2) with one atom failing to be detected, or one atom in $|e\rangle$ and the other in $|g\rangle$ ($M_e M_g$) with or without error detections and with one atom undetected, or finally two atoms in $|g\rangle$ (M_g^2) with one atom escaping detection and one detection error. The final density matrices corresponding to all these actual sample compositions must be averaged, weighted by the occurrence probabilities of the corresponding events, which are obtained by a straightforward but tedious Bayesian inference analysis.[37]

The state estimator must also take care of the $344\,\mu s$ time interval between the interaction of an atomic sample with C and its detection in D. When a sample is finally detected, four óther ones have already interacted with C and are on their way towards D. State estimation takes into account the yet unread measurements performed by these samples.[38]

All the state estimation computations should be performed in a time shorter than the time interval T_a between atomic samples. We perform them with a carefully optimized code in a truncated Fock state basis, with 8 elements for $n_t \leq 2$ ($|0\rangle$ to $|7\rangle$) and 9 for $n_t = 3, 4$. Moreover, since the initial amplitude is real, the initial density operator is real in the Fock state basis, as well as the target. We may thus restrict the amplitude α to real values at each step, and hence the estimated density operator to real matrix elements.

The controller adjusts the value of the real injected field amplitude α in order to minimize the distance $d(n_t, \rho_\alpha)$ of the estimated density matrix after translation, $\rho_\alpha = D(\alpha)\rho D(-\alpha)$, with the target state $|n_t\rangle\langle n_t|$. A simple distance definition is $d = 1 - F_t(\rho_\alpha)$ where $F_t(\rho_\alpha) = \text{Tr}(\rho_\alpha|n_t\rangle\langle n_t|)$ is the fidelity with respect to the target state. However, this distance reaches its maximum value, 1, for all states orthogonal to $|n_t\rangle$. It does not provide clear information on the actual difference between the estimated state and the target. We use instead[37]:

$$d(n_t, \rho_\alpha) = 1 - \text{Tr}(\Lambda^{(n_t)}\rho_\alpha), \tag{18}$$

where $\Lambda^{(n_t)}$ is a diagonal matrix in the Fock state basis, with $\langle n_t|\Lambda^{(n_t)}|n_t\rangle = 1$, so that the distance cancels in the target state. The diagonal elements of this matrix are chosen so that d has a global minimum on the target state and local maxima on all other Fock states. They are further fine-tuned by optimizing the feedback convergence speed in numerical simulations.

An explicit optimization of the amplitude α is far too time-consuming to be performed at each iteration of the loop. In order to make computations faster, we use a second-order approximation of the displacement operators:

$$\rho_\alpha = D(\alpha)\rho D(-\alpha) \approx \rho - \alpha[\rho, a^\dagger - a] + \frac{\alpha^2}{2}[[\rho, a^\dagger - a], a^\dagger - a], \qquad (19)$$

leading to

$$d(n_t, \rho_\alpha) \approx d(n_t, \rho) - a_1\alpha - a_2\frac{\alpha^2}{2}, \qquad (20)$$

with $a_1 = \mathrm{Tr}[\Lambda^{(n_t)}, a^\dagger - a]\rho$ and $a_2 = \mathrm{Tr}[[\Lambda^{(n_t)}, a^\dagger - a], a^\dagger - a]\rho$.

Each step should reduce d and hence $(a_1\alpha + a_2\alpha^2/2)$ must be large and positive. If $a_2 < 0$, $d(n_t, \rho_\alpha)$ is concave and the optimum is achieved for $\alpha = -a_1/a_2$. Since we have used a second order approximation in α, we moreover limit the amplitude modulus to $|\alpha| < \alpha_{\max} = 0.1$. For mere experimental convenience, we arbitrarily set to zero all amplitudes whose modulus is lower than 10^{-3}. When $a_2 \geq 0$, d is a convex function of α and we can thus expect to be far from its global minimum. We thus apply the largest allowed displacement with the proper sign: $\alpha = \alpha_{\max}$ for positive a_1 and $\alpha = -\alpha_{\max}$ for non-positive a_1.

The controller K sets the amplitude α by pulsing the continuous wave source S with the switch A. The injected amplitude modulus is a precisely calibrated linear function of the injection time t_i. Typically, the injection of the maximum amplitude $\alpha_{\max}$ is performed in $63\,\mu$s, a time shorter than the interval T_a between samples (note that the injection pulse overlaps with the independent calculation by K of the amplitude for the next iteration of the feedback loop). The sign of α is controlled by the programmable $\{0, \pi\}$ phase-shifter Φ.

Figure 9 presents one feedback run lasting 160 ms, aiming at the preparation of a two-photon state ($n_t = 2$). Initially, the distance to the target is large and the feedback active, with repeated injections. After a while, the cavity state converges towards the target. The probability $P(2)$ for having two photons reaches a high value, in the 80% range and the estimated density matrix is quite close to that of a Fock state. We have thus prepared on demand the prescribed non-classical state.

After a while, a quantum jump towards $n = 1$ occurs. The feedback activity resumes, until the target state is restored. We observe, in frame (e) that the field has no defined phase immediately after the jump, since the density matrix is diagonal. During recovery, the injections create transient non-diagonal elements, which are required to pull the state back towards the target and which finally vanish when convergence is achieved.

We have measured the average convergence time towards the target state. The feedback procedure leads to a Fock state preparation noticeably faster than a 'fail

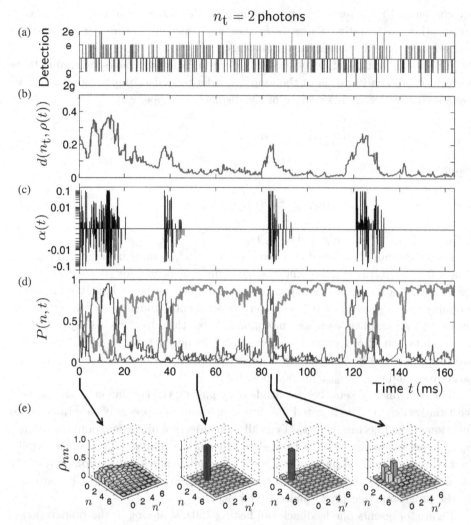

Fig. 9.   One feedback run lasting 160 ms with $|n_t\rangle = |2\rangle$). (a) Atomic detection events (the two-atoms detections appear as double length bars). (b) Distance $d(n_t, \rho)$ between the estimated density matrix and the target state. (c) Applied injection amplitude (in Log scale, as $\mathrm{sgn}(\alpha)\log(|\alpha|)$. (d) Photon number distribution with $P(2)$ in thick line, $P(n < 2)$ in thin light gray and $P(n > 2)$ in thin dark gray line. (e) Four snapshots of the estimated density matrix, at time $t = 0$ (initial coherent state), at a time when convergence is realized, immediately after a quantum jump and finally during quantum jump recovery. The precise times are indicated by arrows from the above time scale. Reprinted by permission from Macmillan Publishers Ltd: Nature.[37]

and resume' approach based on QND measurement (we perform a QND measurement of n, restart the sequence if n is found to be different from n_t, until we finally find $n = n_t$). For $n_t = 3$, 63% of the feedback trajectories have reached $P(3) > 80\%$ after 50 ms only, whereas 250 ms are required for the 'fail and resume' brute force approach. We have also studied the dynamics of the recovery after a quantum jump, from $n = n_t = 3$ down to $n = 2$. The estimator needs about 3 ms (7 detected atoms) to 'realize' that a jump has occurred and the target state restoration lasts for an additional 8 ms.

We estimate the fidelity of the feedback procedure by interrupting the sequence and realizing an independent measurement of the photon number distribution with QND atoms, using a fast procedure outlined in.[24] Figure 10 presents the obtained result for n_t from 1 to 4 [frames (a) to (d) respectively]. We use for these data two feedback interruption criteria. For the light gray histograms, we interrupt the feedback when the probability for having n_t photons in the estimated density matrix, $\langle n_t | \rho | n_t \rangle$, reaches 80% for three consecutive atomic samples. We observe that the state estimation is correct, since the independently measured $P(n_t)$ is then close to 80%.

For the dark gray histograms, we interrupt the feedback sequence after a fixed amount of time (164 ms), large enough for reaching a steady-state on the average. The fidelity with respect to the target state is lower than for the light gray histograms, because the field resides for a fraction of time in states with $n \neq n_t$ due

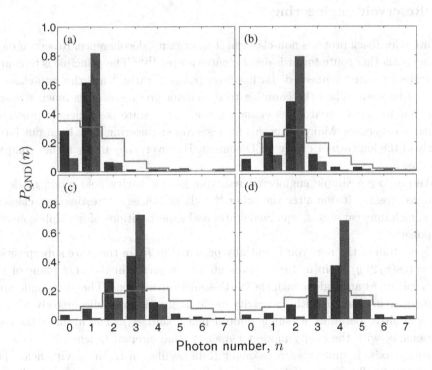

Fig. 10. Photon number distributions after feedback operation for n_t=1 to 4 [frames (a) to (d)]. The dark gray histogram is obtained when the feedback operation is interrupted at a fixed 164 ms time. If the feedback is interrupted when the estimated probability for having n_t photons is larger than 80%, we get the light gray histograms. Finally, the line presents the measured photon number distribution in the initial coherent field. Reprinted by permission from Macmillan Publishers Ltd: Nature.[37]

to relaxation-induced quantum jumps. The photon number distribution is however clearly non-classical, with $P(1) > 60\%$ for $n_t = 1$, $P(3) \approx 45\%$ for $n_t = 3$. This shows that our feedback procedure does protect the non-classical Fock states against decoherence.

We have implemented a real-time quantum feedback operating in the steady-state. The slow pace of the Rydberg atoms cavity QED experiments is a considerable asset, since it allows K to perform complex state estimation procedures in the time interval between two measurements of the cavity field. This would be much more difficult in optical or circuit QED experiments, in which the time scale is typically 1000 times faster. We have recently realized a new feedback experiment with quantum actuators made up of Rydberg atoms interacting resonantly with C, and delivering (or subtracting) single photons.[40] The quantum actuators are well matched to the quantum nature of the photon number jumps and provide a more efficient compensation of decoherence events. We have been able to stabilize photon numbers up to 7 with a high fidelity.

7.   Reservoir engineering

Quantum feedback protects non-classical states against decoherence. Reservoir engineering is another route towards decoherence control.[41, 42] The principle is to couple the system to an 'engineered' bath whose pointer states[13] are the non-classical states of interest. When the coupling to the engineered reservoir is much stronger than that to the standard loss channels, the target state is efficiently protected against decoherence. Many proposals for reservoir engineering have been put forth, mostly in the ion trap or cavity QED context. However, they imply rather complex operations.

We propose a simple engineered reservoir for the cavity field mode made up of atoms crossing it one after the other.[43, 44] It stabilizes interesting non-classical states, including cat states, squeezed states and superpositions of multiple coherent components.

Each atom of the reservoir is initially prepared in R_1 in the state superposition $|u_a\rangle = \cos(u/2)|g\rangle + \sin(u/2)|e\rangle$, represented by a vector in the xOz plane of the Bloch sphere, at an angle u with the north-south vertical axis. The atom undergoes a composite interaction with the cavity mode. It first interact dispersively with it, with a positive detuning δ (single-photon phase shift ϕ_0). The atom is then set in resonance with the cavity mode for a short time interval t_r when it is close to the cavity axis. It undergoes a resonant Rabi oscillation in the cavity field. The detuning is finally set to $-\delta$ for a second dispersive interaction, corresponding to the phase shift $-\phi_0$, opposite to that of the first dispersive interaction. The atom is finally discarded. The Stark effect produced by an electric field applied across the cavity mirror makes it possible to control easily in real time the atomic transition frequency and to achieve this sequence of operations.

Let us first get an insight into the reservoir's operation by considering only the resonant interaction, described by the evolution operator $U_r(\Theta)$:

$$U_r(\Theta) = |g\rangle\langle g|\cos\frac{\Theta\sqrt{N}}{2} + |e\rangle\langle e|\cos\frac{\Theta\sqrt{N+1}}{2}$$
$$- |e\rangle\langle g|a\frac{\sin\Theta\sqrt{N}/2}{\sqrt{N}} + |g\rangle\langle e|\frac{\sin\Theta\sqrt{N}/2}{\sqrt{N}}a^\dagger,$$

where $\Theta = \Omega_0 t_r$. If $u = 0$ (atoms sent in $|g\rangle$), the resonant reservoir obviously stabilizes the vacuum state. The resonant atoms in their lower state absorb any initial field present in C. We use, by the way, this operation routinely to empty rapidly the cavity at the end of all experimental sequences.

When the atom is injected with a small u value, $0 < u \ll 1$, we realize in C a micromaser pumped below population inversion.[45] The balance between atomic emission and absorption stabilizes a nearly coherent state $|\alpha\rangle$. The real amplitude α is given by $\alpha = 2u/\Theta$ in the $\Theta \ll 1$ limit.

This steady state value can be understood in simple terms. When it is reached, the resonant interaction corresponds to a rotation of the Bloch vector by an angle $-\Theta\alpha$ in the xOz plane of the Bloch sphere. The Bloch vector, initially at an angle u with the vertical axis evolves first towards the south pole, corresponding to an atomic emission, and continues its rotation (atomic absorption) until it finally reaches an angle $u - \alpha\Theta = -u$ with the vertical axis. The atom exits the cavity with the same average energy as it had entering it. The cavity energy is unchanged and the field amplitude is stable.

Numerical simulations of the atom-cavity interaction show that this picture remains valid even when u and Θ are of the order of 1. Of course, stabilizing a coherent state is not utterly interesting (a classical source does it).

Let us now consider the full interaction, described by the evolution operator $U = U_d^\dagger(\phi_0)U_r U_d(\phi_0)$, where:

$$U_d(\phi_0) = |g\rangle\langle g|e^{-i\phi_0 N/2} + |e\rangle\langle e|e^{i\phi_0(N+\mathbb{1})/2}, \tag{21}$$

is the dispersive evolution operator corresponding to the phase shift per photon ϕ_0. After some algebra,[43] making an intensive use of the relation $af(N) = f(N+\mathbb{1})a$, we cast the total evolution operator under the form:

$$U = e^{-ih_0(N)}U_r e^{ih_0(N)} \tag{22}$$

with:

$$h_0(N) = -\phi_0 N(N+1)/2 \tag{23}$$

The resonant interaction of a single atom with the mode leaves the coherent state $|\alpha\rangle$ unchanged. It is thus clear from the above equations that the composite interaction leaves invariant the state $U_0|\alpha\rangle$, where $U_0 = \exp(-ih_0(N))$ is the evolution operator in the fictitious Hamiltonian h_0 for a unit time interval.

The pointer state of the composite interaction results thus from the action on the coherent state $|\alpha\rangle$ of a photon-number conserving Hamiltonian including a term proportional to N^2. This corresponds to the propagation of a coherent state through a Kerr medium, with an index of refraction containing a term proportional to the intensity. The Kerr effect has been widely analysed in theoretical literature as a way to produce non-classical fields.

Along its propagation through the Kerr medium, an initial coherent state $|\alpha\rangle$ is first distorted into a squeezed state, with quantum fluctuations on one quadrature reduced below the initial ones. Since the photon number distribution is conserved, the squeezing is limited, and the state gradually evolves into a 'banana-shaped' state spreading in the phase plane along the circle with radius $|\alpha|$. When the spreading extends over the whole circle, quantum interference phenomena appear, which are conspicuous on a plot of the Husimi-Q or of the Wigner function of the field.[1] For definite interaction times, the state evolves into a quantum superposition of 6, 5, 4, 3 and finally 2 coherent components with opposite phases, a cat state of the field. By tuning the phase shift ϕ_0, we can thus stabilize with the composite reservoir any of those highly non-classical states produced by the fictitious Kerr Hamiltonian h_0.

In order to confirm this insight, we have performed numerical simulations, using the parameters of our actual CQED experiment. In particular, we include the competition of the engineered reservoir with the standard cavity losses, with $T_c = 65$ ms and $n_{th} = 0.05$. The dispersive and resonant interactions are computed exactly, integrating the Jaynes and Cummings Hamiltonian [Eq. (2)] with the help of the quantum optics package for MATLAB.[46]

In these computations, the atomic samples are prepared at regular time intervals T_a with velocity v such that they are separated by $3w$ along the beam. They thus interact separately with C. They contain 0.3 atoms on average and we neglect the effect of two-atom samples. The resonant and dispersive interaction periods are symmetric with respect to the time when the samples cross the cavity axis.

Figure 11(a) presents the Wigner function $W(\xi)$ associated to the cavity field state ρ_{200} after its interaction with 200 atomic samples for $v = 70$ m/s, $T_a = 257\,\mu$s, $\Delta = 2.2\Omega_0$, $t_r = 5\,\mu$s, $\Theta \approx \pi/2$, $u = 0.45\pi$. The (irrelevant) initial cavity state is the vacuum. We get a state quite similar to a Schrödinger cat (see Section 5), with an average photon number $\bar{n} = 2.72$ and $\mathcal{D}^2 = 10.9$, where the mesoscopicity parameter $\mathcal{D}$ is, as before, the distance in phase space between the classical components.

The purity $P = Tr(\rho_{200}^2)$ is 51%. We estimate the fidelity $F = Tr[\rho_{200}\rho_{cat}]$ of this state with respect to an ideal cat, optimized by adjusting in the reference state ρ_{cat} the phase and amplitude of the coherent components and their relative quantum phase. We get $F = 69\%$. We have checked that cavity relaxation is the main cause of imperfection, F being 98% with an ideal cavity.

Figure 11(b) presents as a solid line the fidelity F of the prepared state with respect to the ideal cat state as a function of atomic sample number (i.e. as a function of time). The transient reflects the competition between the fast build-up of the state, the fidelity raising over a few atomic samples only, and the decoherence, whose

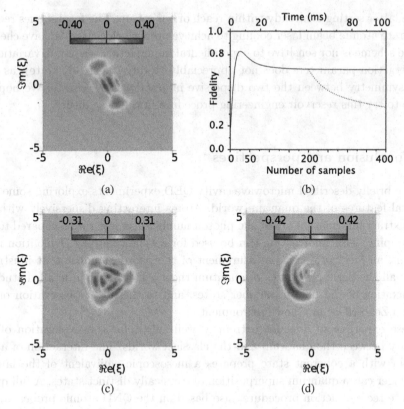

Fig. 11.   Non-classical state stabilization. (a) Wigner function of the cavity field after interaction with 200 atoms. Parameters (see text) are optimized to produced a state close to a two-component cat state. b) Solid line: fidelity of the generated state against the closest cat as a function of number of interactions (bottom axis) or of time (upper axis). Dashed line: reservoir is switched off after 200 interactions. (c) and (d) Wigner functions of stabilized cavity fields close to a three-component cat and to a 'banana state' respectively. From Ref. 43.

time scale $T_d = T_c/(2\bar{n})$, becomes relevant once a large average photon number $\bar{n}$ has been produced. The steady-state fidelity is reached after $\simeq 100$ samples. The dashed line presents F when we switch off the reservoir after 200 interactions. It initially drops rapidly, illustrating the efficient protection offered by the engineered reservoir.

For a slightly larger detuning, $\Delta = 3.7\,\Omega_0$ (all other parameters unchanged), we obtain a superposition of three components [Fig. 11(c)], with $\bar{n} = 2.70$ photons, $P = 56\%$, and $F = 73\%$ with respect to the closest ideal three-coherent-component superposition. Figure 11(d) presents $W(\xi)$ obtained after 200 samples with $v = 150$ m/s, $t_i = 120\,\mu s$, $t_r = 5\,\mu s$ i.e. $\Theta \approx \pi/2$, $u = \pi/2$ and $\Delta = 7\,\Omega_0$. These parameters correspond to a weak Kerr interaction, leading to the production of a 'banana' state, with interesting negativities in its Wigner function. The field has $\bar{n} = 3.52$ and $P = 91\%$.

The latter setting is already within reach of our set-up. The former ones require a moderate atomic beam laser cooling to achieve proper velocities. We have checked that the scheme is not sensitive to experimental imperfections (a few % variation of the interaction parameters does not appreciably modify the steady state), as long as the symmetry between the two dispersive interactions is accurate. We hope to be able to try this reservoir engineering procedure in the near future.

8.   Conclusion and perspectives

We have briefly described microwave cavity QED experiments exploring some fundamental features of the quantum world. Atoms interacting dispersively with the cavity extract information about the photon number at a fast rate compared to the field damping. This information can be used for a Quantum Non Demolition measurement, an ideal projective measurement of the photon number that illustrates directly all the basic postulates of quantum theory. It leads to a detailed study of the relaxation of the photon number states, and to the direct observation of the quantum Zeno effect in a new environment.

These experiments are also extremely well suited for the exploration of the boundary between the quantum and the classical worlds. The interaction of a single atom with a coherent state prepares a mesoscopic equivalent of the famous Schrödinger cat, a quantum superposition of classically distinct states. A full quantum state reconstruction procedure, also based on the QND atomic probes, allows us to get a detailed insight into the decoherence of this cat state.

Finally, the slow pace of these experiments and the exquisite controllability of the atomic system make it possible to test strategies to combat decoherence. Quantum feedback uses information extracted by QND probes to deterministically steer the cavity field towards a prescribed Fock state. It also protects these non-classical resources from decoherence by reversing rapidly the adverse effects of quantum jumps. Quantum reservoir engineering opens the way to the steady-state generation of non-classical states.

All these achievements open promising perspectives, which are nicely complementary to those of circuit QED. In order to achieve an even higher degree of control, we are now developing our experiment in two directions.

The first aims at the deterministic preparation of circular Rydberg atoms on demand. The random excitation used so far, leading to a Poisson atom number distribution, is painful for experiments involving a fixed number of atoms, such as those on quantum information.[17]

The dipole blockade mechanism[47] leads to a deterministic preparation of individual Rydberg atoms. We plan to use a dense sample of $\simeq 100$ cold atoms, with a micrometer-sized extension. A resonant laser excites a transition from the ground state to a low angular momentum Rydberg level ($60S$ state). Once the first atom has been excited in the Rydberg state, it shifts all the others far out of resonance with the laser, due to the extremely strong dipole-dipole interaction between the very excited Rydberg levels. Numerical simulations show that the probability for exciting two atoms at the same time is low, in the fraction of a % range.

The strong confinement required is obtained with an atom-chip device, which should moreover operate at cryogenic temperatures. We have thus developed a superconducting atom chip experiment.[48] This context is quite appropriate for the preparation of a small Bose-Einstein condensate.[49] We are investigating the dipole blockade mechanism. The prepared Rydberg state could easily be transferred in the circular levels and used on-chip,[50] possibly coupled to a stripline microwave resonator, or launched in high-Q cavities using the Stark effect for the acceleration at a macroscopic velocity.

In the present set-up, the atom-cavity interaction time is limited to about $100\,\mu s$ by the thermal velocity of the atomic beam. This is an asset when we need to extract information rapidly out of the cavity, but it is a severe limitation for other experiments. We are thus developing a new set-up, represented in Fig. 12, where the atoms will be prepared inside the cavity itself, out of a slow atomic beam in a fountain arrangement. Atoms excited near their turning point interact with the mode for times in the millisecond range, only limited by their free fall through the mode's waist. The limited level lifetime makes it mandatory to perform the field ionization detection also in the cavity structure.

Fig. 12. Experimental set-up under construction, featuring a slow atomic beam in a fountain arrangement to prepare circular atoms nearly at rest in a high-Q cavity (only one mirror is shown). The electrodes around the cavity are used for the circular state preparation and field-ionisation detection. A fast horizontal atomic beam with its Ramsey zones and detectors is used for cavity diagnostic.

With these long interaction times, we could generate large cat states, containing up to a few tens of photons with high fidelities and monitor their decoherence. We could realize quantum random walks for the phase of the cavity field, driven by a single atom.[51]

Finally, the atom-cavity interaction time is long enough to resolve the anharmonicity of the dressed-levels ladder and to address selectively a transition corresponding to one precise photon number. We have shown recently that we could realize Quantum Zeno Dynamics (QZD) in this context.[52, 53]

QZD[54] generalizes the quantum Zeno effect to measurements of an observable with degenerate eigenspaces. Under such frequently repeated measurements, the system evolution is confined in one of these subspaces, and proceeds under the restriction of the Hamiltonian in the subspace.

Frequent interrogation of a photon-number selective transition in the dressed levels implements such a dynamics, restricting the evolution to photon numbers smaller or larger than the addressed one. This leads to non-trivial dynamics and to the efficient generation of non-classical states.

Combining these interrogations with global displacements, we proposed 'phase space tweezers', able to pick out a single coherent component in a complex cat-like superposition and to move it at will, independently from the others. Moreover, these tweezers can be adapted to prepare such superpositions rapidly from the initial vacuum state, a rather fascinating perspective.

References

1. S. Haroche and J.-M. Raimond, *Exploring the quantum.* (Oxford University Press, Oxford, 2006).
2. M. A. Nielsen and I. L. Chuang, *Quantum Computation and Quantum Information.* (Cambridge University Press, Cambridge, 2000).
3. J.-M. Raimond. Exploring the quantum world with photons trapped in cavities and Rydberg atoms. In eds. M. Devoret, B. Huard, and R. Schoelkopf, *Quantum machines, Les Houches Schools proceedings.* Oxford University Press, (In Press).
4. E. M. Purcell, Spontaneous emission probabilities at radio frequencies, *Phys. Rev.* **69**, 681, (1946).
5. P. Goy, J.-M. Raimond, M. Gross, and S. Haroche, Observation of cavity-enhanced single atom spontaneous emission, *Phys. Rev. Lett.* **50**, 1903, (1983).
6. G. Gabrielse and H. Dehmet, Observation of inhibited spontaneous emission, *Phys. Rev. Lett.* **55**, 67, (1985).
7. R. G. Hulet, E. S. Hilfer, and D. Kleppner, Inhibited spontaneous emission by a Rydberg atom, *Phys. Rev. Lett.* **55**, 2137, (1985).
8. D. Meschede, H. Walther, and G. Müller, One-atom maser, *Phys. Rev. Lett.* **54**, 551, (1985).
9. R. J. Thompson, G. Rempe, and H. J. Kimble, Observation of normal-mode splitting for an atom in an optical cavity, *Phys. Rev. Lett.* **68**, 1132, (1992).
10. J. P. Reithmaier, G. Sek, A. Löffler, G. Hofmann, S. Kuhn, S. Reitzenstein, L. V. Keldysh, V. D. Kulakovskii, T. L. Reinecke, and A. Forchel, Strong coupling

in a single quantum dot-semiconductor microcavity system, *Nature (London)*. **432**, 197, (2004).

11. A. Badolato, K. Hennessy, M. Atatüre, J. Dreiser, E. Hu, P. M. Petroff, and A. Imamoglu, Determinstic coupling of single quantum dots to single nanocavity modes, *Science*. **308**, 1158, (2005).

12. E. Schrödinger, Die gegenwärtige Situation in der Quantenmechanik, *Naturwissenschaften*. **23**, 807, 823, 844, (1935).

13. W. H. Zurek, Decoherence, einselection, and the quantum origins of the classical, *Rev. Mod. Phys.* **75**, 715, (2003).

14. J.-M. Raimond, M. Brune, and S. Haroche, Manipulating quantum entanglement with atoms and photons in a cavity, *Rev. Mod. Phys.* **73**, 565, (2001).

15. S. Kuhr, S. Gleyzes, C. Guerlin, J. Bernu, U. B. Hoff, S. Deléglise, S. Osnaghi, M. Brune, J.-M. Raimond, S. Haroche, E. Jacques, P. Bosland, and B. Visentin, Ultrahigh finesse fabry-pérot superconducting resonator, *Appl. Phys. Lett.* **90**, 164101, (2007).

16. E. T. Jaynes and F. W. Cummings, Comparison of quantum and semiclassical radiation theories with application to the beam maser, *Proc. IEEE.* **51**, 89, (1963).

17. A. Rauschenbeutel, G. Nogues, S. Osnaghi, P. Bertet, M. Brune, J.-M. Raimond, and S. Haroche, Step-by-step engineered multiparticle entanglement, *Science*. **288**, 2024, (2000).

18. G. Lindblad, On the generators of quantum dynamical semigroups, *Commun. Math. Phys.* **48**, 119, (1976).

19. V. B. Braginsky and Y. I. Vorontosov, Quantum-mechanical limitations in macroscopic experiments and modern experimental technique, *Usp. Fiz. Nauk.* **114**, 41, (1974). [Sov. Phys. Usp., **17**, 644 (1975)].

20. P. Grangier, J. A. Levenson, and J.-P. Poizat, Quantum non-demolition measurements in optics, *Nature (London)*. **396**, 537, (1998).

21. S. Gleyzes, S. Kuhr, C. Guerlin, J. Bernu, S. Deléglise, U. B. Hoff, M. Brune, J.-M. Raimond, and S. Haroche, Quantum jumps of light recording the birth and death of a photon in a cavity, *Nature (London)*. **446**, 297, (2007).

22. C. Guerlin, J. Bernu, S. Deléglise, C. Sayrin, S. Gleyzes, S. Kuhr, M. Brune, J.-M. Raimond, and S. Haroche, Progessive field state collapse and quantum non-demolition photon counting, *Nature (London)*. **448**, 889, (2007).

23. M. Bauer and D. Bernard, Convergence of repeated quantum non-demolition measurements and wave function collapse, *Phys. Rev. A.* **84**, 044103, (2011).

24. M. Brune, J. Bernu, C. Guerlin, S. Deléglise, C. Sayrin, S. Gleyzes, S. Kuhr, I. Dotsenko, J.-M. Raimond, and S. Haroche, Process tomography of field damping and measurement of fock states lifetimes by quantum non demolition photon counting in a cavity, *Phys. Rev. Lett.* **101**, 240402, (2008).

25. H. Wang, M. Hofheinz, M. Ansmann, R. C. Bialczak, E. Lucero, M. Neeley, A. D. O'Connell, D. Sank, J. Wenner, A. N. Cleland, and J. M. Martinis, Measurement of the decay of fock states in a superconducting quantum circuit, *Phys. Rev. Lett.* **101**(24), 240401, (2008).

26. B. Misra and E. Sudarshan, Zenos paradox in quantum theory, *J. Math. Phys.* **18**, 756, (1977).

27. W. Itano, D. Heinzen, J. Bollinger, and D. Wineland, Quantum zeno effect, *Phys. Rev. A.* **41**, 2295, (1990).

28. J. Bernu, Deléglise, C. Sayrin, S. Kuhr, I. Dotsenko, M. Brune, J.-M. Raimond, and S. Haroche, Freezing coherent field growth in a cavity by the quantum zeno effect, *Phys. Rev. Lett.* **101**, 180402, (2008).

29. M. Brune, E. Hagley, J. Dreyer, X. Maître, A. Maali, C. Wunderlich, J.-M. Raimond, and S. Haroche, Observing the progressive decoherence of the "meter" in a quantum measurement, *Phys. Rev. Lett.* **77**(24), 4887–4890, (1996).

30. S. Deléglise, I. Dotsenko, C. Sayrin, J. Bernu, M. Brune, J.-M. Raimond, and S. Haroche, Reconstruction of non-classical cavity field states with snapshots of their decoherence, *Nature (London).* **455**, 510, (2008).

31. V. Buzek and G. Drobny, Quantum tomography via the maxent principle, *J. Mod. Opt.* **47**, 2823, (2000).

32. D. Leibfried, D. M. Meekhof, B. E. King, C. Monroe, W. M. Itano, and D. J. Wineland, Experimental determination of the motional quantum state of a trapped atom, *Phys. Rev. Lett.* **77**, 4281, (1996).

33. M. Hofheinz, H. Wang, M. Ansmann, R. Bialczak, E. Lucero, M. Neely, A. O'Connel, D. Sank, J. Wenner, J. Martinis, and A. Cleland, Synthesizing arbitrary quantum states in a superconducting resonator, *Nature (London).* **459**, 546–549, (2009).

34. S. Adler and A. Bassi, Is quantum theory exact, *Science.* **325**, 275–276, (2009).

35. H. Wiseman, Quantum theory of continuous feedback, *Phys. Rev. A.* **49**, 2133–2150, (1994).

36. A. Doherty, S. Habib, K. Jacobs, H. Mabuchi, and S. Tan, Quantum feedback control and classical control theory, *Phys. Rev. A.* **62**, 012105, (2000).

37. C. Sayrin, I. Dotsenko, X. Zhou, B. Peaudecerf, T. Rybarczyk, S. Gleyzes, P. Rouchon, M. Mirrahimi, H. Amini, M. Brune, J.-M. Raimond, and S. Haroche, Real-time quantum feedback prepares and stabilizes photon number states, *Nature (London).* **477**, 73, (2011).

38. I. Dotsenko, M. Mirrahimi, M. Brune, S. Haroche, J.-M. Raimond, and P. Rouchon, Quantum feedback by discrete quantum nondemolition measurements: towards on-demand generation of photon-number states, *Phys. Rev. A.* **80**, 013805, (2009).

39. J.-M. Geremia, Deterministic and nondestructively verifiable preparation of photon number states, *Phys. Rev. Lett.* **97**, 073601, (2006).

40. X. Zhou, I. Dotsenko, B. Peaudecerf, T. Rybarczyk, C. Sayrin, S. Gleyzes, J.-M. Raimond, M. Brune, and S. Haroche, Field locked to a fock state by quantum feedback with single photon corrections, *Phys. Rev. Lett.* **108**, 243602 (Jun, 2012).

41. J. Poyatos, J. Cirac, and P. Zoller, *Phys. Rev. Lett.* **77**, 4728, (1996).

42. A. R. R. Carvalho, P. Milman, R. L. de Matos Filho, and L. Davidovich, Decoherence, pointer engineering, and quantum state protection, *Phys. Rev. Lett.* **86**(22), 4988–4991, (2001).

43. A. Sarlette, J.-M. Raimond, M. Brune, and P. Rouchon, Stabilization of non-classical states of the radiation field in a cavity by reservoir engineering, *Phys. Rev. Lett.* **107**, 010402, (2011).

44. A. Sarlette, Z. Leghtas, M. Brune, J.-M. Raimond, and P. Rouchon, Stabilization of nonclassical states of one and two-mode radiation fields by reservoir engineering, *Phys. Rev. A.* **86**, 012114, (2012).

45. J. J. Slosser, P. Meystre, and S. L. Braunstein, Harmonic oscillator driven by a quantum current, *Phys. Rev. Lett.* **63**(9), 934–937, (1989).

46. S. Tan, A computational toolbox for quantum and atomic optics, *J. Opt. B.* **1**, 424, (1999).

47. M. D. Lukin, M. Fleischhauer, R. Cote, L. M. Duan, D. Jaksch, J. I. Cirac, and P. Zoller, Dipole blockade and quantum information processing in mesoscopic atomic ensembles, *Phys. Rev. Lett.* **87**(3), 037901, (2001).

48. T. Nirrengarten, A. Qarry, C. Roux, A. Emmert, G. Nogues, M. Brune, J.-M. Raimond, and S. Haroche, realization of a superconducting atom chip, *Phys. Rev. Lett.* **97**, 200405, (2006).

49. C. Roux, A. Emmert, A. Lupascu, T. Nirrengarten, G. Nogues, M. Brune, J.-M. Raimond, and S. Haroche, Bose einstein condensation on a superconducting atom chip, *EPL.* **81**, 56004, (2008).

50. P. Hyafil, J. Mozley, A. Perrin, J. Tailleur, G. Nogues, M. Brune, J.-M. Raimond, and S. Haroche, Coherence-preserving trap architecture for long-term control of giant Rydberg atoms, *Phys. Rev. Lett.* **93**, 103001, (2004).

51. B. C. Sanders, S. D. Bartlett, B. Tregenna, and P. L. Knight, Quantum quincunx in cavity quantum electrodynamics, *Phys. Rev. A.* **67**(4), 042305, (2003).

52. J.-M. Raimond, C. Sayrin, S. Gleyzes, I. Dotsenko, M. Brune, S. Haroche, P. Facchi, and S. Pascazio, Phase space tweezers for tailoring cavity fields by quantum Zeno dynamics, *Phys. Rev. Lett.* **105**, 213601, (2010).

53. J.-M. Raimond, P. Facchi, B. Peaudecerf, S. Pascazio, C. Sayrin, I. Dotsenko, S. Gleyzes, M. Brune, and S. Haroche, Quantum Zeno dynamics fo a field in a cavity, *Phys. Rev. A.* **86**, 032120, (2012).

54. P. Facchi and S. Pascazio, Quantum zeno subspaces, *Phys. Rev. Lett.* **89**(8), 080401, (2002).

Chapter 2

Exciton-Polaritons in Bulk Semiconductors and in Confined Electron and Photon Systems

Lucio Claudio Andreani

Physics Department, University of Pavia
Via Bassi 6, I-27100 Pavia, Italy
lucio.andreani@unipv.it

In these lecture notes I shall review excitons and polaritons in bulk semiconductors and in systems with electron and photon confinement. Emphasis will be on a proper definition of exciton-photon coupling, on the effects of reduced electron and photon dimensionality, and on the conditions for the occurrence of the strong-coupling regime of radiation-matter interaction. In addition to bulk semiconductors, I shall treat exciton-photon interaction in two-dimensional systems like quantum well excitons in planar microcavities and in waveguide-embedded photonic crystals. I shall also consider quantum dots in three-dimensional cavities, comparing the strong-coupling regime in zero dimensions with exciton-polariton physics in higher dimensions.

1. Introduction

Exciton-polaritons are the mixed modes arising from interaction of excitons with the retarded electromagnetic field. Excitons are the lowest-lying elementary excitations of the electronic systems in a semiconductor, and their coupling with photons leads to elementary excitations that are half light, half matter — the polaritons. The semiclassical theory of polaritons was first developed by Huang[1] and Born and Huang[2] (actually in the context of *phonon*-polaritons, i.e., of long-wavelength lattice vibrations in polar crystals). The quantum-mechanical theory of *exciton*-polaritons was formulated by Fano,[3] Hopfield[4] and Agranovich.[5] The concept of polaritons turned out to be a very profound and fruitful one in order to interpret optical experiments in direct-gap semiconductors in the region just below the fundamental gap. While excitons and polaritons in bulk semiconductors are essentially low-temperature phenomena, the development of nanostructure physics led to exciting developments related to a reduction of dimensionality of exciton and photon states. Quasi-two dimensional (2D) excitons arising from electron and hole states confined in quantum wells (QWs) are more stable than in the bulk, due to the increased exciton binding energy and oscillator strength. While quasi-2D QW excitons coupled to 3D photons are intrinsically radiative and do not give rise to stationary polariton states, they lead to robust exciton-polariton excitations when photon states are also confined in planar microcavities (MCs). Cavity polaritons in the solid state were first demonstrated by Weisbuch *et al.*[6] and they have been the playground for a number of investigations and amazing discoveries, *in primis* stimulated Bosonic scattering and amplification[7] as well as Bose-Einstein condensation.[8] These phenomena are only possible when exciton and photon states have a quasi-2D character, as the polariton dispersion has an energy minimum in the optical region. A similar situation occurs when QW excitons are embedded in two-dimensional, waveguide photonic crystals, as it has been recently demonstrated.[9] Thus, the *dimensionality of exciton and photon states* plays a key role in determining the nature of exciton-photon interaction and for the resulting linear and nonlinear phenomena.

In these lecture notes I shall review the development of exciton-polariton concept in bulk semiconductors and in systems with 2D electron and photon states. The following systems are treated: bulk semiconductors, QW excitons and planar microcavities, photonic crystals. Then I shall mention the issue of strong exciton-photon coupling in zero dimensions (0D), i.e., when excitons in quantum dots interact with photon confined in three-dimensional microcavities. While the excitations of such system are sometimes referred to as "polaritons", this is actually a misnomer: historical developments show clearly that the concept of polariton is closely related to wavevector conservation, i.e., to the existence of extended excitations that propagate across the crystal. Also, the quantum statistical properties of strongly coupled systems in 0D are very different from those of 3D and 2D polaritons, as the low-lying excitonic excitations in 0D have a Fermionic (rather than a Bosonic) character and the strong coupling regime in 0D is described by a Jaynes-Cummings model with strong intrinsic nonlinearities even at low excitation level. Nevertheless, it is interesting to compare the exciton-photon coupling in 0D with the analogous coupling in 2D and 3D. Such a comparison is done in Section 4, where the conditions for the occurrence of the strong coupling regime as a function of dimensionality are also discussed and summarized. Section 5 contains concluding remarks.

As polariton physics has been studied for more than 60 years, the present lecture notes have to be very concise in each Section and highly selective in the reference list. Their aim is to provide the main concepts together with an entry point to previous reviews and specific literature, while giving a flavor of historical developments and benchmark achievements in the field, from a personal perspective.

2.   Excitons and polaritons in bulk semiconductors

Excitons are excited states of the electronic systems of an insulating crystal that go beyond the one-electron approximation, as they represent bound states of electron-hole pairs. A basic prerequisite before getting involved into exciton physics is the study of the optical properties of solids within the one-electron approximation, i.e., the theory of the frequency-dependent dielectric function and of interband transitions. The optical properties of semiconductors are covered in a number of textbooks, see e.g. Refs. 10, 11. The necessary background material for the present lectures is also summarized in previous lecture notes, see Ref. 12.

Note on electromagnetic units — Most of the old (and less old) literature on polaritons employs cgs or Gaussian units. I shall use Gaussian units in these lecture notes, but some of the formulas will be written using the factor (ϵ_0 is the vacuum permittivity)

$$\kappa_{\text{SI}} = \frac{1}{4\pi\epsilon_0}. \tag{1}$$

This factor should be kept (dropped) to read the expression in SI (Gaussian) units.

We shall be especially concerned with the *oscillator strength* of an optical transition, which is a dimensionless quantity with a precise classical and quantum-mechanical definition. Within the classical Lorentz oscillator model, the oscillator strength f_j is the number of oscillators at frequency ω_j with damping γ_j and the dielectric function is expressed as

$$\epsilon(\omega) = 1 + \kappa_{\text{SI}} \frac{4\pi e^2}{m_0 V} \sum_j \frac{f_j}{\omega_j^2 - \omega^2 - i\gamma_j\omega}, \tag{2}$$

where e is the proton charge, m_0 is the free electron mass and V is the crystal volume. The oscillator strengths obey the sum rule $\sum_j f_j = N$, where N is the total number of electrons in the crystals: this is equivalent to the plasma sum rule following from Kramers-Kronig relations. Quantum mechanically, the oscillator strength of a transition from an initial state i to a final state f for light polarization $\hat{e}$ is defined by

$$f_{i \to f}^{\hat{e}} = \frac{2}{m_0 \hbar \omega} |\langle \Psi_f | \hat{e} \cdot \mathbf{p} | \Psi_i \rangle|^2, \tag{3}$$

where $\hbar\omega = E_f - E_i$ is the transition energy and $\mathbf{p} = -i\nabla$ is the electron momentum operator (for many-body states, it should be understood as $\sum_j \mathbf{p}_j$, where the sum extends over all electrons j). The sum rule becomes $\sum_f f_{i \to f}^{\hat{e}} = N$. Within perturbation theory, the oscillator strength integrated over a frequency interval $[\omega_1, \omega_2]$ is related to the integral of the absorption coefficient $\alpha(\omega)$ by

$$\int_{\omega_1}^{\omega_2} \alpha(\omega)d\omega = \kappa_{\text{SI}} \frac{2\pi^2 e^2}{n m_0 c} \frac{1}{V} f_{\text{tot}}, \tag{4}$$

where n is the refractive index. Thus, the oscillator strength per unit volume for any transition — e.g., an excitonic transition — can be measured from the integral of the absorption coefficient over the transition line.

2.1. Wannier-Mott excitons, oscillator strength, exchange interaction

Excitons in semiconductors are usually Wannier-Mott type or *shallow*, i.e., their binding energy is small and their radius is much larger than the interatomic spacing: as such, they can be described by a two-particle effective mass equation. We shall outline the derivation of the effective mass equation,[13–16] as it is the starting point for generalizing the description to systems of reduced dimensionality.

We start from the N-electron Hamiltonian of the crystal. In the Hartree-Fock approximation, the crystal ground state is given by a Slater determinant in which the electrons occupy all Bloch states in the valence band:

$$\Psi_0(\mathbf{r}_1, \mathbf{r}_2, \ldots, \mathbf{r}_N) = \det\{\psi_{v\mathbf{k}_1}(\mathbf{r}_1), \psi_{v\mathbf{k}_2}(\mathbf{r}_2), \ldots, \psi_{v\mathbf{k}_N}(\mathbf{r}_N)\}, \tag{5}$$

where the determinantal operator gives the required antisymmetry properties to the many-body Fermionic wavefunction. Spin coordinates σ are understood in the electron coordinate $\mathbf{r}$. Excited electron-hole pair states are given by

$$\Psi_{c\mathbf{k}_c, v\mathbf{k}_v}(\mathbf{r}_1, \mathbf{r}_2, \ldots, \mathbf{r}_N) = \mathcal{A}\{\psi_{v\mathbf{k}_1}(\mathbf{r}_1), \ldots, \widehat{\psi_{v\mathbf{k}_v}}(\mathbf{r}_i), \psi_{c\mathbf{k}_c}(\mathbf{r}_i), \ldots, \psi_{v\mathbf{k}_N}(\mathbf{r}_N)\}, \tag{6}$$

where the valence Bloch function $\psi_{v\mathbf{k}_v}$ has been replaced by the conduction Bloch function $\psi_{c\mathbf{k}_c}$. In order to find exciton levels we restrict ourselves to the subspace of states of the form (6). The exciton wavefunction is represented as

$$\Psi_{\text{exc}} = \sum_{\mathbf{k}_c, \mathbf{k}_v} A(\mathbf{k}_c, \mathbf{k}_v) \Psi_{c\mathbf{k}_c, v\mathbf{k}_v}, \tag{7}$$

where $A(\mathbf{k}_c, \mathbf{k}_v)$ is a k-space envelope function. The exciton wavevector $\mathbf{k}_{\text{ex}} = \mathbf{k}_c - \mathbf{k}_v$ is a conserved quantum number. By taking the matrix elements of the N-electron Hamiltonian between basis states (6) the following equation is obtained:

$$(E_c(\mathbf{k}_c) - E_v(\mathbf{k}_v) - E)A(\mathbf{k}_c, \mathbf{k}_v) + \sum_{\mathbf{k}'_c, \mathbf{k}'_v} (J + V_C)A(\mathbf{k}'_c, \mathbf{k}'_v) = 0 \tag{8}$$

in terms of the dispersion $E_c(\mathbf{k}_c)$, $E_v(\mathbf{k}_v)$ of conduction and valence bands. The *direct* matrix element J between one-electron states leads to an exchange interaction between electron and hole, while the electron-electron *exchange* matrix element V_C is responsible for electron-hole Coulomb attraction. Equation (8) can be considerably simplified for shallow excitons, which can be treated more conveniently in terms of a real-space envelope function defined as

$$F(\mathbf{r}_e, \mathbf{r}_h) = \frac{1}{\sqrt{V}} \sum_{\mathbf{k}_c, \mathbf{k}_v} A(\mathbf{k}_c, \mathbf{k}_v) e^{i(\mathbf{k}_c \cdot \mathbf{r}_e - \mathbf{k}_v \cdot \mathbf{r}_h)}. \tag{9}$$

The corresponding Schrödinger equation takes the form

$$\left[E_c(-i\nabla_e) - E_v(-i\nabla_h) - \frac{e^2}{\epsilon_b |\mathbf{r}_e - \mathbf{r}_h|} + J_{cv}\delta(\mathbf{r}_e - \mathbf{r}_h) - E \right] F(\mathbf{r}_e, \mathbf{r}_h) = 0, \tag{10}$$

where the electron-hole Coulomb attraction is screened with the appropriate background dielectric constant, and J_{cv} is the electron-hole exchange interaction.

An exciton associated with a pair of parabolic bands has energy levels of the form

$$E_n(\mathbf{k}_{ex}) = E_g - \frac{R^*}{n^2} + \frac{\hbar^2 k_{ex}^2}{2M}, \quad n = 1, 2, \ldots \tag{11}$$

where

$$R^* = \frac{\mu e^4}{2\epsilon_b^2 \hbar^2} = \frac{\hbar^2}{2\mu(a_B)^2} \tag{12}$$

is the effective Rydberg, $\mu = (1/m_e^* + 1/m_h^*)^{-1}$ is the reduced effective mass and $M = m_e^* + m_h^*$ is the total effective mass. The exciton radius is given by $a_B = \hbar^2 \epsilon_b/(\mu e^2)$ and is usually much larger than the hydrogenic Bohr radius, since semiconductors have large dielectric constants and small effective masses. This result justifies *a posteriori* the use of approximations leading to Eq. (10).

The oscillator strength of the transition from the crystal ground state to the exciton state — briefly referred to as the *exciton oscillator strength* — is evaluated in the effective mass approximation as

$$f_{\hat{e}} = g_s \frac{2|\hat{e} \cdot \mathbf{p}_{cv}|^2}{m_0 \hbar \omega} \left| \int F(\mathbf{r}_e, \mathbf{r}_h) \delta(\mathbf{r}_e - \mathbf{r}_h) d\mathbf{r}_e \, d\mathbf{r}_h \right|^2$$

$$= g_s \frac{2|\hat{e} \cdot \mathbf{p}_{cv}|^2}{m_0 \hbar \omega} \left| \int F(\mathbf{r} = 0, \mathbf{R}) d\mathbf{R} \right|^2, \tag{13}$$

where $\mathbf{p}_{cv}$ is the interband momentum matrix element and in the second line the envelope function is expressed in terms of the relative and center-of-mass variables $(\mathbf{r}, \mathbf{R})$. The factor g_s is twice the singlet component of the exciton state and it will be referred to as the *spin-orbit factor*. Apart from the spin-orbit factor which is a number of order unity, the oscillator strength depends on the envelope function at zero electron-hole separation, times a prefactor

$$f_0 \equiv \frac{2|\hat{e} \cdot \mathbf{p}_{cv}|^2}{m_0 \hbar \omega} \tag{14}$$

which represents the interband oscillator strength (without spin). Using the Kane model, the interband oscillator strength can be expressed as $f_0 \simeq m_0/m_e^*$, where m_e^* is the effective mass at the conduction band minimum. In GaAs we have $m_e^* \simeq 0.067 m_0$ and $f_0 \simeq 15$.

Expression (13) is generally valid for Wannier-Mott excitons (free or bound) and it is the starting point for discussing the behavior of the oscillator strength in systems of low dimensionality like quantum wells, wires and dots. For free excitons in the bulk, conservation of the exciton wavevector $\mathbf{k}_{ex}$ implies

$F(\mathbf{r}, \mathbf{R}) = V^{-1/2} e^{i\mathbf{k}_{ex} \cdot \mathbf{R}} F(\mathbf{r})$ and we obtain

$$f_{\hat{e}} = g_s \frac{2|\hat{e} \cdot \mathbf{p}_{cv}|^2}{m_0 \hbar \omega} V |F_n(0)|^2 \delta_{\mathbf{k}_{ex}, 0}, \tag{15}$$

For hydrogenic s-states (the only ones with $F_n(0) \neq 0$), we have

$$|F_n(0)|^2 = \frac{1}{\pi n^3 (a_B)^3}. \tag{16}$$

Equation (15) shows that the dimensionless oscillator strength is proportional to the crystal volume, as the exciton center-of-mass wavefunction extends over the whole crystal. The meaningful quantity is the oscillator strength per unit volume, which for an allowed transition decreases as n^{-3}. The excitonic oscillator strength per unit volume can be related to the absorption coefficient $\alpha(\omega)$ integrated over the absorpion peak, as given by Eq. (4). However, this holds true only when polariton effects can be neglected, as we shall see later in this lecture.

Exchange interaction. The electron-hole exchange interaction is usually small and can be treated as a first-order perturbation to the solutions of the effective-mass equation. It is important, however, since it splits the exciton states with different symmetries with respect to the crystal point group.

The evaluation of the electron-hole exchange term J in Eq. (8) can be done in a Wannier-function formulation[12,15] or in k-space.[17,18] In the latter scheme, the products of the periodic parts of the Bloch functions are expanded in plane waves with reciprocal lattice vectors, and the exchange matrix element is expressed as

$$J(\mathbf{k}, \mathbf{k}') = \frac{1}{V} \sum_{\mathbf{G}} D_{c v \mathbf{k}_c \mathbf{k}_v}(\mathbf{G}) D_{v c \mathbf{k}'_v \mathbf{k}'_c}(-\mathbf{G}) \frac{4\pi e^2}{|\mathbf{G} + \mathbf{k}_{ex}|^2} \equiv J^{ana} + J^{non-ana}, \tag{17}$$

where

$$D_{c v \mathbf{k}_c \mathbf{k}_v}(\mathbf{G}) = \int u^*_{c \mathbf{k}_c}(\mathbf{r}) u_{v \mathbf{k}_v}(\mathbf{r}) e^{-i\mathbf{G} \cdot \mathbf{r}} d\mathbf{r} \tag{18}$$

and $u_{c\mathbf{k}}, u_{v\mathbf{k}}$ are the periodic parts of the Bloch functions. In expression (17), the sum of the $\mathbf{G} \neq 0$ terms gives a contribution J^{ana} which has a well-defined limit as $\mathbf{k}_{ex} \to 0$, and therefore is called the *analytic* part. The $\mathbf{G} = 0$ term, instead, has different limits as $\mathbf{k}_{ex} \to 0$ depending on the direction of $\mathbf{k}_{ex}$ with respect to the exciton dipole moment: it is therefore called the *nonanalytic* part. The nonanalytic part can be evaluated by expanding the periodic parts of the Bloch functions in $\mathbf{k} \cdot \mathbf{p}$ theory around the band extremum (taken at $\mathbf{k} = 0$), with the result

$$J^{non-ana}(\mathbf{k}_{ex}) = \frac{4\pi e^2}{V k_{ex}^2} \langle u_c | \mathbf{k}_{ex} \cdot \mathbf{r} | u_v \rangle \langle u_v | \mathbf{k}_{ex} \cdot \mathbf{r} | u_c \rangle. \tag{19}$$

For $\mathbf{k}_{ex} \to 0$, the nonanalytic part has a limiting value $4\pi e^2 |\langle u_c | \mathbf{r} | u_v \rangle|^2 / V$ for an exciton wavevector parallel to the dipole moment (longitudinal exciton), and a vanishing limiting value for an exciton wavevector perpendicular to the dipole moment (transverse exciton). It is possible to show that inclusion of the analytic part of the exchange interaction corresponds to taking into account local-field effects, while inclusion of the nonanalytic part corresponds to taking into account the depolarization field in the dielectric response.[19]

In the effective mass approximation, both terms J^{ana}, $J^{non-ana}$ are independent of $(\mathbf{k}, \mathbf{k}')$. The corresponding term in the effective-mass equation (10) is $g_s V(J^{ana} + J^{non-ana})\delta(\mathbf{r})\delta(\mathbf{r}')$, where g_s is the same spin-orbit factor which appears in the oscillator strength. When the Schrödinger equation for the exciton is projected onto a pair of bands, the effect of exchange coupling to the other bands results in a screening of the nonanalytic part of the exchange interaction with the high-frequency dielectric constant ϵ_∞.[18,20] The splitting between longitudinal and transverse excitons is then calculated to be

$$\Delta E_{LT} = \frac{4\pi|\langle u_c|e\mathbf{r}|u_c\rangle|^2}{\epsilon_\infty} g_s|F(0)|^2. \tag{20}$$

The LT splitting is proportional to the oscillator strength per unit volume according to the relation

$$\Delta E_{LT} = \kappa_{SI}\frac{2\pi}{\epsilon_\infty}\frac{\hbar e^2}{m_0\omega_0}\frac{f}{V}. \tag{21}$$

A schematic representation of the effect of the exchange interaction on the ground-state exciton in zincblende semiconductors is shown in Fig. 1. The analytic part of the exchange interaction separates the optically active state Γ_5, which is a linear combination of singlet and triplet excitons, from the forbidden states which are pure triplets. In addition, the nonanalytic part of the exchange interaction splits the Γ_5 states into a longitudinal and a transverse exciton, with degeneracies one and two respectively.

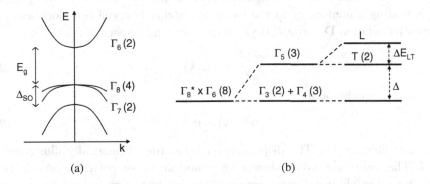

(a) (b)

Fig. 1. (a) Schematic representation of the uppermost valence bands and the lowest conduction band in direct-gap zincblende semiconductors. (b) Schematic representation of the effect of the exchange interaction on the ground-state exciton (Δ is the analytic part of the exchange splitting). Numbers in brackets indicate the degeneracy of the state.

2.2.   Exciton-polaritons: semiclassical theory

The splitting between longitudinal and transverse excitons cannot hold for the true eigenstates in the limit $\mathbf{k}_{ex} \rightarrow 0$, as dipole-active states must be threefold degenerate at $\mathbf{k}_{ex} = 0$. The solution of the dilemma lies in the effects of retardation, which cannot be neglected for $k_{ex} < k_0 = n\omega_0/c$. For wavevectors of the order of the wavevector of light in the crystal, interaction between the excitons and the transverse electromagnetic field modifies the dispersion relation through the formation of stationary states called *excitonic polaritons*.

The classical theory of polaritons was first developed by Kun Huang [1,2] in the context of lattice vibrations in polar crystals, leading to quasiparticles which are now called *phonon-polaritons*.[21] Here we present a formulation suitable for *exciton-polaritons*, which takes into account spatial dispersion following from exciton center-of-mass motion. Exciton-polaritons arise from the solution of Maxwell equations, with a constitutive relation between excitonic polarization and electric field described by the equation

$$\frac{1}{\omega_0^2}\ddot{\mathbf{P}} - \frac{\hbar}{M\omega_0}\nabla^2\mathbf{P} + \mathbf{P} = \beta\mathbf{E}. \tag{22}$$

The polarizability β is related to the oscillator strength per unit volume by

$$\beta = \frac{e^2}{m_0\omega_0^2}\frac{f}{V}. \tag{23}$$

Taking into account the frequency-independent contribution ϵ_∞ arising from all other electronic resonances of the crystals, the dielectric function is given by

$$\epsilon(\omega, \mathbf{k}) = \epsilon_\infty + \kappa_{\text{SI}}\frac{4\pi\beta\omega_0^2}{\omega_k^2 - \omega^2}, \tag{24}$$

where $\hbar\omega_k = \hbar\omega_0 + \hbar^2 k^2/(2M)$ is the exciton energy. Notice that the dielectric function depends on wavevector in addition to frequency, i.e., it is *spatially dispersive* indicating a nonlocality in the medium. Solving Maxwell equations with the constitutive relation $\mathbf{D} = \epsilon(\omega, k)\mathbf{E}$ yields the following dispersion relations:

$$\frac{c^2 k^2}{\omega^2} = \epsilon(\omega, k) \tag{25}$$

for transverse modes, and

$$\epsilon(\omega, k) = 0 \tag{26}$$

for longitudinal modes. The dispersion relations are schematically illustrated in Fig. 2. The transverse modes show a lower and an upper polariton branch, both of which are twofold degenerate because of the two transverse polarizations. Lower and upper polaritons anticross close to the transverse frequency ω_0. The longitudinal mode is a purely electrostatic solution and it is not modified by retardation. At $\mathbf{k}_{ex} \rightarrow 0$, the two transverse upper polariton modes are degenerate with the

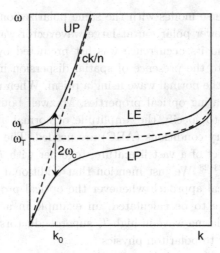

Fig. 2.   Schematic representation of the dispersion of the upper and lower polaritons and of the longitudinal exciton (solid lines). The dispersion of the independent photon and transverse exciton modes is also shown (dashed lines).

longitudinal exciton, as required by cubic symmetry. The electrostatic shift of the longitudinal mode is given by

$$\hbar\omega_{LT} = \hbar\omega_0 \left(1 + \kappa_{SI} \frac{4\pi\beta}{\epsilon_\infty}\right)^{1/2} - \hbar\omega_0 \simeq \kappa_{SI} \frac{2\pi\hbar e^2}{\epsilon_\infty m_0 \omega_0} \frac{f}{V} \qquad (27)$$

and it is seen to coincide with the LT splitting (21). In other words, the electrostatic shift of the longitudinal exciton calculated from the nonanalytic part of the exchange interaction coincides with that calculated from the solution of Maxwell equations for the longitudinal mode. Thus the dielectric function (24) of the exciton must *not* contain the exchange interaction, otherwise the effect would be counted twice.

It should be emphasized that the LT splitting (21) is the *electrostatic splitting* between L and T excitons at large wavevectors ($k \gg k_0$, when retardation effects can be neglected) but it should not be confused with the *polariton splitting* at the crossing point. The latter is written as $2\hbar\omega_c$, where the polariton coupling ω_c is calculated as

$$\omega_c = \sqrt{\frac{\omega_0\omega_{LT}}{2}} = \sqrt{\kappa_{SI} \frac{\pi e^2}{\epsilon_\infty m_0} \frac{f}{V}}. \qquad (28)$$

The polariton coupling is much larger than the LT splitting: in GaAs, for example, $\hbar\omega_{LT} = 0.08\,\text{meV}$ while $\hbar\omega_c = 7.8\,\text{meV}$. Even if the polariton coupling is quite large (even larger than the exciton binding energy), polariton effects may not be easy to observe, as discussed later in this lecture.

Because of exciton center-of-mass motion with the total mass M, the polariton dispersion turns upwards at large wavevectors. For $\omega > \omega_L$, this results in the presence of two transverse modes with the same polarization but with a different refractive index. The lower polariton at large wavevector was traditionally called the *additional wave* and its occurrence was fist predicted by Pekar.[22] The additional wave is related to the presence of spatial dispersion in the medium and it can be separated from the normal wave using a prism. When using the bulk dielectric function for calculating optical properties, Maxwell boundary conditions are not sufficient in order to specify the amplitude of normal and additional waves, and additional boundary conditions (ABC) are needed. The so-called ABC problem has been the subject of a vast literature, together with ABC-free theories, as reviewed elsewhere.[12, 23–28] We just mention that additional boundary conditions are introduced in a local approach whenever the optical properties of a spatially dispersive medium have to be calculated: an example in a rather different context is the Josephson plasma wave in high-T_c superconductors,[29, 30] which bears an interesting resemblance to polariton physics.

The dispersion of exciton-polaritons has been experimentally verified in a number of optical experiments on good-quality semiconductor crystals at low temperature. We refer to previous reviews[12, 25, 31–33] as entry points to specific literature. Many experiments on polaritons have been performed on materials with a large exciton binding energy, like CuCl or II-VI semiconductors (CdS, CdSe, ZnSe and alike). It is worth mentioning that polariton effects have been especially hard to demonstrate in GaAs, because of the very small exciton binding energy (4 meV): two key papers in this respect are those of Sell *et al.*[34] where polariton effects were first measured in reflectance and photoluminescence, and by Ulbrich and Weisbuch[35] where resonant Brillouin scattering by exciton-polaritons was demonstrated.

2.3.   *Exciton-polaritons: quantum theory*

Quantum theory of polaritons. A quantum theory of polaritons consists of setting up a second-quantized Hamiltonian which describes excitons and photons with their mutual interaction. The approximation of keeping only quadratic terms yields a Hamiltonian which can be diagonalized exactly, and which (for an infinite crystal) describes stationary states. Terms beyond the quadratic approximation, as well as terms associated to the exciton-phonon interaction, describe the damping of polariton states. The exciton-photon Hamiltonian can be obtained either from a microscopic model for the exciton[4, 5] with the radiation-matter Hamiltonian $H = (\mathbf{p} + e\mathbf{A}/c)^2/2m_0$, where $\mathbf{A}$ is the vector potential, or from second quantization of Maxwell equations plus the equation of motion (22) for the excitonic polarization. The resulting Hamiltonian is

$$H = \sum_k \left[\hbar v k \left(a_k^\dagger a_k + \frac{1}{2} \right) + \hbar \omega_k \left(b_k^\dagger b_k + \frac{1}{2} \right) \right.$$

$$\left. + iC_k \left(a_k^\dagger + a_{-k} \right)\left(b_k - b_{-k}^\dagger \right) + D_k \left(a_k^\dagger + a_{-k} \right)\left(a_k + a_{-k}^\dagger \right) \right], \qquad (29)$$

where $k = (\mathbf{k}, \lambda)$ is a combined index which includes wavevector and polarization vector for the two transverse modes, $a_k^\dagger, a_k$ ($b_k^\dagger, b_k$) are Bose operators for the photon (exciton), $v = c/\sqrt{\epsilon_\infty}$ is the speed of light in the crystal, $\hbar\omega_k = \hbar\omega_0 + \frac{\hbar^2 k^2}{2m}$ is the exciton energy including spatial dispersion, and

$$C_k = \hbar\omega_0 \left(\frac{\pi\beta\omega_k}{vk\epsilon_\infty}\right)^{1/2}, \quad D_k = \hbar\omega_0 \frac{\pi\beta\omega_k}{vk\epsilon_\infty}. \tag{30}$$

The normal modes of the quadratic Hamiltonian (30) can be found by an operator transformation[3, 4] which is known in this context as the *Hopfield transformation*. An analogous transformation in the BCS theory of superconductivity is called the *Bogoljubov transformation*.[36] New operators $\alpha_{kl}, l = 1, 2$ are introduced, which are related to the exciton and photon operators by the linear relation

$$\begin{bmatrix} \alpha_{k1} \\ \alpha_{k2} \\ \alpha_{-k1}^\dagger \\ \alpha_{-k2}^\dagger \end{bmatrix} = \begin{bmatrix} W_1 & X_1 & Y_1 & Z_1 \\ W_2 & X_2 & Y_2 & Z_2 \\ Y_1^* & Z_1^* & W_1^* & X_1^* \\ Y_2^* & Z_2^* & W_2^* & X_2^* \end{bmatrix} \begin{bmatrix} a_k \\ b_k \\ a_{-k}^\dagger \\ b_{-k}^\dagger \end{bmatrix}. \tag{31}$$

The new operators are determined by the condition that they satisfy Bose commutation relations, and that the Hamiltonian becomes diagonal, i.e.,

$$[\alpha_{kl}, H] = \hbar\Omega_l(k)\alpha_{kl}. \tag{32}$$

This leads to the following secular equation for the eigenfrequencies Ω_{kl}:

$$\Omega_k^4 - \left(v^2 k^2 + \omega_k^2 + \kappa_{\mathrm{SI}} \frac{4\pi\beta}{\epsilon_\infty}\omega_0^2\right)\Omega_k^2 + v^2 k^2 \omega_k^2 = 0. \tag{33}$$

This can be seen to be identical with the classical dispersion relation (25). The coefficients of the transformation are given e.g. in Refs. 4, 33.

Notice that imposing the condition (32) for the operators, as was first done by Hopfield,[4] is an especially smart way for diagonalizing the Bosonic Hamiltonian as it leads to linear equations for the coefficients. The Hopfield procedure can be generalized to more complex situations with an arbitrary number of Bosonic operators interacting via a quadratic Hamiltonian, and it yields a linear eigenvalue problem (non-hermitian, but with real solutions for the eigenvalues) that can be solved by standard numerical methods. We shall see examples of this procedure in Subsection 3.5 (for photonic crystal polaritons) and 4.1 (for polaritons in pillar microcavities).

The above diagonalization procedure gives the polariton *operators*, but not the polariton *states*. In particular, the polariton vacuum defined by the condition

$$\alpha_{kl}|0'\rangle = 0 \tag{34}$$

does not coincide with the classical vacuum.[4] It is shown in Ref. 37 that the conditions (34) generate recursive equations, which allow an exact determination of the polariton vacuum in terms of photon and exciton states. The polariton states can be obtained by applying the polariton creation operators on the new vacuum. In general, the polariton states are not simply given by a linear combination of one exciton and one photon states, but contain instead additional components like three photons, two excitons plus one photon, etc. While these nonlinear components cannot be separated from the polariton states in normal experiments, it has been shown recently that the polariton vacuum can be dynamically tuned producing correlated photon pairs via quantum-electrodynamical (QED) phenomenona that are a manifestation of a dynamical Casimir effect.[38] Such phenomena are especially prominent for intersubband polaritons, for which the ratio of polariton coupling to resonant frequency can be quite substantial: this situation is known as the *ultra-strong coupling regime*.[38]

Two-level model, polariton coupling. Exciton-photon coupling is often analyzed within a two-oscillator model. This can be derived from the Hamiltonian (29) by (i) restricting oneself to a given wavevector k, since different wavevectors are decoupled in a bulk semiconductor due to translational invariance; (ii) neglecting the last term, which arises from the A^2 interaction; (iii) neglecting anti-resonant terms of the type $a^\dagger b^\dagger$ or ab, which is known as the *rotating-wave approximation*. With these simplifications, the Hamiltonian becomes a 2×2 matrix of the form

$$H = \begin{bmatrix} \hbar v k & \hbar g \\ \hbar g & \hbar \omega_k \end{bmatrix}. \tag{35}$$

The polariton coupling is given by the off-diagonal element $\hbar g$, which coincides with C_k in Eq. (30). The polariton coupling can also be viewed as the matrix element of the linear interaction Hamiltonian, $H_I = (e/m_0 c)\mathbf{A} \cdot \mathbf{p}$, using the second-quantized form for the vector potential:

$$\hbar g = \langle \text{exciton}|H_I|\text{photon}\rangle = \left(\kappa_{SI} \frac{\pi}{\epsilon_\infty} \frac{k_0}{k} \frac{e^2 \hbar^2}{m_0} \frac{f}{V} \right)^{1/2} = \left(\frac{k_0}{k} \right)^{1/2} \hbar \omega_c, \tag{36}$$

where ω_c has been defined in Eq. (28). *We see that the quantum-mechanical polariton coupling coincides with the classical one at the crossing point $k = k_0$*. The quantum-mechanical coupling diverges for $k \to 0$, which is a manifestation of the *infrared catastrophe* in quantum electrodynamics. In practice, the two-oscillator model can be used close to the crossing point, but not for $k \to 0$, when the full Hamiltonian (29) must be solved. Still, we shall see that the quantum-mechanical definition (36) of the polariton coupling is quite convenient for generalization to systems of lower dimensionality.

Criteria for relevance of polariton effects. The criteria for polariton effects being or not being relevant for specific experiments can be discussed by generalizing Eq. (28) to the presence of a damping term Γ:

$$\frac{c^2 k^2}{\omega^2} = \epsilon(\omega, k) = \epsilon_\infty + \kappa_{SI} \frac{4\pi \beta \omega_0^2}{\omega_k^2 - \omega^2 - i\Gamma \omega}. \tag{37}$$

As shown long ago by Tait,[39] Eq. (37) has two kinds of solutions, which are found either at fixed wavevector (*quasiparticle solutions*) or at fixed frequency (*forced harmonic solutions*).

In the first case, which applies to nonlinear optical experiments like two-photon absorption in which the polariton wavevector is fixed, the polariton splitting remains as long as

$$\Gamma < \omega_c. \tag{38}$$

This is a criterion of *temporal coherence*, as spatial coherence between exciton and photon is provided by the condition of fixed wavevector. In practice, the dephasing time must be much smaller than $(\omega_c)^{-1} \simeq 40$ fs for GaAs: under these conditions the polariton is *very stable*.

In the second case in which the frequency is fixed, like e.g. in absorption or reflectivity experiments, the polariton splitting occurs as long as

$$\Gamma < \Gamma_c = \left(\frac{8\omega_{LT}}{Mc^2}\right)^{1/2} \omega_0. \tag{39}$$

This is now a criterion of spatial coherence, as the exciton-photon coupling has to remain coherent while the polariton propagates in the medium in the presence of damping. This criterion is equivalent $l \gg \lambda$, where $l = v_g/\Gamma$ is the mean free path and λ is the light wavelength. The criterion of spatial coherence is much more difficult to satisfy: to give an example, $\hbar\Gamma_c \simeq 0.1$ meV for GaAs. When $\Gamma \gg \Gamma_c$ and spatial coherence applies, as is often the case (except in very pure samples at low temperature), light-matter coupling is in the so-called *coherence-volume regime*, in which the exciton has a coherence volume $\propto l^3$. Only when the exciton coherence length exceeds the light wavelength in all directions does the exciton-photon coupling lead to exciton-polariton states with a vacuum Rabi splitting. This remark explains why polariton effects in a material with very shallow excitons, like GaAs, are so difficult to observe.

Absorption and luminescence, bound excitons. In the absence of dissipative processes arising from electron-phonon interaction, exciton-polaritons do not give rise to electromagnetic absorption. A photon incident on the crystal produces a polariton (i.e., it propagates while oscillating between exciton and photon states: see Fig. 3), then it transforms back into a photon when going out of the crystal. Thus, polariton absorption must vanish at zero temperature, and it must decrease with the damping Γ of Eq. (37): this has been shown by experiments performed on many semiconductors, which indicate that the integrated absorption decreases with decreasing temperature. On the other hand, Eq. (4) implies that the integrated absorption of the exciton depends only on the oscillator strength per unit volume, as it has been shown explicitly when spatial dispersion is absent.[40] Indeed, the decrease of integrated absorption is due to spatial dispersion, which

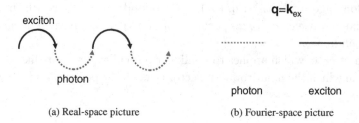

(a) Real-space picture   (b) Fourier-space picture

Fig. 3.   Schematic representation of the polariton picture. (a) Real space: polariton propagation occurs with a continuous and reversible exchange of energy between exciton and photon states. (b) Fourier space: because of momentum conservation, an exciton with a given wavevector interacts with one photon with the same wavevector.

invalidates the usual derivation of the plasma sum rule: see Ref. 25 for a review of this topic.

The polariton framework is also crucial in order to describe exciton luminescence at low temperature. When the polariton effect sets in, the exciton is not expected to decay radiatively: because of wavevector conservation, an exciton with a given wavevector $\mathbf{k}$ interacts with one photon with the same wavevector $\mathbf{q} = \mathbf{k}$ and the same polarization. The interaction between two discrete states does not give rise to radiative decay, but rather to oscillations, i.e., to a reversible exchange of energy between photon and exciton wavefunctions. This concept, which was emphasized by Hopfield in his seminal work,[4] goes under the name of *vacuum Rabi oscillations*. However, in a crystal of finite size, the polaritons will eventually reach the sample surface and exit the crystal by transforming into photons: this mechanism leads to *polariton luminescence*. The thermalization mechanisms related to polariton luminescence were first discussed by Toyozawa,[41] who introduced the phenomenon of *bottlenecking*: while propagating in the crystal, polaritons accumulate at the knee of the dispersion relation of the LP branch and at the minimum of the UP branch. Indeed, a two-peak luminescence was first observed in very pure GaAs at low temperature, and attributed to LP and UP luminescence.[34] Nevertheless, the mechanism for polariton luminescence is inefficient and low-temperature luminescence is dominated by excitons bound to neutral impurity states, as is clearly shown in the experimental spectra.[34, 42]

For excitons bound to impurities, the situation is quite different since crystal momentum is not conserved. Thus there is a density of states for radiative decay, given by the usual three-dimensional density of states of photons. The radiative lifetime of bound excitons is described by the atomic-like formula

$$\tau = \frac{3m_0 c^3}{2n e^2 \omega^2} \frac{1}{f}, \qquad (40)$$

and is inversely proportional to the oscillator strength. Excitons bound to impurities are characterized by the so-called *giant oscillator strength*.[43, 44] Assuming that the internal electron-hole wavefunction is not modified, and that the wavefunction of the center-of-mass motion is that of a bound state with a radius l,

$$F(\mathbf{R}, \mathbf{r}) = F_{CM}(\mathbf{R}) f_{rel}(\mathbf{r}) = \frac{e^{-R/l}}{\sqrt{2\pi l}R} \frac{e^{-r/a_B^*}}{\sqrt{\pi(a_B^*)^3}}, \qquad (41)$$

the oscillator strength is calculated to be (dropping the spin-orbit factor for simplicity)

$$f_\epsilon = \frac{2}{m_0 \hbar\omega} |\langle u_c | \epsilon \cdot \mathbf{p} | u_v \rangle|^2 \left| \int F(\mathbf{R}, 0) d\mathbf{R} \right|^2 = 8\pi l^3 \frac{f_{free}}{V} \simeq \frac{m_0}{m^*} \frac{8\pi l^3}{\pi(a_B^*)^3}. \qquad (42)$$

In CdS, the confinement length of excitons bound to impurities is $l \sim 15$ Å, which leads to an oscillator strength of order unity. The corresponding lifetime is $\tau \sim 0.5 - 1$ ns, as observed experimentally.[44] The "giant oscillator strength" explains the fact that bound excitons are easily observed in absorption, even for low impurity concentration.

Notice that the oscillator strength is larger when the attractive potential is smaller, because the exciton is less localized and l is larger. Equation (42) is sometimes interpreted by saying that the oscillator strength of the bound exciton is enhanced over the oscillator strength *per molecule* of the free exciton by a factor $8\pi l^3/\Omega$, where Ω is the unit-cell volume: this is actually the origin of the term "giant oscillator strength". This interpretation is not quite appropriate for Wannier excitons: since the electron-hole separation is much larger than the lattice spacing, the "oscillator strength per molecule" cannot be related to a property of the constituents of the crystal. The dimensionless oscillator strength is proportional to the crystal volume, see Eq. (15), and is infinite in the thermodynamic limit.

It is interesting to discuss the radiative lifetime of the exciton in terms of its "coherence volume" — although this concept is not precisely defined, and it should be used with caution. Loosely speaking, the scenario for radiative decay of the exciton depends on the coherence length l of the center of mass. When the exciton is weakly bound, or when the mean free path is small compared to the wavelength of light, the coherence volume is $\propto l^3$ and the exciton decays radiatively with a decay rate which is also $\propto l^3$. However when $l > \lambda$ polariton effects sets in, radiation-matter interaction leads to a reversible exchange of energy between exciton and photon, and "radiative decay" becomes inefficient as it occurs within the polariton framework.

3. Excitons and polaritons in two dimensions

A *quantum well* is a thin layer of a semiconductor surrounded by thick layers of another semiconductor with a larger band gap. The alignment of the bulk band structures determines the band discontinuity, or band offset, between conduction (or valence) band edges. The variation of the band edge from one material to the other gives rise to quantum confinement effects. Electronic states in heterostructures (quantum wells, superlattices, and systems of lower dimensionality like quantum wires and dots) can be treated in a phenomenological yet accurate way by the envelope-function method. In this framework, electrons and holes are described by an effective mass equation (possibly a multiband one in the case of band degeneracy) with a confining potential determined by the band offset. A quantum well is characterized by a square-well potential, leading to the particle-in-a-box problem of quantum mechanics. The motion along the growth direction (taken as z) is quantized, but the energy levels have a two-dimensional dispersion $E_n(\mathbf{k}_\parallel)$ as a function of in-plane wavevector $\mathbf{k}_\parallel$ and are called *subbands*. The electronic states and optical transitions in heterostructures are treated in a number of textbooks.[11,45,46]

A *semiconductor microcavity* is a dielectric multilayer structure consisting of a cavity region embedded between two distributed Bragg reflectors (DBRs) acting as dielectric mirrors. It acts as a Fabry-Pérot cavity, in the sense of giving rise to a resonant Fabry-Pérot mode within the stop band of the DBRs. For a high number of pairs in the DBRs, leading to a high mirror reflectivity, the quality factor of the Fabry-Pérot resonance can be quite high — up to 10^5 and higher, with present epitaxial technologies. The Fabry-Pérot mode has a dispersion with the incident angle, which can be viewed as a dispersion of the photons confined in the cavity region. Photon states in a semiconductor microcavity are quasi-2D, as they are confined along the cavity axis (taken again as z) and are free along the planes. Thus, photons confined in semiconductor microcavities have the same dimensionality of electronic (or excitonic) states in QWs.

The interaction of QW excitons with 3D or with 2D photons is the subject of this Section. In addition to planar microcavities, which are known since more than 20 years to be a rich playground for the study of exciton-polariton physics, we also consider 2D waveguide-embedded photonic crystals, which have been recently shown to support another variant of exciton-polariton states.

3.1. *Exciton and polariton effects in quantum wells*

Excitons in quantum wells are described to a first approximation by the following hamiltonian (E_g is the band gap of the well material and $V_e(z_e)$, $V_h(z_h)$ are square-well confining potentials):

$$H = E_g - \frac{\hbar^2 \nabla_e^2}{2m_e^*} - \frac{\hbar^2 \nabla_h^2}{2m_h^*} + V_e(z_e) + V_h(z_h) - \frac{e^2}{\epsilon_b |\mathbf{r}_e - \mathbf{r}_h|}. \tag{43}$$

Due to conservation of the in-plane exciton wavevector $\mathbf{k}_{ex}$, the real-space envelope function factorizes as $F(\mathbf{r}_e, \mathbf{r}_h) = e^{i\mathbf{k}_{ex} \cdot \mathbf{R}_\parallel} F(\rho, z_e, z_h)$ where ρ is the in-plane relative coordinate. The oscillator strength of a quantum-well exciton is proportional to the sample surface S: only the oscillator strength per unit area has a physical meaning. Specifying Eq. (13), the expression is:

$$\frac{f_{\hat{\epsilon}}}{S} = g_s \frac{2|\hat{\epsilon} \cdot \mathbf{p}_{cv}|^2}{m_0 \hbar \omega} \left| \int F(\rho = 0, z, z) dz \right|^2, \tag{44}$$

where g_s again is a spin-orbit factor. The oscillator strength per unit area can be measured from the absorption probability (a dimensionless quantity) integrated over the excitonic peak through a relation similar to Eq. (4).[12]

The interplay between quantum confinement and electron-hole attraction gives rise to various physical regimes, which are discussed at length in Refs. 12, 25. Briefly, the most common regime for QW excitons is when the well width L is of the order of (or slightly larger than) the exciton Bohr radius. In this situation, which is called *electron-hole confinement regime*, the binding energy and the oscillator strength per unit area increase as the well width is reduced, due to a reduction of the 2D exciton radius. This behavior, which increases the stability of the exciton compared to the bulk, persists as long as electron and hole states remain confined in the well region, i.e., when the well thickness is not too small. In the opposite case, when the well width is much larger than the Bohr radius, the internal electron-hole wavefunction is essentially the 3D one, but the center of mass wavefunction is confined in the thin layer of width L. This is called *center-of-mass confinement regime*, and it leads to an oscillator strength per unit area which increases linearly with the well width. Thus, the oscillator strength per unit area must have a *minimum* at a given well thickness, which turns out to be of the order of 2.5 exciton radii.[47] Yet another regime of weak center-of-mass localization occur for very narrow QWs[48] or for excitons bound to monolayer impurity planes.[49]

Similar concepts hold when the dimensionality is reduced in other spatial directions: for example, the oscillator strength of QW excitons localized to interface fluctuations has a minimum as a function of defect size and it increases with the size when the dimension of the island is much larger than a Bohr radius, due to lateral quantization of the center-of-mass motion.[50] The regime of center-of-mass confinement is an example of a *coherent-volume regime*: this applies for confinement of the exciton along any given spatial direction, when the localization length is much larger than the exciton radius but much smaller than the wavelength of light in the sample.

For a detailed discussion of exciton and polariton quantization in thin films we refer again to previous reviews.[12, 25] Basically, polariton interference for large film thickness — including the anomalous features resulting from the additional wave and interference between LP and UP — goes over to exciton center-of-mass quantization in thinner films. For recent work in this context see, e.g., Ref. 51.

An appropriate theoretical framework for treating the above-mentioned phenomena (polariton and exciton interference in thin films, exciton radiative decay, formation of evanescent non-radiative states) is the nonlocal susceptibility formalism. Indeed, due to the lack of translational invariance along the growth direction, the dielectric response associated with the exciton is intrinsically nonlocal. The constitutive relation can be taken of the form

$$\mathbf{D}(z) = \epsilon_\infty \mathbf{E}(z) + 4\pi \int \chi(z, z') \mathbf{E}(z') dz', \tag{45}$$

where the nonlocal susceptibility $\chi(z, z')$ is given in linear response theory as:[52, 53]

$$\chi(z, z') = \sum_\lambda \chi_\lambda(\omega, k_{ex}) \rho_\lambda(z) \rho_\lambda(z'), \tag{46}$$

where λ are quantum numbers of the excited states of the crystal. We consider only the resonant term of the excitonic state, and incorporate the contribution of all nonresonant terms in the background dielectric constant. For excitons in thin layers, we have

$$\chi_\lambda(\omega, k_{ex}) = \frac{g_\lambda |\langle u_c | e\mathbf{r} | u_v \rangle|^2}{\hbar(\omega_k - \omega - i\gamma_{ex})}, \tag{47}$$

$$\rho_\lambda(z) = F_\lambda(\rho = 0, z, z), \tag{48}$$

where $F_\lambda(\rho, z_e, z_h)$ is the exciton envelope function and γ_{ex} is a *nonradiative* broadening. Thus the semiclassical description of polariton effects in thin layers consists of the following three steps:[54–56]

(1) Determine the excitonic levels $\hbar\omega_\lambda$ and wavefunctions $F_\lambda(\rho, z_e, z_h)$ by solving the Schrödinger equation for the exciton;
(2) Identify the resonant terms and calculate the nonlocal susceptibility (47);
(3) Solve Maxwell equations with the constitutive relation (45).

The formalism is intrisically ABC-free, since no-escape boundary conditions are imposed for electrons and hole separately. It can be applied for both thin films and quantum wells, provided the quantum numbers λ of the excited states are correctly identified and, possibly, a sum over nearby resonant levels is taken. More elaborate and powerful ABC-free schemes are the coherent-wave approach[57] or semiconductor Maxwell-Bloch equations.[28, 58, 59]

We shall now focus on the radiative properties of 2D excitons. It was first shown by Agranovich and Dubovskii,[60] and later by several authors,[61–66] that 2D excitons are subject to an intrinsic radiative decay due to the lack of wavevector conservation along the confinement direction. An exciton with parallel wavevector $\mathbf{k}_{ex} \equiv \mathbf{k}_\parallel$ interacts with photons with wavevectors $\mathbf{q} = (\mathbf{q}_\parallel, q_z)$, where the parallel wavevector $\mathbf{q}_\parallel = \mathbf{k}_\parallel$ is conserved, while the z-component q_z can take any value. Thus, the 2D exciton interacts with a quasi-1D density of photon states, given by

$$\rho(\mathbf{k}_\parallel, \omega) \equiv \sum_{k_z} \delta\left(\hbar\omega - \frac{\hbar c}{n}\sqrt{k_\parallel^2 + k_z^2}\right) = \frac{V}{\pi S}\left(\frac{n}{\hbar c}\right)^2 \frac{\hbar\omega}{\sqrt{k_0^2 - k_\parallel^2}}\theta(k_0 - k_\parallel). \tag{49}$$

Modes that lie above the light line in the $(\omega, \mathbf{k}_\parallel)$ plane are subject to radiative decay with the 1D density of states (49). On the other hand, modes that lie below the light line are evanescent and non-radiative: they resemble surface modes, as they can only be excited by coupling with a grating[67] or else with a prism in an attenuated total reflection configuration. For a schematic representation see Fig. 4.

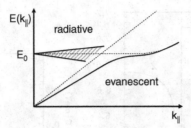

Fig. 4. Radiative properties of 2D excitons: states above the light line are intrinsically radiative, while those below the light line gives rise to stationary, evanescent modes. Adapted from Ref. 60.

Exciton radiative decay and formation of nonradiative QW polaritons are discussed in previous reviews.[12,25] The radiative widths of T, L, and Z-polarized QW exciton modes can be calculated by Fermi's golden rule and are given by:

$$\Gamma_T(k_\parallel) = \kappa_{\mathrm{SI}} \frac{2\pi}{n} \frac{e^2}{m_0 c} \frac{f_{xy}}{S} \frac{k_0}{k_z},$$

$$\Gamma_L(k_\parallel) = \kappa_{\mathrm{SI}} \frac{2\pi}{n} \frac{e^2}{m_0 c} \frac{f_{xy}}{S} \frac{k_z}{k_0}, \tag{50}$$

$$\Gamma_Z(k_\parallel) = \kappa_{\mathrm{SI}} \frac{2\pi}{n} \frac{e^2}{m_0 c} \frac{f_z}{S} \frac{k_\parallel^2}{k_0 k_z}$$

with $k_z = \sqrt{k_0^2 - k_\parallel^2}$ (the factor of two was missing in Ref. 64). The radiative width obviously vanishes for nonradiative excitons with $k_\parallel > k_0$. The oscillator strengths per unit area f_{xy}, f_z depend on the polarization vector. The decay rate of the exciton at $\mathbf{k}_\parallel = 0$ is given by $2\Gamma_0$, where

$$\Gamma_0 = \kappa_{\mathrm{SI}} \frac{\pi}{n} \frac{e^2}{m_0 c} \frac{f_{xy}}{S}. \tag{51}$$

It should be noted that the intrinsic radiative lifetime of QW excitons at $\mathbf{k}_\parallel = 0$ is of the order of 10 ps in GaAs, i.e., very short.[64,66] This lifetime can be observed under *resonant* excitation, in order that a coherent exciton population with a specified in-plane wavevector is initially excited.[65,68] Under nonresonant excitation, the exciton population is thermalized, leading to an effective lifetime which increases linearly with temperature,[69] see e.g. Ref. 42 for a review.

It follows from the previous considerations that polariton effects for QW excitons should be viewed in a rather different way compared to the bulk. We can generally define polariton effects as those resulting from the interaction of excitons with the transverse (retarded) part of the electromagnetic field. The frequency shift can be real (leading to exciton-polariton dispersion, as in the bulk), or complex: in this case, the imaginary part of the frequency shift is related to the radiative width describing radiative decay of the exciton. An example for QW excitons is shown in Fig. 5. In this figure (like in the previous Fig. 4), the exciton dispersion in the

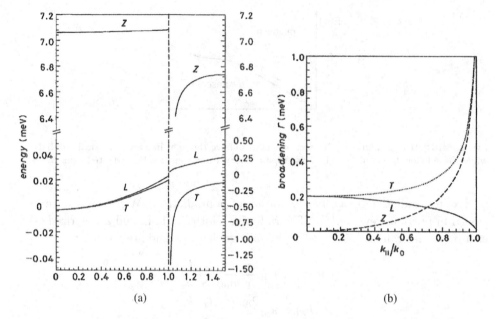

(a)   (b)

Fig. 5.   Polariton effect on excitons in a 20 A wide CuCl quantum well: (a) dispersion of radiative and nonradiative modes (the dashed line is the light dispersion), (b) lifetime broadening of radiative modes. Adapted from Ref. 63.

radiative region is obtained by perturbation theory. Nonperturbative phenomena taking place when the exciton dispersion approaches the light line are treated in Ref. 70.

If the QW is described by a local dielectric layer, its dielectric function should be taken of the form

$$\epsilon(\omega) = \epsilon_\infty \left(1 + \frac{\tilde{\omega}_{LT}}{\omega_{ex} - \omega - i\gamma_{ex}} \right) \tag{52}$$

It has been shown[71] that the effective LT splitting is given by

$$\tilde{\omega}_{LT} = \frac{2\Gamma_0}{k_0 L_w} = \kappa_{SI} \frac{2\pi}{\epsilon_\infty} \frac{e^2}{m_0 \omega_0} \frac{f_{xy}}{S L_w}. \tag{53}$$

This formula is analogous to Eq. (21), but in this case it does not correspond to the exchange splitting: indeed, it has been shown in Ref. 62 that the actual LT exchange splitting of QW excitons vanishes linearly in $\mathbf{k}_{ex}$ when $\mathbf{k}_{ex} \to 0$. Moreover, the local model reproduces the results of the more fundamental nonlocal approach only if $qd \ll 1$, where $q = \sqrt{\epsilon(\omega)}\omega/c$. In practice, the local model is approximately valid only when $\gamma_{ex} \gg \tilde{\omega}_{LT}$. When the nonlocal model is used, the normal-incidence reflectivity and transmission amplitudes of a single QW are found as

$$r_{QW}(\omega) = -\frac{i\Gamma_0}{\Delta + i\Gamma_0}, \quad t_{QW}(\omega) = 1 + r_{QW}(\omega) \tag{54}$$

where

$$\Delta = \omega - \omega_{ex} - \Delta\omega_{ex} + i\gamma_{ex}. \tag{55}$$

Here $\Delta\omega_{ex}$ is a radiative shift of the exciton energy, which is usually small and can be neglected, while γ_{ex} is the *nonradiative* exciton broadening. The reflectivity has a Lorentzian lineshape with total width $\Gamma_0 + \gamma_{ex}$, where Γ_0 is the radiative exciton broadening: in other words, the radiative decay rate originates directly from the solution of Maxwell equations with retardation, and for this reason it should not be put in the nonlocal susceptibility (46). The result (54) is important as it can be directly used in a transfer-matrix formulation which incorporates the nonlocal dielectric response of the quantum well exciton: this topic will be treated later in this Section.

3.2. Radiative recombination of excitons in 2D, 1D, 0D

When the exciton is confined along one, two or three spatial directions, its dispersion as a function of center-of-mass wavevector becomes 2D, 1D, 0D respectively. In all these cases, the confined exciton interacts with 3D photons and it has a density of states for radiative decay. It is then interesting to compare the results for intrinsic radiative decay of the exciton as a function of dimensionality. In the case of 2D excitons (quantum wells) the radiative decay rate is given by:[64, 66]

$$\Gamma_{2D} = \kappa_{SI}\frac{2\pi}{n}\frac{e^2}{m_0 c}\frac{f_{xy}}{S} \quad \text{(quantum wells)}, \tag{56}$$

where f_{xy}/S is the oscillator strength per unit area. In the case of 1D exciton (quantum wires) is has been shown that:[72]

$$\Gamma_{1D} = \kappa_{SI}\pi\frac{e^2}{m_0 c}\frac{\omega}{c}\frac{f}{L} \quad \text{(quantum wires)}, \tag{57}$$

where f/L is the oscillator strength per unit length. Finally, for 0D excitons (quantum dots), the atomic-like formula (40) gives

$$\Gamma_{0D} = \kappa_{SI}\frac{2n}{3}\frac{e^2}{m_0 c}\frac{\omega^2}{c^2}f \quad \text{(quantum dots)}, \tag{58}$$

where now the oscillator strength is a dimensionless number. Roughly speaking, we can see that $\Gamma_{2D} \approx (\lambda/2\pi l)^2\Gamma_{0D}$ and $\Gamma_{1D} \approx (\lambda/2\pi l)\Gamma_{0D}$, where l is the well width or the wire width, respectively. With typical number for GaAs, we obtain $\Gamma_{2D} \sim 13$ ps, $\Gamma_{1D} \sim 130$ ps, $\Gamma_{0D} \sim 1.3$ ns. In other words, the shortest intrinsic radiative lifetime occurs for 2D excitons, as the exciton wavefunction has a coherent interaction with the photons along two spatial directions; the case of 1D excitons is intermediate, as coherent exciton-photon interaction is along one direction only; in the case of 0D exciton, confinement along the three spatial directions is such that the exciton interacts with the photon only in a volume given by quantum confinement, thus yielding a radiative decay rate and a lifetime that are comparable with atomic ones. *Surprisingly, radiative decay of the free exciton is fastest in 2D and it gets slower in 1D and 0D.*

The fast intrinsic radiative decay of the 2D exciton is likely the explanation for the observation made by Weisbuch *et al.*[73] that free exciton recombination dominates the photoluminescence (PL) in GaAs/GaAlAs QWs, while bulk PL is dominated by impurity-bound excitons, even if the GaAs layer contains a comparable amount of impurities. A nice example is shown in Fig. 6. The point is that radiative recombination in the bulk is inefficient and slow, as it is due to exciton-polaritons which propagate up to the sample surface, while radiative recombination in QWs in intrinsically due to coupling of 2D excitons with 3D photons.

Under many circumstances, time-resolved PL follows from a thermalized exciton distribution, which in nondegenerate conditions is simply a Boltzmann distribution function over the exciton center-of-mass dispersion. Averaging the decay rate over the thermal distribution gives the following results:

$$\langle \Gamma_{2D}(T) \rangle \propto T \quad \text{(quantum wells)}, \tag{59}$$

$$\langle \Gamma_{1D}(T) \rangle \propto T^{1/2} \quad \text{(quantum wires)}, \tag{60}$$

$$\langle \Gamma_{0D}(T) \rangle \propto T^0 \quad \text{(quantum dots)}. \tag{61}$$

It is tempting to generalize the above equations to 3D and to write $\langle \Gamma_{3D}(T) \rangle \propto T^{3/2}$; however, this relation cannot hold in the polariton regime, because there is no intrinsic radiative decay of the exciton. It is expected to hold in the coherence-volume

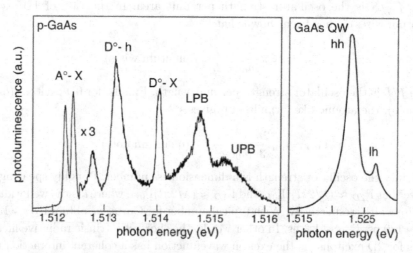

Fig. 6.   (left) Photoluminescence of a high-purity epitaxial GaAs layer, showing free exciton polariton peaks (LPB and UPB), excitons bound to neutral donor (D^0-X) and to neutral acceptors (A^0-X), recombination of holes with neutral donors (D^0-h) (R.G. Ulbrich and C. Weisbuch, 1976, unpublished); (right): Photoluminescence of a GaAs/GaAlAs multiquantum well sample,[73] displaying only heavy and light-hole free-exciton peaks. From Ref. 42.

regime, in which the exciton coherence length is too short to give rise to coherent polariton coupling, and radiative decay of the exciton is recovered.

3.3. *Photon confinement in planar microcavities*

In order to have planar solid-state microcavities one needs mirrors with high reflectivity and low losses. In the optical region, multilayer dielectric mirrors, called *Distributed Bragg Reflectors (DBRs)*, are usually employed. A DBR is a periodic repetition of a two-layer unit structure with different refractive indices n_1, n_2 for the two layers.[74, 75] The electromagnetic problem for such a structure is analogous to the one-dimensional Kronig-Penney model for electrons. The presence of a refractive index contrast in the medium of period d opens a gap in the photonic dispersion relation $\omega(k)$ when the Bragg condition $k = m\pi/d$ is satisfied, which corresponds to the wavevector k being at the boundary of the Brillouin zones. A *photonic gap* is a forbidden energy region, in which photons cannot propagate: for a finite structure, this means that the reflectivity is large inside the gap and it becomes exponentially close to unity when the number of periods is increased. DBRs have in fact regions of large reflectivity, called *stop bands*, separated by regions called side-bands in which the reflectivity is small and oscillating. For a $\lambda/4$ DBR (i.e., when the thicknesses L_1, L_2 of two layers satisfy the condition $k_1 L_1 = k_2 L_2 = \lambda/2$ at the operating wavelength) the reflectivity at normal incidence is $R = 1 - 4(n_L/n_H)^{2N}$, where n_L (n_H) is the lower (higher) of the refractive indices and N is the number of periods. The reflectivity is large within a frequency region which depends on the refractive index contrast, and within an angular cone whose width also depends on the contrast: indeed, the center of the stop band has an angular dispersion which (for small contrast and small internal angle θ) goes approximately like $\omega_s(\theta) = \omega_s(0)/\cos\theta$.

A planar Fabry-Pérot microcavity is an epitaxial structure in which a central cavity region of width L_c is surrounded by two mirrors. If the reflection amplitude of the mirrors is assumed to be real and independent of frequency (an approximation which may be valid for metallic, but not for dielectric mirrors) the transmission coefficient is found to have resonances when the Fabry-Pérot condition $k_c Lc = m\pi$ is satisfied: this corresponds to the cavity width being an integer multiple of half wavelengths. The Fabry-Pérot condition is modified for semiconductor microcavities with dielectric mirrors, due to the fact that the reflection amplitude of a DBR is complex and frequency-dependent. The Fabry-Pérot resonance can be viewed as a quasi-bound state of light which is analog to the quantized level of an electron in a potential well (or, more precisely, to the quasi-bound state corresponding to the tunneling resonance in a double-barrier configuration). The transmission resonance has a Lorentzian lineshape characterized by a frequency ω_c and a linewidth γ_c: the quality factor of the Fabry-Pérot mode is defined as $Q_c = \omega_c/\gamma_c$ and is proportional to $(1 - R)^{-1}$, where R is the mirror reflectivity. Like the center of the stop band, the resonance frequency ω_c has an angular dispersion like $\omega_c(\theta) = \omega_c(0)/\cos\theta$ (for small index contrast and internal angle): this can be derived by writing the frequency in terms of wavevector components and noting that the wavevector along the growth direction is discretized by the Fabry-Pérot condition. The dispersion of the Fabry-Pérot mode is sometimes referred to as "heavy photons", since the optical mode has a quadratic dispersion for small in-plane wavevector and can be characterized by an "effective mass". The photon mass is easily estimated as $m_{ph} = \pi \hbar n_c / (cL_c)$ and is of the order of $10^{-5} m_0$.

Planar microcavities can be employed to modify the emission pattern of a dipole.[76,77] The change of the emission pattern is the basis for the operation of Vertical-Cavity Surface Emitting Lasers (VCSELs). On the other hand the emission lifetime undergoes small modifications in a planar microcavity.[78] In order to strongly modify the radiative lifetime, full 3D photon confinement is needed.

3.4. *Polaritons in planar microcavities*

A very interesting phenomenon which takes place in planar microcavities with embedded quantum wells is the strong-coupling regime of radiation-matter coupling. This manifests itself as a splitting between coupled exciton-photon modes at resonance, as was first shown in the pioneering paper by Weisbuch *et al.*[6] The experiment is done by measuring the reflectivity at different positions on the sample surface: since the sample thickness is nonuniform due to inhomogeneous epitaxial growth, the frequency of the Fabry-Pérot mode depends on the position, thus the detuning between the optical mode and the exciton resonance can be changed at will. At zero detuning, two dips are observed in reflectivity with a splitting of a few meV. If the dip positions are plotted as a function of detuning, a clear anticrossing behavior is seen with a minimum separation at resonance. The energy separation at resonance is called *vacuum-field Rabi splitting*: it represents the splitting between mixed modes arising from the interaction between quantum confined excitons in the quantum wells and the Fabry-Pérot mode of the cavity. The mixed exciton-cavity modes are the 2D analog of exciton-polaritons in bulk semiconductors and are called *cavity polaritons*. For specific reviews see Refs. 79–82.

The semiclassical theory of cavity polaritons can be developed by simple extension of usual transfer-matrix theory which is employed to describe the Bragg reflectors.[74,75] The additional ingredient which is needed is the transfer matrix of a quantum well, as obtained from the nonlocal susceptibility description of Subsection 3.1. The formalism was developed in Refs. 83, 84 and is reviewed in Ref. 86. For a finite angle of incidence, the transfer matrix must be calculated for both TE (s) and TM (p) polarizations. For TE polarization, the transfer matrix which "jumps over" a single QW of width L_w is

$$T_{QW} = \frac{1}{\Delta_T} \begin{bmatrix} e^{ik_z L_w}(\Delta_T - i\gamma_T) & -i\gamma_T \\ i\gamma_T & e^{-ik_z L_w}(\Delta_T + i\gamma_T) \end{bmatrix}, \tag{62}$$

where $k_z = \sqrt{\epsilon_\infty \omega^2 / c^2 - k_{ex}^2}$. The expressions for Δ_T and γ_T in the long-wavelength approximation are $\Delta_T = \omega - \omega_{ex} + i\gamma_{ex}$, $\gamma_T = \Gamma_0 k_0 / k_z$, where ω_{ex} is the exciton energy including spatial dispersion, γ_{ex} is the nonradiative exciton broadening, and Γ_0 defined in Eq. (51) is the radiative decay rate of the exciton amplitude in a single QW at $\mathbf{k}_{ex} = 0$. Notice that the decay rate of the exciton probability is given by $\Gamma_T = 2\gamma_T$, see Eq. (50). The subscript "T" refers to the T-mode, which is the optically active mode for TE polarization.

We now consider a semiconductor microcavity with one (or more) QW excitons in the cavity region, in positions corresponding to antinodes of the electromagnetic field. Once the transfer matrix of the whole cavity structure is found, the eigenmodes are determined by the poles of the transmission coefficient, i.e., by the roots of the equation $T_{22}(\omega) = 0$ in the complex plane. By expanding close to resonance region, the following equation is found:[84]

$$(\omega - \omega_{ex} + i\gamma_{ex})(\omega - \omega_c + i\gamma_c) = |g|^2, \tag{63}$$

where ω_c is the cavity-mode frequency, $\gamma_c = c(1 - \sqrt{R})/(n_c L_{eff}\sqrt{R})$ is the cavity mode linewidth related to the Q-factor by $Q_c = \omega_c/\gamma_c$, and

$$g = \left(\kappa_{SI}\frac{2\pi e^2}{\epsilon_\infty m_0 L_{eff}}\frac{f_{xy}}{S}\right)^{1/2} \tag{64}$$

is the coupling constant of the exciton-photon interaction depending on the oscillator strength per unit area f_{xy}/S of the QW exciton (if there are several quantum wells, the oscillator strength has to be multiplied by an effective number of wells n_{eff} which takes into account their relative positions with respect to the e.m. field profile.[84]) In Eq. (64) $L_{eff} = L_c + L_{DBR}$ is an effective cavity length that depends on the physical length L_c and on a penetration depth in the dielectric mirrors L_{DBR}. Equation (63), which is often introduced phenomenologically as a two-oscillator model, has two solutions:

$$\omega = \frac{\omega_{ex} + \omega_c - i(\gamma_{ex} + \gamma_c)}{2} \pm \sqrt{|g|^2 + \frac{1}{4}(\omega_{ex} - \omega_c - i(\gamma_{ex} - \gamma_c))^2}. \tag{65}$$

At resonance ($\omega_{ex} = \omega_c$), we see that a splitting in the real part of the frequency occurs provided

$$g > \frac{|\gamma_{ex} - \gamma_c|}{2}. \tag{66}$$

When this condition is satisfied, the exciton-photon system is in a strong-coupling regime, with the formation of cavity polaritons. An example of crossover from weak to strong coupling as a function of mirror reflectivity is shown in Fig. 7. Notice that γ_{ex}, γ_c are *half-widths at half maximum* (HWHM). It often happens that one of the two linewidths dominates, nevertheless Eq. (66) implies the interesting possibility that QW excitons and cavity photons can be in a strong coupling regime even with large, but nearly identical linewidths. In this case, exchange of energy between QW exciton and cavity photon states takes place in a coherent and reversible way, even if both states (and the resulting amplitude of Rabi oscillations) are strongly damped.

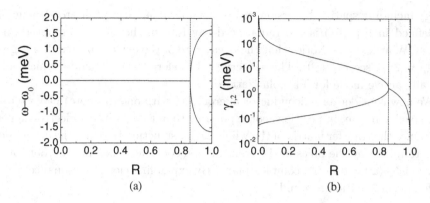

Fig. 7.   Crossover from weak to strong coupling for the λ-cavity structure of Ref. 6 assuming the following parameters: resonance condition ($\omega_{\text{ex}} = \omega_{\text{c}} \equiv \omega_0$), $\Gamma_0 = 0.032$ meV (corresponding to one QW), $\gamma_{\text{ex}} = 0.05$ meV, $n_{\text{c}} = 3.17$, $L_{\text{eff}} = 1.5\,\mu$m: (a) real parts, (b) imaginary parts of Eq. (65). The vertical dashed lines denote the crossover point. Adapted from Ref. 84.

A quantum-mechanical treatment of cavity polaritons gives the same results as the semiclassical theory.[84]

The dispersion of cavity polaritons is found within the same formalism and from the same equations (63)–(65), provided the exciton and cavity frequencies are interpreted as dependent on the wavevector in the 2D plane: $\omega_{\text{ex}} = \omega_{\text{ex}}(\mathbf{k}_{\parallel})$, $\omega_{\text{c}} = \omega_{\text{c}}(\mathbf{k}_{\parallel})$. Notice that the exciton resonance is usually the ground-state heavy-hole exciton in the QW, which is polarized in the plane, thus only TE-polarization needs to be kept. The dispersion of cavity polaritons can be measured by angle-resolved reflectivity or photoluminescence.[79,86,87] In analogy with bulk polaritons, we see that wavevector conservation plays a crucial role: a QW exciton with a given in-plane wavevector and polarization interacts with a cavity photon with the same wavevector and polarization, thus giving rise to a coupled system that behaves like a two-oscillator model of two interacting bosons.

It is interesting to compare the exciton-light coupling (64) for QW excitons in planar microcavities with the analogous quantity in the bulk, i.e., ω_{c} of Eq. (28) or the quantum coupling g of Eq. (36), which is half of the polariton splitting in the absence of damping. Basically, the oscillator strength per unit volume $\frac{f}{V}$ of the bulk exciton is replaced by $\frac{f_{xy}}{SL_{\text{eff}}}$, since exciton-light coupling occurs only across the QW thickness. The additional factor of two in the numerator of Eq. (64) is due to multiple reflections from the QW and it is a specific feature of semiconductor microcavities. The polariton splitting is *reduced* in microcavities with respect to the bulk, even when the cavity contains several quantum wells (typical values are 16 meV for bulk GaAs compared to a few meV for GaAs-based microcavities). Nevertheless cavity polaritons are much easier to observe than bulk polaritons and are sometimes observed even at room temperature: this is because there is no polariton motion along the cavity axis, therefore the criterion of spatial coherence (see Section 1.3) does not come into play and only the much less stringent criterion of temporal coherence embodied in the relation $V > |\gamma_{ex} - \gamma_c|/2$ applies. Similar remarks can be made for polaritons in so called bulk microcavities,[85] in which the excitonic resonance is that of the whole cavity medium.

The bosonic nature of cavity polaritons has been demonstrated by stimulated scattering and amplification of the emission in the strong coupling regime.[7] Bose-Einstein condensations of cavity polaritons has been realized a few years later:[8] this ranks among the most important discoveries in condensed matter physics in the present Millennium. It should be kept in mind that Bose-Einstein condensation in atomic gases takes place at temperatures of the order of nano-Kelvin, and that Bose-Einstein condensation of excitons has always remained an elusive phenomenon, a sort of "Holy Graal" of semiconductor physics. Bose-Einstein condensation of cavity polaritons is made possible by the fact that radiation-matter interaction leads to stable exciton-polariton states, and that the dispersion of cavity polaritons has an energy minimum due to photon confinement along the microcavity axis. The small photon mass gives rise to a condensation temperature higher than a few degree Kelvin. This area of research is very active at present and is unravelling new challenging phenomena related to Bose-Einstein condensates in solids. The reader interested in this exciting field is referred to recent textbooks.[88,89]

3.5. *Polaritons in waveguide-embedded photonic crystals*

Photonic crystals realized in planar waveguides, commonly known as photonic crystal (PhC) slabs,[90-94] are at the heart of current interest in periodic dielectric structures thanks to their capability to control propagation of light in all spatial directions. This is achieved by combining Bragg reflection due to a 2D lattice in the slab plane with total internal reflection in the vertical direction. In particular, semiconductors are very suited for the realization of PhC slabs, due to their high refractive index yielding good confinement properties and to the availability of mature processing technologies. The lithographic definition of the 2D lattice allows introducing line and point defects, which behave as linear waveguides and nanocavities, respectively. In this Subsection we show that the interaction of PhC slab modes with QW excitons that are embedded at the center of the slab gives rise again to quasi-stationary exciton-polariton states.

A crucial issue related to PhC slabs is that of losses, especially due to scattering (diffraction) out of the slab plane. Photonic modes whose dispersion lies above the cladding light line(s) in the $k - \omega$ plane are subject to intrinsic losses, as they are coupled to leaky modes of the slab, and are usually called quasi-guided. On the other hand, modes lying below the cladding light line(s) in the $k\omega$ plane are lossless, or truly guided, in an ideal structure without disorder, and are subject only to extrinsic losses due to fabrication imperfections.

Solving Maxwell equations for a PhC slab structure is a complicated numerical task, especially for what concerns quasi-guided modes and their diffraction losses. Here we adopt an approximate method, which has been named Guided-Mode Expansion (GME).[95] The basic idea of the GME method is to represent the electromagnetic field in a finite basis set consisting of the guided modes of an effective homogeneous waveguide. This method is especially suited for formulating the exciton-polariton problem, as we shall see. We start from the second-order equation for the magnetic field in a source-free dielectric medium and for harmonic time dependence:

$$\nabla \times \left[\frac{1}{\epsilon(\mathbf{r})} \nabla \times \mathbf{H} \right] = \frac{\omega^2}{c^2} \mathbf{H}, \tag{67}$$

where $\epsilon(\mathbf{r})$ is the spatially dependent dielectric constant. Due to translational invariance in the slab (xy) plane implying Bloch-Floquet theorem, the magnetic field can be expanded on a basis in which planar and vertical coordinates are factorized

$$\mathbf{H}_\mathbf{k}(\mathbf{r}) = \sum_\mathbf{G} \sum_\alpha c_{\mathbf{k}+\mathbf{G},\alpha} \mathbf{h}_{\mathbf{k}+\mathbf{G},\alpha}(z) e^{i(\mathbf{k}+\mathbf{G}) \cdot \boldsymbol{\rho}}, \tag{68}$$

where $\mathbf{r} = (\boldsymbol{\rho}, z)$, $\mathbf{k}$ is the in-plane Bloch vector in the first Brillouin zone (BZ), $\mathbf{G}$ are reciprocal lattice vectors, and the functions $\mathbf{h}_{\mathbf{k}+\mathbf{G},\alpha}(z)$ $(\alpha = 1, 2, \ldots)$ are the (discrete) guided modes of the effective planar waveguide with an average dielectric constant $\bar{\epsilon}_j$ in each layer $j = 1, 2, 3$, calculated from the air fraction of the given photonic lattice (in this Subsection, for simplicity, we indicate the wavevectors in the plane by $\mathbf{k}$, $\mathbf{G}$ instead of the more precise notation $\mathbf{k}_\parallel$, $\mathbf{G}_\parallel$). Thus, Eq. (68) is reduced to a linear eigenvalue problem

$$\sum_{\mathbf{G}'} \sum_{\alpha'} \mathcal{H}_{\mathbf{k}+\mathbf{G},\mathbf{k}+\mathbf{G}'}^{\alpha,\alpha'} c_{\mathbf{k}+\mathbf{G}',\alpha'} = \frac{\omega^2}{c^2} c_{\mathbf{k}+\mathbf{G},\alpha} \tag{69}$$

where the matrix $\mathcal{H}$ is given by

$$\mathcal{H}_{\mathbf{k}+\mathbf{G},\mathbf{k}+\mathbf{G}'}^{\alpha,\alpha'} = \int \frac{1}{\epsilon(\mathbf{r})} (\nabla \times \mathbf{h}_{\mathbf{k}+\mathbf{G},\alpha}^*(\mathbf{r})) \cdot (\nabla \times \mathbf{h}_{\mathbf{k}+\mathbf{G}',\alpha'}(\mathbf{r})) d\mathbf{r}. \tag{70}$$

This formulation of the electromagnetic problem has strong analogies with the quantum-mechanical treatment of electrons, with the Hermitian matrix $\mathcal{H}$ playing the role of a Hamiltonian. The properties of the specific photonic lattice enter via the Fourier transform of the inverse dielectric constant in each layer. The eigenvalue problem (69) is solved by numerical diagonalization and the resulting photonic modes are classified according to their band index, n, and their in-plane Bloch vector $\mathbf{k}$. Once the magnetic field is calculated, the electric field is obtained from

$$\mathbf{E}_\mathbf{k}(\mathbf{r}) = \frac{ic}{\omega\epsilon(\mathbf{r})} \nabla \times \mathbf{H}_\mathbf{k}(\mathbf{r}). \tag{71}$$

The fields $\mathbf{E}_{\mathbf{k}n}(\mathbf{r})$ and $\mathbf{H}_{\mathbf{k}n}(\mathbf{r})$ calculated by the GME approach satisfy the orthonormality conditions

$$\int \epsilon(\mathbf{r})\mathbf{E}_{\mathbf{k}n}^*(\mathbf{r}) \cdot \mathbf{E}_{\mathbf{k}'n'}(\mathbf{r})d\mathbf{r} = \delta_{\mathbf{k},\mathbf{k}'}\delta_{n,n'}, \tag{72}$$

$$\int \mathbf{H}_{\mathbf{k}n}^*(\mathbf{r}) \cdot \mathbf{H}_{\mathbf{k}'n'}(\mathbf{r})d\mathbf{r} = \delta_{\mathbf{k},\mathbf{k}'}\delta_{n,n'}, \tag{73}$$

thus they constitute a very convenient set for the second-quantized formulation to be discussed below.

The basis set consisting of the guided modes of the effective waveguide is orthonormal, but not complete, since leaky modes are not included. Coupling to leaky modes produces a second-order shift of the mode frequency, which is usually small and is neglected in the GME method. When the guided modes are folded in the first Brillouin zone, many of them fall above the light line and become quasi-guided. Indeed, first-order coupling to leaky modes at the same frequency leads to a radiative decay, which can be calculated by time-dependent perturbation theory (like in Fermi Golden Rule for quantum mechanics).[95] This formulation is very suited for calculating intrinsic as well as disorder-induced extrinsic losses.[96,97]

We now treat exciton-polaritons in photonic crystal slabs.[98] We consider a PhC slab with a quantum well (QW) grown in the middle of the core layer, as illustrated in Fig. 8(a). The ground-state QW exciton is usually a heavy-hole exciton, with in-plane polarization of the transition dipole, and is able to interact with TE-like ($\sigma_{xy} = +1$, even) modes of the PhC slab. In the following we take the specific case of a square lattice of holes with lattice constant a, whose photonic bands for TE-like modes are plotted in Fig. 8(b) in dimensionless units. When the QW exciton is

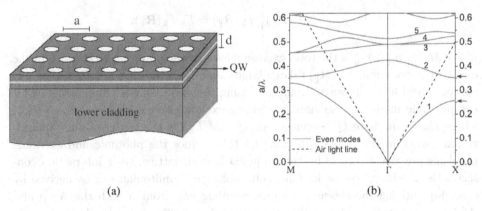

(a)   (b)

Fig. 8. (a) Schematic view of a 2D photonic crystal slab with core thickness d and lattice constant a, with an embedded single quantum well at the midplane of the core layer. (b) Photonic mode dispersion (even modes, $\sigma_{xy} = +1$) for a square lattice in a high-index ($\epsilon = 11.76$) photonic crystal membrane with $d/a = 0.3$, $r/a = 0.34$. The first few modes are labelled by a band number. Dashed lines represent the light dispersion in air. From Ref. 102.

resonant with truly-guided photonic modes below the light line (e.g., at the frequencies indicated by the two arrows in Fig. 8(b)) it forms stationary modes which we call *guided photonic crystal polaritons*. On the other hand, when the exciton is resonant with quasi-guided photonic modes above the light line, the exciton-photon interaction can be in a weak or in a strong-coupling regime: when the exciton-photon coupling is larger than the intrinsic photonic mode linewidth, exciton-polariton states are formed, which we name *radiative photonic crystal polaritons*. Those states are especially interesting, since they can be probed by incident light from the surface of the sample in a reflectance or transmission experiment. Indeed, strong exciton-light coupling has been reported in photonic crystal slabs filled with organic excitons[99] and the semiclassical theory of the optical response has been developed using a scattering-matrix formalism, where the excitonic resonance is treated by a Lorentz oscillator.[100, 101]

Here we formulate a theory of radiation-matter interaction, which is based on a full quantization of the exciton and photon fields. In a non-homogeneous dielectric medium, the vector potential can be chosen to satisfy the generalized Coulomb gauge $\nabla \cdot (\epsilon(\mathbf{r})\mathbf{A}(\mathbf{r}, t)) = 0$ and it is expanded in normal modes whose normalization condition is related to Eq. (72) for the electric field. Thus, the classical fields calculated by the GME method (neglecting the QW dielectric discontinuity and prior to including the exciton contribution) can be conveniently used as normal modes for second quantization. The exciton field can also be quantized by introducing operators $\hat{b}_{\mathbf{k}\nu}^\dagger$ and $\hat{b}_{\mathbf{k}\nu}$, which satisfy Bose commutation relations in the limit of weak excitation, using quantum number $\mathbf{k}, \nu$ analogous to those of the photon modes. Indeed, the free motion of the exciton center-of-mass in the QW plane is restricted to the dielectric region and it can be described by a 2D Schrödinger equation for the envelope function in an effective potential $V(\mathbf{R}_{\|})$:

$$\left[-\frac{\hbar^2 \nabla^2}{2M_{\text{ex}}} + V(\mathbf{R}_{\|}) \right] F_{\mathbf{k}}(\mathbf{R}_{\|}) = E_{\mathbf{k}} F_{\mathbf{k}}(\mathbf{R}_{\|}), \tag{74}$$

where $M_{\text{ex}} = m_e^* + m_h^*$ is the total exciton mass and the potential $V(\mathbf{R}_{\|}) = 0$ in the dielectric regions, while $V(\mathbf{R}_{\|})$ takes a large value V_∞ in the air holes. Equation (74) can be solved by plane-wave expansion, using the same Fourier components as for the photonic modes in the slab, leading to exciton energies $E_{\mathbf{k}\nu}^{(\text{ex})} = E_{\text{ex}} + E_{\mathbf{k}\nu}$, where E_{ex} is the bare QW exciton energy and $E_{\mathbf{k}\nu}$ is the center-of-mass quantization energy in the in-plane potential $V(\mathbf{R}_{\|})$. Since the photonic and excitonic problems are characterized by the same 2D Bravais lattice, their interaction conserves the 2D Bloch vector $\mathbf{k}$. The exciton-photon Hamiltonian can be derived by second-quantizing the classical minimal-coupling hamiltonian with the $\mathbf{A} \cdot \mathbf{p}$ and $\mathbf{A}^2$ interaction terms. The resulting quantum Hamiltonian takes the form

$$\hat{H} = \sum_{\mathbf{k},n} \hbar\omega_{\mathbf{k}n} \left(\hat{a}_{\mathbf{k}n}^\dagger \hat{a}_{\mathbf{k}n} + \frac{1}{2} \right) + \sum_{\mathbf{k},\nu} E_{\mathbf{k}\nu}^{(\text{ex})} \hat{b}_{\mathbf{k}\nu}^\dagger \hat{b}_{\mathbf{k}\nu}$$

$$+ i \sum_{k,n,\nu} C_{kn\nu}(\hat{a}_{kn} + \hat{a}^{\dagger}_{-kn})(\hat{b}_{-k\nu} - \hat{b}^{\dagger}_{k\nu})$$

$$+ \sum_{k,\nu,n,n'} D_{k\nu nn'}(\hat{a}_{-kn} + \hat{a}^{\dagger}_{kn})(\hat{a}_{kn'} + \hat{a}^{\dagger}_{-kn'}), \tag{75}$$

with the coupling coefficients being given by

$$C_{kn\nu} = E^{(ex)}_{k\nu} \left(\frac{2\pi e^2 \hbar \omega_{kn}}{\hbar^2 c^2} \right)^{1/2} \left\langle \Psi^{(ex)}_{k\nu} \middle| \sum_j \mathbf{A}_{kn}(\mathbf{r}_j) \cdot \mathbf{r}_j \middle| 0 \right\rangle, \tag{76}$$

while $D_{k\nu nn'} = C^*_{kn\nu} C_{kn'\nu}/E^{(ex)}_{k\nu}$. Equation (75) has the same structure of the Hopfield Hamiltonian (29) for bulk exciton-polaritons,[4,5] however the interaction takes place between photons and excitons with the same Bloch vector $\mathbf{k}$ but with all pairs of quantum numbers n, ν. The coefficients $C_{kn\nu}$ can be expressed in terms of the oscillator strength per unit area of the QW exciton, f/S, and of the overlap integral between electric field and exciton center-of-mass wavefunction in the 2D plane as

$$C_{kn\nu} \simeq -i \left(\kappa_{SI} \frac{\pi \hbar^2 e^2}{m_0} \frac{f}{S} \right)^{1/2} \int \hat{\mathbf{e}} \cdot \mathbf{E}_{kn}(\boldsymbol{\rho}, z_{QW}) F^*_{k\nu}(\boldsymbol{\rho}) d\boldsymbol{\rho} \tag{77}$$

Thus, all parameters of the quantum Hamiltonian are determined from a classical calculation of the electric field eigenmodes, through the GME method, and of the exciton envelope function confined in the dielectric regions. The Hamiltonian (75) can be diagonalized by a generalized Hopfield transformation, which leads to a non-hermitian eigenvalue problem of dimension $2(N_{max} + M_{max}) \times 2(N_{max} + M_{max})$ at each Bloch vector $\mathbf{k}$, where N_{max} and M_{max} are the number of photon and exciton modes kept in the expansion, respectively. Numerical solution of this eigenvalue problem yields finally the eigenfrequencies of the mixed exciton-photon modes. It should be noted that the dampings of photon and exciton states can be taken into account phenomenologically in the formalism by means of an imaginary part of their frequencies in the diagonal terms of Eq. (75). In particular, the intrinsic photon linewidth plays a crucial role in the theory: even if the exciton linewidth is assumed to be small (which requires very high-quality samples at low temperature), and for an ideal sample without disorder, quasi-guided PhC slab modes have an intrinsic linewidth arising from out-of-plane diffraction. Thus, the occurrence of radiative PhC polaritons depends on the relative size of the exciton-photon interaction (quantified by the coupling coefficients $C_{kn\nu}$) versus the photonic mode linewidth $\text{Im}(\omega_{kn})$, which is also calculated by the GME method.

Radiative photonic crystal polaritons have been recently observed in a sample consisting containing three $In_{0.05}Ga_{0.95}As/GaAs$ quantum wells embedded in the core of a $GaAs/Al_{0.8}Ga_{0.2}As$ slab waveguide.[9] The top layer was patterned by inductively coupled plasma etching with a square lattice of circular air holes: areas with different lattice parameters $a = 245, 250, 255,$ and 260 nm were defined. Only the top layer was patterned, i.e., patterning was only partial and did not reach the QW regions: this is a key aspect for success of the experiment, as the quality — and linewidth — of the QW excitons was unaffected by processing. Still, the periodic corrugation is adequate in order to yield a dispersion folding on the guided modes of the slab waveguide within the first Brillouin zone, making them radiative around normal incidence. Such folding depends on the lattice parameter a, which gives the possibility of engineering the dispersion relation by changing the lattice parameter.

Two kinds of experiments have been undertaken, namely resonant light scattering with cross-polarized light as well as angle-resolved photoluminescence. For suitable parameters, both measurements indicate the occurrence of a polariton splitting with a clear anticrossing when the dispersion of photonic modes is close to resonance with the exciton state. The results are in fair agreement with the theoretical description reported above. A characteristic feature of photonic crystal polaritons is that there are a number of exciton states which remain in the weak-coupling regime and bright. The reason is the leaky character of the PhC slab waveguides, which allow coupling of exciton states with far-field radiation modes Bright excitons tend to dominate resonant scattering and (especially) photoluminescence spectra, yet the formation of polariton states is clearly recognized. More recently, photonic crystal polaritons coupled to L3 PhC cavities have been shown to lead to polariton lasing with very low threshold.[103]

Photonic crystal polaritons result from the coupling of quasi-2D QW excitons to the photonic modes of a PhC slab, which are also quasi-2D, as the slab modes are (partially) confined along the vertical direction. They combine features of bulk polaritons and of QW excitons, namely, they are at the same time in a strong coupling regime and (partially) radiative.

To conclude this Section, we summarize the general features of exciton-photon interaction in systems of different dimensionalities, as illustrated in Fig. 9. Stationary exciton-polariton states occur when the dimensionality of photon states is equal to that of the excitons: this happens in the bulk (3D versus 3D) and with QW excitons in microcavities or in PhC slabs (2D versus 2D). Stationary polaritons are formed even when the dimensionality of photon states is smaller then that of the excitons: this is the situation with so-called bulk microcavities, i.e., 3D excitons interacting with 2D photons.[85] On the other hand, when the dimensionality of photons is larger than that of the exciton, an intrinsic radiative decay occurs: this is the case of 2D (or 1D, 0D) excitons interacting with 3D photons. The reason is that a single exciton interacts with a continuum of photons, which makes it impossible to have reversible energy exchange and gives rise instead to irreversible decay.

4.   From three to zero dimensions

In this Section we briefly consider the phenomena associated with full confinement of the electromagnetic field in three dimensions, i.e., when photon states become

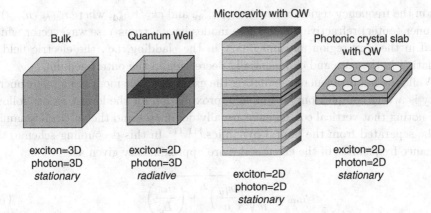

Fig. 9. Effect of dimensionality on the interaction between excitons and photons. Stationary polariton states arise only when the dimensionality of photon states is equal to (or, possibly, smaller than) that of the excitons. In the opposite case, the exciton is subject to radiative decay and no stationary polaritons are formed (except for nonradiative states below the light line, which are evanescent and do not couple to an external electromagnetic field).

zero-dimensional (0D). We first recall the case of QW excitons in pillar microcavities: in this situation the exciton states are 2D, while photon states are 0D. We then consider quantum dots in 3D microcavities, i.e., 0D excitons interacting with 0D photons. The strong-coupling regime in 0D has been demonstrated since 2004 in different systems, namely quantum dots in pillar microcavities,[104] QW excitons bound to monolayer fluctuations (interface islands) in microdisks,[105] and quantum dots in photonic crystal cavities.[106, 107] Although strongly coupled systems in 0D should not be called "polaritons" (as there is no wavevector conservation, and the excitations are not bosons), it is instructive to compare the coupling constants and the conditions for strong coupling with those of 3D and 2D systems. A review of cavity-QED experiments with self-assembled quantum dots in the weak-coupling regime is presented in Ref. 108.

4.1. *Polaritons in pillar microcavities*

A pillar microcavity, or micropillar, is obtained from a planar microcavity by lateral etching to form a dielectric cylinder with radius of the order of a micron. Photonic confinement in such a structure is given by two mechanisms: vertical confinement is provided by the dielectric mirrors of the planar microcavity, while lateral confinement is given by total internal reflection produced by the discontinuity of the average refractive index. Indeed, for III-V microcavities based on GaAs/AlGaAs or InP/InGaAs the refractive indices of the constituent materials are close to each other and of the order of 3 to 3.5; an infinite pillar can be viewed as a cylindrical waveguide[109, 110] where the core has the average refractive index of the dielectric and the cladding is air. For a given wavevector k_z along the pillar axis, guided modes exist in the frequency region between ck_z/n_{core} and ck_z/n_{clad}, where n_{core} (n_{clad}) is the inner (outer) refractive index; these modes have a transverse wavevector which is real in the core region and imaginary in the cladding, i.e., the electric field is oscillating in the core and exponentially decreasing in the outer medium.

While the full problem of calculating the exact photonic modes in a pillar micro-cavity is a very complex one, a simple approximation for the lowest modes follows from noting that vertical confinement usually dominates and the vertical dynamics can be separated from the lateral dynamics.[111, 112] In this decoupling scheme, the resonance frequencies of the eigenmodes are approximately given as

$$\omega_{lnm} = \frac{c}{n}\sqrt{\left(\frac{x_{ln}}{a}\right)^2 + \left(\frac{m\pi}{L_c}\right)^2} \tag{78}$$

where a is the radius of the pillar, L_c is the length of the planar microcavity, and x_{ln} is a zero of a Bessel function of order l: the pillar microcavity acts as a resonant cavity in the optical region, with discretized vertical and transverse wavevector components. The eigenfrequencies (78), or the expressions which result from more realistic calculations, are those of a *photonic dot*.

The energies of photonic modes in three-dimensional microcavities can be measured from the photoluminescence peaks when the cavity contains a broad-band emitter, since the emission spectrum is concentrated at the energies of cavity modes (a weak continuum emission due to coupling with leaky modes of the waveguide is also present). Often the emitter consists in a planar array of self-assembled InAs quantum dots embedded in a GaAs cavity, as was first demonstrated in Ref. 111 Analogous results have been obtained in Ref. 113 with quantum well excitons in rectangular microcavities. Currently, pillar microcavities with Q factors higher than ~150000 have been realized.[114, 115]

When the cavity contains one or several quantum wells, the interaction between 2D QW excitons and 0D cavity modes can be described by choosing suitable basis functions for the exciton states. While momentum eigenstates (i.e., plane waves with well defined in-plane center of mass wavevector) are the natural eigenstates in a planar cavity, for a pillar microcavity it is convenient to consider linear combinations of center-of-mass wavefunctions which have the same spatial and polarization dependence as the cavity mode. A detailed procedure is presented in Ref. 112 and it allows formulating a second-quantized Hamiltonian with quadratic coupling between exciton and photon states with different quantum numbers. This can be diagonalized by the Hopfield procedure, as in the case of PhC polaritons. The main results can summarized as follows: when a cavity mode is in resonance with the exciton a Rabi splitting occurs. The splitting is close to that of the planar cavity for large pillar radius, but it decreases for a radius smaller than ~1 μm due to leakage of the cavity modes and reduction of overlap with the exciton states. This result is consistent with the observation of a constant Rabi splitting for micro-pillars with radius >1.25 μm.[116] Even when no resonance condition occurs, exciton levels with different quantum numbers are split by radiative coupling with the corresponding cavity modes, although the splitting is very small and difficult to observe.

To summarize, quantum wells in pillar microcavities can give rise to exciton-polariton states, or (from another point of view) to lateral confinement of 2D cavity polaritons. The polariton splitting decreases with respect to the planar microcavity, as overlap between exciton and photon states is reduced by the spatial mismatch of the respective mode profiles. Quantum dots embedded in pillar microcavities can give rise to a Purcell effect, i.e., to a reduction of the exciton radiative lifetime because of modification of the vacuum e.m. field fluctuations. In the next Subsection we shall see that this effect can evolve into a strong-coupling regime in 0D, when the exciton-photon coupling is sufficiently large.

4.2. Quantum dots in three-dimensional microcavities: Strong coupling of 0D excitons with 0D photons

In this Subsection we give a short theoretical account of the problem of 0D excitons coupled to 0D photons, in order to make proper comparison with the previous treatment of polaritons in 3D and 2D. We are especially interested in comparing the statistical character of the excitations, as well as the coupling constant of the interaction for different dimensionalities.

We now consider an exciton resonance in a *single* quantum dot (or, more generally, a two-level system) which interacts with confined photons in a 3D microcavity. While QW excitons are described by bosonic operators, a single quantum dot transition has a fermionic character and can be described the Jaynes-Cummings model. If the QD can be modeled as a two-level system in interaction with a single cavity mode, the hamiltonian is

$$H = \hbar\omega_0\hat{\sigma}_3 + \hbar\omega_\mu \left(\hat{a}_\mu^\dagger \hat{a}_\mu + \frac{1}{2}\right) + i\hbar g(\hat{\sigma}_- \hat{a}_\mu^\dagger - \hat{\sigma}_+ \hat{a}_\mu), \qquad (79)$$

where $\hat{\sigma}_+$, $\hat{\sigma}_-$, $\hat{\sigma}_3$ are pseudo-spin operators for the two-level system with ground (excited) state $|g\rangle$ ($|e\rangle$) and $a_\mu^\dagger, a_\mu$ are creation/destruction operators for the cavity mode μ. The coupling constant $g = \langle \mathbf{d} \cdot \mathbf{E} \rangle/\hbar$ of the quantum dot-cavity interaction is

$$g = \left(\kappa_{SI} \frac{\pi e^2 f}{\epsilon_r m_0 V_\mu(\mathbf{r}_1)}\right)^{1/2}, \qquad (80)$$

where $\kappa_{SI} = 1/(4\pi\epsilon_0)$ for SI units, ϵ_∞ is the high-frequency relative permittivity, m_0 is the free-electron mass, f is the oscillator strength of the transition, and $V_\mu(\mathbf{r}_1)$ is the mode volume evaluated at the QD position defined as

$$\frac{1}{V_\mu(\mathbf{r}_1)} = \frac{\epsilon(\mathbf{r}_1)|\mathbf{E}_\mu(\mathbf{r}_1)|^2}{\int \epsilon(\mathbf{r})|\mathbf{E}_\mu(\mathbf{r})|^2 d\mathbf{r}}. \qquad (81)$$

When interaction between the two-level system and the radiation field in a 3D microcavity is treated in perturbation theory, we recover the change of spontaneous emission rate known as *Purcell effect*.[117] If the emitter has a well-defined frequency and is in resonance with a single cavity mode μ with Q-factor $Q_\mu = \omega_\mu/\gamma_\mu$, its emission rate calculated by standard perturbation theory is $\gamma_{SE} = F_\mu \gamma_0$, where γ_0 is the emission rate without the cavity and

$$F_\mu = \frac{3}{4\pi^2} \left(\frac{\lambda}{n}\right)^2 \frac{Q_\mu}{V_\mu} \tag{82}$$

is called the Purcell factor. The mode volume is calculated at the position $\mathbf{r}_1$ of the emitter; it is usually of the order of the cavity volume. The enhancement of the spontaneous emission rate is maximized if the cavity volume is small, the quality factor is large and the emitter is placed at a maximum of the electric field with the proper polarization. The enhanced spontaneous emission was first observed for self-assembled InAs quantum dots in micropillars[118–120] as well as microdisks[121] and later in photonic crystal cavities.[122] Typical measured values of the Purcell factor range from 5–10 to no more than 100, due to effects of spatial and spectral averaging.[108]

Treating Hamiltonian (79) beyond perturbation theory, its basis states consist of a ground state $|g, 0\rangle$ and of a ladder of doublets $|e, n_\mu\rangle, |g, n_\mu + 1\rangle, n_\mu = 0, 1, \ldots$, where n_μ is the number of photons in the mode. In the resonance case $\omega_0 = \omega_\mu$ each doublet gives rise to dressed states split by $2\hbar g \sqrt{n_\mu + 1}$. For $n_\mu = 0$ one has a vacuum-field Rabi splitting, while for $n_\mu \gg 1$ the physical behavior becomes that of the dynamical Stark effect with the classical Rabi splitting.

However, the Hamiltonian treatment does not take into account the finite linewidths γ_{ex}, γ_μ of the quantum dot transition and of the cavity mode — i.e., it does not yield the *conditions* for achieving strong coupling in the presence of damping. To derive these conditions, we follow the formalism of Ref. 50, which builds on previous work[123] and has been subject to generalization and debate in the last few years.[124, 125] Dampings are treated by a master equation for the density matrix:

$$\frac{d\hat{\rho}}{dt} = \frac{1}{i\hbar}[H, \hat{\rho}] + \frac{\gamma_{ex}}{2}(2\hat{\sigma}_-\hat{\rho}\hat{\sigma}_+ - \hat{\sigma}_+\hat{\sigma}_-\hat{\rho} - \hat{\rho}\hat{\sigma}_+\hat{\sigma}_-)$$
$$+ \frac{\gamma_\mu}{2}(2\hat{a}_\mu\hat{\rho}\hat{a}_\mu^\dagger - \hat{a}_\mu^\dagger\hat{a}_\mu\hat{\rho} - \hat{\rho}\hat{a}_\mu^\dagger\hat{a}_\mu). \tag{83}$$

Within this formalism, the dynamics of the quantum dot-cavity system can be calculated analytically in the case of weak excitation (i.e., when only the ground state and the lowest excited doublet are kept). We assume the quantum dot to be in resonance with the cavity mode, i.e., the QD has to be not only *spatially* but also *spectrally* resonant. The luminescence spectrum is then found to have maxima at complex frequencies given by

$$\Omega_\pm = \omega_0 - \frac{i}{4}(\gamma_{ex} + \gamma_\mu) \pm \sqrt{g^2 - \left(\frac{\gamma_{ex} - \gamma_\mu}{4}\right)^2}. \tag{84}$$

This equation is similar to (65) for planar microcavities, noting that here we are taking γ_{ex}, γ_μ to be the *full-widths at half maximum* (FWHM). For $g > |\gamma_{ex} - \gamma_\mu|/4$ there is a splitting in the real part: this is the vacuum-field Rabi splitting. Typically the coupling constant $g \sim 50\text{--}150\,\mu eV$ and the excitonic linewidth can be made much smaller, of the order of a few μeV: thus the strong-coupling condition can be put in the form $g > \gamma_\mu/4 = \omega_\mu/(4Q_\mu)$. In practice, achieving the strong-coupling regime requires having large oscillator strength, small mode volume, and high Q-factor. The figure of merit scales as $f^{1/2}Q_\mu/V_\mu$.

Notice that in the weak-coupling regime, the square root is purely imaginary and the exciton-like root has a modified radiative linewidth: this is nothing else than the increase in spontaneous emission rate, i.e., the Purcell effect. Indeed, expanding Eq. (84) in the weak-coupling regime gives the exciton decay rate as $\Gamma = 4g^2/\gamma_\mu$, which — eliminating the oscillator strength through the atomic-like formula (40) — gives back the Purcell enhancement factor (82). An example is given in Fig. 10 below.

The condition for being in strong-coupling regime is formally similar to that for quantum well excitons in planar microcavities, however the physical system has very different behavior since the quantum well exciton is a boson for weak excitation while the quantum dot transition is a fermion — in the sense that any two-particle excited level cannot be occupied more than once. The difference manifests itself in the behavior for increasing excitation intensity: the Rabi splitting of quantum well excitons does not depend on excitation level for weak enough excitation, while the Rabi splitting for a quantum dot transition increases as $(n_\mu + 1)^{1/2}$. Starting from the weak-coupling regime, the luminescence spectra in the presence of a laser beam resonant with the quantum dot transition evolve from a Lorentzian lineshape at low excitation intensity to a three-peak spectrum at high excitation[125, 126] and go over to the Mollow triplet in the classical limit.[127]

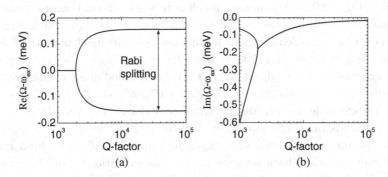

Fig. 10. (a) Real and (b) imaginary parts of the energy shifts (Eq. (84)) for the coupling between a two-level system and a cavity mode, as a function of cavity Q-factor. Parameters are appropriate to a InAs quantum dot in a GaAs photonic crystal nanocavity: $\lambda = 950\,nm$, $n = 3.54$, $\gamma_{ex} = 50\,\mu eV$, $f = 10.7$, mode volume $V_\mu = 1.1 \cdot 10^{-2}\,(\mu m)^2$. From Ref. 136.

The two most common types of quantum dots that can be incorporated in semi-conductor microcavities (micropillars, microdisks, or photonic crystal nanocavities) are InAs self-assembled QDs in GaAs and excitons localized to interface fluctuations in QWs. For the former, the dimensions are usually such that the lowest (1s-1s) transition is in a strong confinement regime, i.e., the excitonic effect gives only an energy shift to the transition and the oscillator strength *increases* with the dot radius.[128] For small spherical dots with infinite barriers for electrons and holes, the oscillator strength evaluated from Eq. (13) is estimated as

$$f = g_s \frac{2|\langle \psi_c | \hat{\epsilon} \cdot \mathbf{p} | \psi_v \rangle|^2}{m_0 \hbar \omega} = g_s \frac{2|p_{cv}|^2}{m_0 \hbar \omega} \left| \int c(\mathbf{r}) v(\mathbf{r}) d\mathbf{r} \right|^2 \simeq g_s f_0, \qquad (85)$$

where $f_0 \simeq E_P/E_g \simeq m_0/m_e^*$ is the interband oscillator strength in the bulk ($E_P = 2|p_{cv}|^2/m_0$ is the Kane energy): thus, in the limit of perfect 0D confinement, the oscillator strength tends to that of the interband transition in the bulk, taking into account the spin-orbit factor. For InAs QDs in GaAs the Kane energy $E_p \sim 18\,\mathrm{eV}$, the transition energy $E_g \sim 1.3\,\mathrm{eV}$, the spin-orbit factor $g_s = 4/3$, thus $f = (4/3)f_0 \sim 18.5$. For real quantum dots with pyramidal shape, there is no spherical symmetry and the spin-orbit factor g_s becomes become 1 instead of 4/3; the overlap integral $\int c(\mathbf{r}) v(\mathbf{r}) d\mathbf{r}$ is slightly less than unity. As a result, the oscillator strength $f < f_0$. The oscillator strength can be measured from absorption by single quantum dots[129] or from the radiative lifetime,[118, 130, 131] which is around $\tau = 1.3\,\mathrm{ns}$. A consistent estimation from all these measurements is $f \simeq 11$. With typical Q/V factors in micropillar and microdisk cavities, the oscillator strength is too small to achieve strong coupling.

In order to increase the oscillator strength one has to employ larger dots. The first report of strong coupling regime with QDs in pillar microcavities[104] used "natural" $In_{0.3}Ga_{0.7}As$ dots whose dimensions are nearly one order of magnitude larger than those of conventional InAs QDs produced by Stranski-Krastanow growth. Strong coupling is achieved with the following parameters $f \sim 50$, $V_\mu \sim 0.3\,\mu m^3$, $g \sim 45\,\mu eV$, $Q_\mu \sim 8000$. Another approach is to use excitons bound to monolayer fluctuations in quantum wells, which are shallower and with a much larger size of the center of mass: calculations as a function of confinement potential[50] lead to oscillator strengths higher than 100. Indeed, strong coupling has been observed with localized excitons in $GaAs/Al_{0.33}Ga_{0.67}As$ QWs embedded in microdisk cavities.[105] The parameters are as follows: $f \sim 100$, $V_\mu \sim 0.07\,\mu m^3$, $g \sim 200\,\mu eV$, $Q_\mu \sim 12000$.

As a last example we consider photonic crystal cavities. Very large Q-factors have been demonstrated in nanocavities realized in PhC slabs,[132–134] especially in silicon at telecom wavelengths, where Q-factors up to $2 \cdot 10^6$ have been reached. High Q-factors have also been achieved in GaAs nanocavities around 1.5 μm wavelength,[135] but for GaAs nanocavities around 900 nm wavelength the Q-factors are usually smaller, due to the smaller period and lower quality of interfaces at the hole sidewalls. Typically, L3 cavities in GaAs slabs have Q-factors of the order of 10000-15000. The crucial advantage of PhC cavities like the L3 cavity is that the mode volume can be very small, of the order of $(\lambda/n)^3$. As an example, in Fig. 10 we plot the solutions of Eq. (84) as a function of Q-factor with the parameters of InAs quantum dots in a L3 PhC cavity: $\lambda = 950$ nm, $n = 3.54$, $\gamma_{\text{ex}} = 50\,\mu\text{eV}$, $f = 10.7$, mode volume $V_\mu = 1.1 \cdot 10^{-2}\,(\mu\text{m})^2$. For small Q-factors, the imaginary part of the exciton-like root (upper curve in Fig. 10(b)) increases as a function of Q, because of the Purcell effect. The crossover from weak to strong coupling is seen to appear at $Q \sim 2000$: at this point the radiative linewidths of exciton- and cavity-like modes join and remain equal to $(\gamma_{\text{ex}} + \gamma_\mu)/2$. This phenomenon, known as *linewidth averaging*, is familiar from cavity polaritons, see Subsection 3.3. For $Q > 2000$ a vacuum Rabi splitting develops in the real part and it saturates for $Q \sim 10000$. Thus, strong coupling in PhC cavities can be achieved with Q-factors of the order of 10^4, thanks to small mode volumes. Notice the similarity with the weak-to-strong coupling crossover in planar microcavities, see Fig. 7. It should also be noticed, however, that the Rabi splitting for QDs in 3D microcavities is about one order of magnitude smaller than for cavity polaritons.

Strong coupling of quantum dots in photonic crystal cavities has been demonstrated in Refs. 106, 107. The experiment of Ref. 107, which was conducted with single quantum dots precisely placed at the center of an L3 cavity, gives evidence of quantum correlations of the emitted light through photon antibunching in the time domain. In simple words, the QD transition cannot occur more than once, unless the QD is "charged" again: this is direct evidence of "Fermionic" statistics, and it is a fundamental proof of different quantum behavior of strong coupling in 0D as compared to polaritons in either 3D or 2D.

4.3. *Exciton-photon coupling in different dimensionalities*

In this Subsection we compare the exciton-photon coupling in systems of different dimensionalities. Table 1 summarizes the expressions for the exciton-photon coupling constant $g = \langle \mathbf{d} \cdot \mathbf{E}/\hbar \rangle$ in 3D (bulk), 2D (QW excitons in planar microcavities), and 0D (QDs in three-dimensional microcavities). Contrary to a common misbelief, the exciton-photon coupling is *decreased* on reducing the dimensionality. While the

Table 1. Exciton-photon coupling in systems of different dimensionalities: 3D = bulk, 2D = quantum wells in planar microcavities, 0D = quantum dots in three-dimensional microcavities. Formulas can be read both in SI units (with the factor $\kappa_{\text{SI}} = 1/(4\pi\epsilon_0)$) or in Gaussian units (dropping the factor κ_{SI}).

System	Formula for exciton-photon coupling	Typical value (GaAs)
3D	$g = \omega_c = \sqrt{\kappa_{\text{SI}} \dfrac{\pi e^2}{\epsilon_\infty m_0} \dfrac{f}{V}} = \sqrt{\dfrac{\omega_0 \omega_{\text{LT}}}{2}}$	$\sim 8\,\text{meV}$
2D	$g = \sqrt{\kappa_{\text{SI}} \dfrac{2\pi e^2}{\epsilon_\infty m_0} \dfrac{f_{xy}}{S L_{\text{eff}}}} = \sqrt{\omega_0 \tilde{\omega}_{\text{LT}} \dfrac{L_{\text{w}}}{L_{\text{eff}}}}$	$\sim 1.5\,\text{meV}$
0D	$g = \sqrt{\kappa_{\text{SI}} \dfrac{\pi e^2}{\epsilon_\infty m_0} \dfrac{f}{V_\mu}}$	$\sim 0.05 - 0.15\,\text{meV}$

Table 2. Oscillator strength per unit volume f/V and effective volume V_{eff} in systems of different dimensionalities: 3D = bulk, 2D = quantum wells in planar microcavities, 0D = quantum dots in three-dimensional microcavities.

System	f/V	V_{eff}	Typical value (GaAs)
3D	$\frac{f}{V} = g_s \frac{f_0}{\pi a_B^{*3}} \simeq g_s \frac{m_0}{m^*} \frac{1}{\pi a_B^{*3}}$	$V_{\text{eff}} = V_{\text{exciton}} = \pi a_B^{*3}$	$\pi \cdot (10\,\text{nm})^3$
2D	$\frac{f}{V} = \frac{f_{xy}}{S} \frac{2}{L_{\text{eff}}} \simeq g_s \frac{m_0}{m^*} \frac{1}{\pi a_{2D}^2}$	$V_{\text{eff}} = S_{2D\,\text{exciton}} \frac{L_{\text{eff}}}{2}$	$\pi \cdot (10\,\text{nm})^2 \cdot 0.5\,\mu\text{m}$
0D	$\frac{f}{V} = \frac{f_0}{V_\mu} \simeq g_s \frac{m_0}{m^*} \frac{1}{V_\mu}$	$V_{\text{eff}} = V_\mu \geq \left(\frac{\lambda}{n}\right)^3$	$> (0.25\,\mu\text{m})^3$

coupling is obviously fixed in 3D, it can to some extent be tuned in 2D and 0D. The exciton-photon coupling can be increased by increasing the oscillator strength (putting more identical QWs in 2D, using larger quantum dots in 0D) or by decreasing the mode volume (decreasing the mirror penetration depth in 2D, decreasing the cavity mode volume in 0D).

To better understand the reasons for the decrease of exciton-photon coupling on reducing the dimensionality, we write the coupling constant as

$$g = \sqrt{\kappa_{\text{SI}} \frac{\pi e^2}{\epsilon_\infty m_0} \frac{f}{V}} \equiv \sqrt{\kappa_{\text{SI}} \frac{\pi e^2}{\epsilon_\infty m_0} \frac{f_0}{V_{\text{eff}}}}, \tag{86}$$

where f_0 is the bulk interband oscillator strength, Eq. (14), and V_{eff} is an effective volume. In Table 2 we report the oscillator strength and the effective volume in 3D, 2D and 0D. In the bulk, thanks to wavevector conservation, photon coupling to the exciton center-of-mass wavefunction is coherent to the maximum possible extent and the effective volume is equal to the exciton volume, i.e., of the order of 10 nm per dimension. On reducing the dimensionality, the effective volume is *increased*: for any direction of confinement, the effective volume is at best of the order of λ/n, i.e., 200 nm or so. This explains why the exciton-photon coupling constant is actually *reduced* on reducing the dimensionality. Nevertheless, control over radiation-matter coupling is improved (even if the interaction is decreased) because of the modifications in the *density of states*. The optimum situation for tailoring radiation-matter interaction is in 0D, as we can exploit the narrow density of states of 0D-confined electronic and photonic resonances.

5. Conclusions

We can conclude these lectures on exciton-polaritons with the following general considerations.

Stationary polaritons can exist only when exciton and photon states have the same dimensionality, or else when the dimensionality of the exciton states is *higher* than that of the photons. The former is the case of the bulk (3D), planar microcavity of photonic crystal slab with quantum wells (2D). The latter case occurs, e.g., in bulk microcavities (3D excitons interacting with 2D photons), in micropillar or photonic crystal cavities with quantum wells (2D excitons with 0D photons).

When the dimensionality of the exciton states is *smaller* than that of the photons, radiative excitons have an intrinsic decay mechanism due to coupling of a single exciton state with a density of photon modes. Typical examples are excitons in quantum wells or quantum wires, but also quantum dot transitions in a bulk material or in planar microcavities. In this case, one should use a more general definition of polariton effects as being the real and imaginary parts of the energy shift due to interaction of the exciton with the retarded electromagnetic field.

The observability of polariton effects (i.e., their robustness against scattering processes and disorder) depends on the experiment being performed through the criteria of temporal and spatial coherence: the former applies when the wavevector is well-defined and real, while the latter applies when the frequency is real and the polariton distribution is nonuniform. Polaritons in planar microcavities are easier to observe than bulk polaritons, although the exciton-photon matrix element is smaller, because there is no propagation along the cavity axis and the relevant criterion is that of temporal coherence.

The concept of *polariton* is closely related to *wavevector conservation*, which implies coherent coupling between exciton and photon. Exciton-polaritons have a dispersion relation along the directions of free motion, and this dispersion determines also their statistical properties: excitons and polaritons in bulk, quantum wells, and quantum wires are approximate bosons, as the eigenstates are delocalized and several states can be populated with no (or little) interaction, to a first approximation. The same holds for excitons in an *ensemble* of quantum dots at low excitation level. On the other hand, excitons in *single* quantum dots with a sufficiently large spacing between levels have different statistical properties, as the Pauli principle due to Fermi statistics applies and each excitation can be populated only once (when spin degrees of freedom are taken into account). The same is true for quantum-dot excitons in 3D microcavities, leading to strongly coupled exciton-photon states in 0D — the "dressed states", in atomic physics language. The different statistical properties of bosonic versus fermionic excitations become manifest only on increasing the excitation intensity — leading to Bose-Einstein condensation for 2D polaritons, and to nonlinear quantum correlations for 0D strongly coupled exciton-photon modes.

Exciton-photon coupling is expressed by a dipole coupling g, which in turn depends on the square root of the oscillator strength per unit volume. The oscillator strength can be increased by decreasing the average electron-hole separation (in the case of e-h confinement) or by increasing the volume occupied by the center of mass (in the case of center-of-mass confinement). The analysis performed in Subsection 4.3 shows that the exciton-photon coupling is *decreased* on decreasing the dimensionality. Nevertheless, better control over radiation-matter interaction is achieved, due to confinement of the motion along the cavity axis (for quantum-well excitons in planar microcavities) or due to narrowing of electronic and photonic density of states (for quantum dots in 3D microcavities). While polaritons in 3D, 2D and 1D are delocalized states with a continuous energy spectrum, it is only by exploiting the narrow quantum dot and cavity lines in 0D that strongly coupled exciton-photon systems resemble the dressed atom-photon states of cavity quantum electrodynamics.

References

1. K. Huang, *Proc. Roy. Soc. A* **208**, 352, (1951).
2. M. Born and K. Huang, *Dynamical Theory of Crystal Lattices* (Clarendon Press, Oxford), 1954.
3. U. Fano, *Phys. Rev.* **103**, 1202, (1956).
4. J. J. Hopfield, *Phys. Rev.* **112**, 1555, (1958).
5. V. M. Agranovich, *J. Exptl. Theoret. Phys.* **37**, 430, (1959) [*Sov. Phys. JETP* **37**, 307, (1960)].
6. C. Weisbuch, M. Nishioka, A. Ishikawa, and Y. Arakawa, *Phys. Rev. Lett.* **69**, 3314, (1992).
7. P. G. Savvidis, J. J. Baumberg, R. M. Stevenson, M. S. Skolnick, D. M. Whittaker, and J. S. Roberts, *Phys. Rev. Lett.* **84**, 1547, (2000).
8. J. Kasprzak, M. Richard, S. Kundermann, A. Baas, P. Jeambrun, J. M. J. Keeling, F. M. Marchetti, M. H. Szymańska, R. André, J. L. Staehli, V. Savona, P. B. Littlewood, B. Deveaud, and Le Si Dang, Nature (London) **443**, 409 (2006).
9. D. Bajoni, D. Gerace, M. Galli, J. Bloch, R. Braive, I. Sagnes, A. Miard, A. Lemaître, M. Patrini, and L. C. Andreani, *Phys. Rev. B* **80**, 201308(R), (2009).
10. C. F. Klingshirn, *Semiconductor Optics*, 3rd edition. (Springer-Verlag, Berlin, 2006).
11. P. Y. Yu and M. Cardona *Fundamentals of Semiconductors*, 4th edition. (Springer-Verlag, Berlin, 2010).
12. L. C. Andreani, in *Confined Electrons and Photons- new Physics and Devices*, edited by E. Burstein and C. Weisbuch (Plenum Press, New York) 1995, p. 57.
13. R. S. Knox, in *Solid State Physics*, suppl. 5, edited by F. Seitz and D. Turnbull (Academic Press, New York, 1963).
14. J. O. Dimmock, in *Semiconductors and Semimetals*, Vol. 3, edited by R. K. Willardson and A. C. Beer (Academic Press, New York, 1967), p. 259.
15. F. Bassani and G. Pastori Parravicini, *Electronic States and Optical Transitions in Solids*. (Pergamon Press, Oxford, 1975).
16. G. Grosso and G. Pastori Parravicini, *Solid State Physics*. (Academic Press, London, 2000).
17. M. M. Denisov and V. P. Makarov, *Phys. Status Solidi (b)* **56**, 9, (1973).
18. U. Rössler and H.-R. Trebin, *Phys. Rev. B* **23**, 1961, (1981).
19. K. Ehara and K. Cho, *Solid State Commun.* **44**, 453, (1982).
20. K. Cho, *Solid State Commun.* **33**, 911, (1980).
21. C. Kittel, Introduction to Solid State Physics, 8th edition (Wiley, New York, 2005).
22. S. I. Pekar, *J. Exptl. Theoret. Phys.* **33**, 1022, (1957) [*Sov. Phys. JETP* **6**, 785, (1958)].
23. V. M. Agranovich and V. L. Ginzburg, *Spatial Dispersion in Crystal Optics and the Theory of Excitons* (Interscience Publ., London, 1966).

24. S. I. Pekar, *Crystal Optics and Additional Light Waves* (Benjamin-Cummings, Menlo Park, California, 1983).
25. L. C. Andreani, in *Electron and Photon Confinement in Semiconductor Nanostructures*, Proc. of the International School of Physics "E. Fermi", Course CL, edited by B. Devaud, A. Quattropani and P. Schwendimann (IOS Press, Amsterdam, 2003), p. 105.
26. K. Henneberger, *Phys. Rev. Lett.* **80**, 2889, (1998).
27. D. F. Nelson and B. Chen, *Phys. Rev. Lett.* **83**, 1263, (1999); R. Zeyher, *ibid.* **83**, 1264, (1999); K. Henneberger, *ibid.* **83**, 1265, (1999).
28. S. Schumacher, G. Czycholl, F. Jahnke, I. Kudyk, H. I. Rckmann, J. Gutowski, A. Gust, G. Alexe and D. Hommel, *Phys. Rev. B* **70**, 235340, (2004).
29. L. N. Bulaevskii, Ch. Helm, A. R. Bishop, and M. P. Maley, *Europhys. Lett.* **58**, 415, (2002).
30. Ch. Helm and L. N. Bulaevskii, *Phys. Rev. B* **66**, 094514, (2002).
31. C. Weisbuch and R. G. Ulbrich, in *Light Scattering in Solids III*, edited by M. Cardona and G. Günterodt, Topics in Applied Physics, Vol. **51** (Springer, Berlin, 1982) p. 207.
32. B. Hönerlage, R. Lévy, J. B. Grun, C. Klingshirn, and K. Bohnert, *Phys. Rep.* **124**, 163, (1985).
33. F. Bassani F. and L. C. Andreani, in *Excited-State Spectroscopy in Solids*, Proc. of the International School of Physics "E. Fermi", Course XCVI, edited by Grassano U. and Terzi N. (Academic Press, Amsterdam) 1987 p. 1.
34. D. D. Sell, S. E. Stokovski, R. Dingle, and J. V. DiLorenzo, *Phys. Rev. B* **7**, 4568, (1973).
35. C. Weisbuch and R. G. Ulbrich, *Phys. Rev. Lett.* **38**, 865, (1977).
36. N. N. Bogoljubov, *Nuovo Cimento* **7**, 794, (1958).
37. A. Quattropani, L. C. Andreani, and F. Bassani, *Il Nuovo Cimento D* **7**, 55, (1986).
38. C. Ciuti, G. Bastard, and I. Carusotto, *Phys. Rev. B* **72**, 115303 (2005).
39. W. C. Tait, *Phys. Rev. B* **5**, 648, (1972).
40. R. Loudon, *J. Phys. A* **3**, 233, (1970).
41. Y. Toyozawa, *Prog. Theor. Phys. Suppl.* **12**, 111, (1959).
42. C. Weisbuch, H. Benisty, and R. Houdré, *J. Lumin.* **85**, 271, (2000)
43. E. I. Rashba and G. E. Gurgenishvili, *Fiz. Tverd. Tela* **4**, 1029, (1982) [*Sov. Phys.- Solid State* **4**, 759, (1962)].
44. C. H. Henry, and K. Nassau, *Phys. Rev. B* **1**, 1628, (1970).
45. G. Bastard, *Wave Mechanics Applied to Semiconductor Heterostructure* (Les Editions de Physique, Paris) 1989.
46. J. H. Davies, *The Physics of Low-Dimensional Semiconductors* (Cambridge University Press) 1998.
47. L. C. Andreani, A. D'Andrea, and R. Del Sole, *Phys. Lett. A* **168**, 451, (1992).
48. R. Iotti and L. C. Andreani, *Phys. Rev. B* **56**, 3922, (1997).
49. R. Iotti, L. C. Andreani, and M. Di Ventra *Phys. Rev. B* **57**, R15072, (1998).
50. L. C. Andreani, G. Panzarini, and J.-M. Gérard, *Phys. Rev. B* **60**, 13276, (1999); *Physica Status Solidi (a)* **178**, 145, (2000).
51. D. Schiumarini, N. Tomassini, L. Pilozzi, and A. DAndrea, *Phys. Rev. B* **82**, 075303 (2010).
52. R. Kubo, *J. Phys. Soc. Jpn.* **12**, 570, (1957).
53. D. A. Dahl and L. J. Sham, *Phys. Rev. B* **16**, 651, (1977).
54. K. Cho, *J. Phys. Soc. Jpn.* **55**, 4113, (1986).
55. A. D'Andrea and R. Del Sole, *Phys. Rev. B* **41**, 1413, (1990).

56. H. Ishihara and K. Cho, *Phys. Rev. B* **41**, 1424, (1990).
57. A. Stahl and I. Balslev, *Electrodynamics of the Semiconductor Band Edge* (Springer, Berlin) 1987.
58. H. C. Schneider, F. Jahnke, S. W. Koch, J. Tignon, T. Hasche, and D. S. Chemla, *Phys. Rev. B* **63**, 045202, (2001).
59. H. Haug and S. W. Koch, *Quantum Theory of the Optical and Electronic Properties of Semiconductors*, 4th ed. (World Scientific, Singapore, 2004).
60. V. M Agranovich and O. A. Dubovskii, *JETP Lett.* **3**, 223, (1966).
61. E. Hanamura, *Phys. Rev. B* **38**, 1228, (1988).
62. L. C. Andreani and F. Bassani, *Phys. Rev. B* **41**, 7536, (1990).
63. F. Tassone, F. Bassani, and L. C. Andreani, *Nuovo Cimento D* **12**, 1673, (1990).
64. L. C. Andreani, F. Tassone and F. Bassani, *Solid State Commun.* **77**, 641, (1991).
65. B. Deveaud, F. Clérot, N. Roy, K. Satzke, B. Sermage, and D. S. Katzer, *Phys. Rev. Lett.* **67**, 2355, (1991).
66. D. S. Citrin, *Phys. Rev. B* **47**, 3832, (1993).
67. M. Kohl, D. Heitmann, P. Grambow, and K. Ploog, *Phys. Rev. B* **37**, 10927, (1988).
68. P. Vledder, A. V. Akimov, J. I. Dijkhuis, J. Kusano, Y. Aoyagi, and T. Sugano, *Phys. Rev. B* **56**, 15282, (1997).
69. J. Martinez-Pastor, A. Vinattieri, L. Carraresi, M. Colocci, Ph. Roussignol, and G. Weimann, *Phys. Rev. B* **47**, 10456, (1993) 10456.
70. C. Creatore, A. L. Ivanov, *Phys. Rev. B* **77**, 075324, (2008).
71. E. L. Ivchenko, *Fiz. Tverd. Tela* **33**, 2388, (1991) [*Sov. Phys. Solid State* **33**, 1344, (1991)].
72. D. S. Citrin, *Phys. Rev. Lett.* **69**, 3393, (1992).
73. C. Weisbuch, R. C. Miller, R. Dingle, A. C. Gossard, and W. Wiegmann, *Solid State Commun.* **37**, 219, (1981).
74. M. Born and E. Wolf, *Principles of Optics* (Pergamon Press, Oxford) 1970.
75. A. Yariv and P. Yeh, *Optical Waves in Crystals* (Wiley, New York) 1984.
76. G. Björk, S. Machida, Y. Yamamoto, and K. Igeta, *Phys. Rev. A* **44**, 669, (1991).
77. H. Yokoyama, *Science* **256**, 66, (1992).
78. G. Björk and Y. Yamamoto, in *Spontaneous Emission and Laser Oscillations in Microcavities*, edited by Yokoyama H. and Ujihara K. (CRC Press, New York) 1995 p. 189.
79. M. S. Skolnick, T. A. Fisher, and D. M. WHittaker, *Semicond. Sci. Technol.* **13**, 645, (1998).
80. V. Savona, C. Piermarocchi, A. Quattropani, P. Schwendimann, and F. Tassone, *Phase Transitions* **68**, 169, (1999).
81. G. Khitrova, H. M. Gibbs, F. Jahnke, M. Kira, and S. W. Koch, *Rev. Mod. Phys.* **71**, 1591, (1999).
82. V. Savona, *J. Phys.: Condens. Matter* **19**, 295208, (2007).
83. L. C. Andreani, *Phys. Lett. A* **192**, 99, (1994).
84. V. Savona, L. C. Andreani, A. Quattropani, and P. Schwendimann, *Solid State Commun.* **93**, 733, (1995).
85. Y. Chen, A. Tredicucci, and F. Bassani, *Phys. Rev. B* **52**, 1800, (1985).
86. G. Panzarini, L. C. Andreani, A. Armitage, D. Baxter, M. S. Skolnick, V. N. Astratov, J. S. Roberts, A. V. Kavokin, M. R. Vladimirova, and M. A. Kaliteevski, *Phys. Rev. B* **59**, 198, (1999); *Physics of the Solid State* **41**, 1223, (1999).
87. R. Houdré, R. P. Stanley, U. Oesterle, M. Ilegems, and C. Weisbuch, *Phys. Rev. B* **73**, 2043, (1994).
88. A. Kavokin, J. J. Baumberg, G. Malpuech, F. P. Laussy, Microcavities *Microcavities*, revised ed. (Oxford University Press, 2011).

89. D. Sanvitto and V. Timofeev, Editors, *Exciton Polaritons in Microcavities: New Frontiers*, Springer Series in Solid State Sciences Vol. **172** (Springer, 2012).

90. K. Sakoda, *Optical Properties of Photonic Crystals* (Springer, Berlin, 2001).

91. S. G. Johnson and J. D. Joannopoulos, *Photonic Crystals: The Road from Theory to Practice* (Kluwer Academic Publishers, Boston, 2002).

92. K. Busch, S. Lölkes, R. B. Wehrspohn, and H. Föll (eds.), *Photonic Crystals: Advances in Design, Fabrication, and Characterization* (Wiley-VCH, Weinheim, 2004).

93. J.-M. Lourtioz, H. Benisty, V. Berger, J.-M. Gérard, D. Maystre, and A. Tchelnokov (eds.), *Photonic Crystals: Towards Nanoscale Photonic Devices* (Springer-Verlag, Berlin, 2005).

94. H. Benisty and C. Weisbuche, *Photonic Crystals*, in Progress in Optics Vol. **49**, edited by E. Wolf (Elsevier, 2006).

95. L. C. Andreani and D. Gerace, *Phys. Rev. B* **73**, 235114, (2006).

96. D. Gerace and L. C. Andreani, *Opt. Express* **13**, 4939, (2005).

97. D. Gerace and L. C. Andreani, *Photon. Nanostruct. Fundam. Appl.* **3**, 120, (2005).

98. D. Gerace and L. C. Andreani, *Phys. Rev. B* **75**, 235325, (2007).

99. T. Fujita, Y. Sato, T. Kuitani, and T. Ishihara, *Phys. Rev. B* **57**, 12428, (1998).

100. A. L. Yablonskii, E. A. Muljarov, N. A. Gippius, S. G. Tikhodeev, T. Fujita, and T. Ishihara, *J. Phys. Soc. Jpn.* **70**, 1137, (2001).

101. L. C. Andreani, D. Gerace, and M. Agio, *Photon. Nanostr. Fundam. Appl.* **2**, 103, (2004).

102. L. C. Andreani and D. Gerace, *Physica Status Solidi (b)* **244**, 3528, (2007).

103. S. Azzini, D. Gerace, M. Galli, I. Sagnes, R. Braive, A. Lemaître, J. Bloch, and D. Bajoni, *Appl. Phys. Lett.* **99**, 111106, (2011).

104. J. P. Reithmaier, G. Sek, A. Löffler, C. Hofmann, S. Kuhn, S. Reitzenstein, L. V. Keldysh, V. D. Kulakovskii, T. L. Reinecke, and A. Forchel, *Nature* **432**, 197, (2004).

105. E. Peter, P. Senellart, D. Martrou, A. Lemaitre, J. Hours, J. M. Gérard, and J. Bloch, *Phys. Rev. Lett.* **95**, 067401, (2005).

106. T. Yoshie, A. Scherer, J. Hendrickson, G. Khitrova, H. M. Gibbs, G. Rupper, C. Ell, O. B. Shchekin, and D. G. Deppe, *Nature* **432**, 2002, (2004).

107. K. Hennessy, A. Badolato, M. Winger, D. Gerace, M. Atatüre, S. Gulde, S. Fält, E. L. Hu, and A. Imamoglu, *Nature* **445**, 896, (2007).

108. J. M. Gérard, *Top. Appl. Phys.* **90**, 269, (2003).

109. J. D. Jackson, *Classical Electrodynamics*, second edition (Wiley, New York) 1975.

110. N. S. Kapany and J. J. Burke, *Optical Waveguides* (Academic Press, New York) 1972.

111. J.-M. Gerard, D. Barrier, J.-Y. Marzin, R. Kuszelewicz, L. Manin, E. Costard, V. Thierry-Mieg, and T. Rivera, *Appl. Phys. Lett.* **69**, 449, (1996).

112. G. Panzarini and L. C. Andreani, *Phys. Rev. B* **60**, 16799, (1999).

113. J. P. Reithmaier, M. Röhner, H. Zull, F. Schäfer, A. Forchel, P. A. Knipp, and T. L. Reinecke, *Phys. Rev. Lett.* **78**, 378, (1997).

114. S. Reitzenstein, C. Hofmann, A. Gorbunov, M. Gorbunov, M. Straub, S. H. Kwon, C. Schneider, A. Löffler, S. Höfling, M. Kamp, and A. Forchel, *Appl. Phys. Lett.* **90**, 251109, (2007).

115. C. Arnold, V. Loo, A. Lemaitre, I. Sagnes, O. Krebs, P. Voisin, P. Senellart, L. Lanco, *Appl. Phys. Lett.* **100**, 111111, (2012).

116. J. Bloch, F. Boeuf, J.-M. Gérard, B. Legrand, J.-Y. Marzin, R. Planel, V. Thierry-Mieg, and E. Costard, *Physica E* **2**, 915, (1998).

117. E.M. Purcell, *Phys. Rev.* **69**, 681, (1946).

118. J.-M. Gérard, B. Sermage, B. Gayral, B. Legrand, E. Costard, and V. Thierry-Mieg, *Phys. Rev. Lett.* **81**, 1110, (1998).

119. M. Bayer, T. L. Reinecke, F. Weidner, A. Larionov, A. McDonald, and A. Forchel, *Phys. Rev. Lett.* **86**, 3168, (2001).

120. G. S. Solomon, M. Pelton, and Y. Yamamoto, *Phys. Rev. Lett.* **86**, 3903, (2001).

121. B. Gayral, J.-M. Gérard, B. Sermage, A. Lemaitre, and C. Dupuis, *Appl. Phys. Lett.* **78**, 2828, (2001).

122. See e.g. A. Kress, F. Hofbauer, N. Reinelt, M. Kaniber, H. J. Krenner, R. Meyer, G. Bohm, and J. J. Finley, *Phys. Rev. B* **71**, 241304, (2005); D. Englund, D. Fattal, E. Waks, G. Solomon, B. Zhang, T. Nakaoka, Y. Arakawa, Y. Yamamoto, and J. Vuckovic, *Phys. Rev. Lett.* **95**, 013904, (2005); W.-Y. Chen, H.-S. Chang, T.-P. Hsieh, J.-I. Chyi, and T.-M. Hsu, *Phys. Rev. Lett.* **96**, 117401, (2006).

123. H. J. Carmichael, R. J. Brecha, M. G. Raizen, H. J. Kimble, and P. R. Rice, *Phys. Rev. A* **40**, 5516, (1989).

124. F. Laussy, E. Del Valle, and C. Tejedor, *Phys. Rev. B* **79**, 235325, (2009).

125. E. Del Valle, F. Laussy, and C. Tejedor, *Phys. Rev. B* **79**, 235326, (2009).

126. L. C. Andreani, in *Radiation-Matter Interaction in Confined Systems*, edited by L. C. Andreani, G. Benedek, and E. Molinari (Società Italiana di Fisica, Bologna) 2002, p. 11.

127. B. R. Mollow, *Phys. Rev.* **188**, 1969, (1969).

128. Y. Kayanuma, *Phys. Rev. B* **38**, 9797, (1988).

129. R. J. Warburton, C. S. Dürr, K. Karrai, J. P. Kotthaus, G. Medeiros-Ribeiro, and P. M. Petroff, *Phys. Rev. Lett.* **79**, 5282, (1997).

130. W. Langbein, P. Borri, U. Woggon, V. Stavarache, D. Reuter, and A. D. Wieck, *Phys. Rev. B* **70**, 033301, (2004).

131. J. Johansen, S. Stobbe, I. S. Nikolaev, T. Lund-Hansen, P. T. Kristensen, J. M. Hvam, W. L. Vos, and P. Lodahl, *Phys. Rev. B* **77**, 073303, (2008).

132. T. Akahane, T. Asano, B.-S. Song, and S. Noda, *Nature* **425**, 944, (2003).

133. M. Notomi, A. Shinya, S. Mitsugi, E. Kuramochi, and H.-Y. Ryu, *Opt. Express* **12**, 1551, (2004).

134. B.-S. Song, S. Noda, T. Asano, and Y. Akahane, *Nature Materials* **4**, 207 (2005).

135. E. Weidner, S. Combrié, N.-V.-Q. Tran, A. De Rossi, J. Nagle, S. Cassette, A. Talneau, and H. Benisty, *Appl. Phys. Lett.* **89**, 221104, (2006)

136. L. C. Andreani, D. Gerace, and M. Agio, *Physica Status Solidi (b)* **242**, 2197, (2005).

Chapter 3

Experimental Circuit QED

Patrice Bertet

Quantronics Group, SPEC/IRAMIS/DSM, CEA Saclay
91191 Gif-sur-Yvette CEDEX, France
patrice.bertet@cea.fr

Superconducting qubits are electrical circuits that behave as artificial two-level systems when cooled at low temperatures; they have allowed the implementation of basic quantum information protocols with up to three entangled qubits. Modern ways of manipulating and reading out these qubits all involve strongly coupling them to superconducting resonators, giving rise to the field of circuit quantum electrodynamics (circuit QED).

Introduction

In building a circuit QED setup, one of the main challenges consists in managing to cool down the superconducting circuit to its quantum ground state, which requires complete decoupling from noise coming from the room-temperature environment, while at the same time being able to measure its final state at the end of the experimental sequence with low-power time-resolved microwave impedance measurements. In these notes, the standard experimental techniques that are used in all circuit QED experiments are presented. The first section introduces useful concepts and notations. The goal of the second section is to present the "standard circuit QED" setup, that we will justify starting from simpler setups, and to briefly discuss the signal-to-noise issues related to qubit readout.

1. Basics of circuit QED

Several comprehensive reviews on circuit QED are available by now.[1] The goal here is only to introduce the basic concepts and notations needed to discuss the specific issues that we want to tackle in these notes.

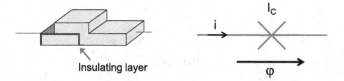

Fig. 1.   A Josephson junction is a tunnel junction between two superconducting electrodes separated by an insulating layer. It is characterized by its critical current I_C, and by the difference of
superconducting phase in the electrodes connected to the junction φ. In Josephson circuits, it is
represented by a cross symbol.

1.1.   *Basics of Josephson junctions and SQUIDs*

Superconducting qubits and quantum circuits are all based on the properties of Josephson junctions. A Josephson junction is a tunnel junction
between two superconducting electrodes, often of the type Aluminum/Aluminum
Oxide/Aluminum (see Fig. 1). Its electrodynamics properties are entirely described
by the two Josephson relations linking the current i and voltage v to φ, φ being the
difference of superconducting phase in the electrodes connected to the junction:

$$i = I_c \sin \varphi \tag{1}$$

$$v = \varphi_0 \frac{d\varphi}{dt} \tag{2}$$

where I_c is the junction critical current, and $\varphi_0 = \hbar/2e$ is the reduced superconducting flux quantum. These two relations imply that

$$v = \frac{L_J}{\cos \varphi} \frac{di}{dt}, \tag{3}$$

with $L_J = \varphi_0/I_C$. From an electrical engineering point of view, a Josephson junc-
tion thus behaves as a very peculiar point-like inductance, with a value $L_J(\varphi) = L_J/\cos\varphi$. Importantly, this inductance is non-linear, since its value depends on
the current i passing through the junction via the first Josephson relation. This
Josephson inductance is both non-linear and entirely non-dissipative, thanks to
superconductivity;[1] it is therefore ideally suited to form the building block of superconducting quantum circuits as explained below. In circuit notation a Josephson
junction is commonly represented as a cross (see Fig. 1). The potential energy associated to the Josephson inductance is $E_J(\varphi) = -E_J \cos \varphi$, with $E_J = \varphi_0 I_C$ the
so-called Josephson energy of the junction.

It is worth noting that the strength of the Josephson inductance non-linearity is
directly linked to I_C. Indeed, the Josephson inductance can be written as $L_J(i) = L_J/\sqrt{1-(i/I_C)^2}$ which can be developed as $L_J(i) = L_J + (L_J/2I_C^2)i^2$. Therefore,
roughly speaking, Josephson junctions can be treated as linear inductances as long
as $i/I_C \ll 1$. Depending on the application, it may be desired to use only the
linear part of the Josepshon inductance, in which case junctions with the largest

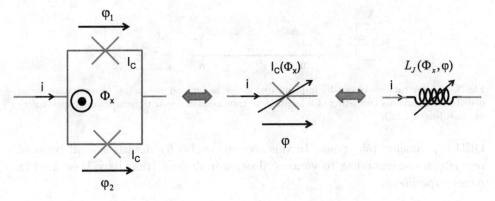

Fig. 2.   A SQUID is a superconducting loop with two Josephson junctions, biased by an external flux Φ_x and a current i. It behaves as a single Josephson junction with tunable critical current $I_C(\Phi_x)$(see text), and thus also as a tunable nonlinear Josephson inductance $L_J(\Phi_x, \varphi)$.

possible critical current should be used; on the opposite, in the applications where its non-linearity is essential, Josephson junctions with small critical currents will be prefered. In some qubit circuits, the non-linearity of the inductance should play a role even when a single quantum of microwave energy is stored in the junction, which corresponds to currents on the order of 10–20 nA; junctions having critical currents of the same order of magnitude will thus be used.

Another essential circuit for superconducting quantum electronics is the SQUID, which is a superconducting loop with two Josephson junctions (see Fig. 2) threaded by an externally applied magnetic fluc Φ_x. In such a loop, provided the loop self-inductance is negligible compared to the Josephson inductance of the junctions, the phases $\varphi_{1,2}$ across each junction verify the phase quantization relation $\varphi_1 - \varphi_2 = \Phi_x/\varphi_0$. Combined with the first Josephson relation applied to each junction, this yields that the SQUID bias current i is related to the average phase $\varphi \equiv (\varphi_1 + \varphi_2)/2$ as $i = I_C(\Phi_x) \cos \varphi$, with $I_C(\Phi_x) = 2I_C|\cos(\Phi_x/\varphi_0)|$. A SQUID with negligible loop inductance is thus entirely equivalent to an effective Josephson junction having a critical current tunable by an external magnetic flux. This also means that a SQUID is a flux-tunable inductance (see Fig. 2). It should however be kept in mind that this flux-tunable inductance will also have a non-linear character for bias currents comparable to $I_C(\Phi_x)$, as explained above.

1.2.   *Josephson resonators in the quantum regime*

Combining the inductive response of Josephson junctions with capacitances, one obtains resonators that inherit the properties of the junctions: non-linearity, flux-tunability, ... while keeping a large quality factor since the Josephson inductance is non-dissipative. If these resonators are cooled to a sufficiently low temperature that there is no thermal noise at their resonance frequency, they are in addition brought in the quantum regime. This is precisely the regime in which circuit

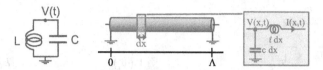

Fig. 3.   Linear resonators can be implemented either as lumped element L, C resonators, or as distributed resonators consisting of a length Λ of transmission line with inductance (capacitance) per unit length $\mathcal{L}$ $(\mathcal{C})$.

QED experiments take place. In this section, we briefly discuss several types of resonators, corresponding to various "Josephson devices" that have been used in recent experiments.

1.2.1.   Linear resonators

We start with the simplest resonator: the linear resonator. It is most simply implemented by a linear inductance L in parallel with a capacitance C (see Fig. 3). As is well known, the Hamiltonian of this circuit is $H/\hbar = \omega_0 a^\dagger a$, $\omega_0 = 1/\sqrt{LC}$ being the resonator frequency, and a $(a^\dagger)$ being the annihilation (creation) operator of the corresponding mode. All the electrodynamic quantities such as current i, voltage v, flux through the inductance Φ, charge on the capacitor Q, are readily expressed as combinations of a and $a^\dagger$:[1]

$$\hat{v} = \frac{Q}{C} = \delta v_0 (a + a^\dagger) \tag{4}$$

$$\hat{i} = \frac{\Phi}{L} = \delta i_0 (-a + a^\dagger) \tag{5}$$

with $\delta v_0 = \omega_0 \sqrt{\frac{\hbar z_0}{2}}$ and $\delta i_0 = i\omega_0 \sqrt{\frac{\hbar}{2z_0}}$, $z_0 = \sqrt{L/C}$ being the resonator characteristic impedance. Our resonators operate at typical frequency around 5–10 GHz, their characteristic impedance being close to 50 Ω. With such figures, the vacuum fluctuations of the voltage and current are of order $\delta v_0 = 2$–$3\,\mu$V and $\delta i_0 = 30$–50 nA.

Linear resonators can also be implemented with a length Λ of transmission line of inductance (capacitance) per unit length $\mathcal{L}$ $(\mathcal{C})$, as shown in Fig. 3. Such distributed resonator supports an infinite number of discrete modes of resonance frequency $\omega_n = (n+1)\pi\bar{c}/\Lambda$, with $\bar{c} = 1/\sqrt{\mathcal{L}\mathcal{C}}$. Restricting ourselves to the fundamental mode $n = 0$, its Hamiltonian is, as previously, $H/\hbar = \omega_0 a^\dagger a$. The expression for currents and voltages are now position dependent and given by

$$\hat{V}(x) = \delta V_0 \cos \frac{\pi x}{\Lambda}(a + a^\dagger) \tag{6}$$

$$\hat{I}(x) = \delta I_0 \sin \frac{\pi x}{\Lambda}(-a + a^\dagger) \tag{7}$$

with $\delta V_0 = \omega_0 \sqrt{\hbar Z_0/\pi}$ and $\delta I_0 = \omega_0 \sqrt{\hbar/\pi Z_0}$, $Z_0 = \sqrt{\mathcal{L}/\mathcal{C}}$ being the transmission line characteristic impedance. A popular choice of transmission line geometry in cQED experiments is the so-called coplanar waveguide (see Fig. 4). The planar

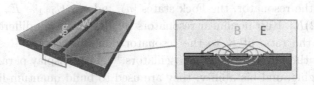

Fig. 4.   A coplanar waveguide (CPW) is a planar transmission line consisting of a central line of width w surrounded by two ground planes at a distance g. The transverse structure of the electric and magnetic fields in the CPW mode are schematically shown in the right panel.

transmission line consists of a central conductor surrounded on two sides by a ground plane.

In addition to their frequency ω_0 and characteristic impedance Z_0, resonators are also characterized by the rate at which the stored energy is damped. This damping rate has two very distinct origins: first, a resonator is usually coupled to measurement transmission lines, used to send microwave signals at the resonator input and/or to retrieve the signals coming out of the resonator. This damping rate is usually called κ_{ext}, corresponding to a quality factor $Q_{ext} = \omega_0/\kappa_{ext}$. The second contribution to energy damping κ_{int} comes from the losses internal to the resonator, due to quasiparticles in the resonator, or to dielectric losses, yielding an internal quality factor Q_{int}. For a given resonator, the value of κ_{ext} and Q_{ext} is chosen by design: when it is essential to measure signals from the resonator with a large bandwidth, one generally designs $\kappa_{ext} \gg \kappa_{int}$ (over-coupled resonator); whereas if it is desired that the field is stored the longest possible in the resonator the reverse $\kappa_{ext} \ll \kappa_{int}$ will be designed (under-coupled resonator). Typical values for Q_{int} range between 10^5 and 10^6 at the single-photon level.[2] This is valid both for linear resonators as well as resonators that include Josephson junctions as described below.

1.2.2.   *Nonlinear resonators*

When a Josephson junction is inserted in such a resonator as shown in Fig. 5, the Josephson inductance nonlinearity affects the resonator dynamics. It is in general sufficient to consider only the first nonlinear term. The Hamiltonian of such Josephson resonator can very generally be written as $H/\hbar = \omega_0 a^\dagger a + (K/2)a^{\dagger 2}a^2$. The second term is called a Kerr nonlinearity; its strength is entirely characterized by the Kerr constant K, whose expression can be found as a function of the circuit parameters.[3] K has a simple physical interpretation if one considers the energy

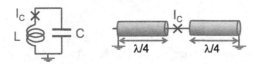

Fig. 5.   Nonlinear resonators can be implemented by inserting a Josephson junction of critical current I_C, either in a lumped element L, C resonator (left panel) or in a distributed resonator (right panel).

eigenstates of the resonator, the Fock states $|n\rangle$: indeed, $E_{n+1} - E_n \equiv \hbar\omega_{n,n+1} = \hbar[\omega_0 + (n+1/2)K]$. Kerr nonlinear resonators (KNR) have very different dynamics, depending on the ratio of K to the resonator damping rate κ. Resonators with $K/\kappa \ll 1$ obey the physics of Duffing oscillators.[4,5] They display parametric amplification, squeezing, and bistability; they are used to build quantum-limited amplifiers as explained later in these notes. In resonators with $K/\kappa \gg 1$ instead, the difference between the transition frequency ω_{01} between states $|0\rangle$ and $|1\rangle$ and ω_{12} between states $|1\rangle$ and $|2\rangle$ is much larger than the width of this transition. As a result, if the resonator is in its ground state and it is driven at or close to ω_{01}, it has the dynamics of an effective two-level system, because level $|2\rangle$ is far off-resonance. This is exactly the basic principle of a superconducting qubit: a resonator which is nonlinear at the single photon level.

1.2.3. *Transmon qubit*

The transmon qubit is a special case of such a Kerr nonlinear resonator. It consists of a small Josephson junction, of critical current I_C and Josephson energy $E_J = \varphi_0 I_C$, in parallel with a capacitor C, corresponding to a so-called Cooper-pair charging energy $E_C = (2e)^2/2C$, as shown in Fig. 6.[6] The linear resonance frequency of such circuit is readily given by $\omega_0 = \sqrt{2E_J E_C}/\hbar$; whereas the Kerr constant is simply expressed as $K = E_C/(4\hbar)$.[6] Typical parameters are $\omega_0 \approx 2\pi 6\,\text{GHz}$, and $K \approx 2\pi 300\,\text{MHz}$ (corresponding to $C \approx 60\,\text{fF}$), whereas typical damping rates are of the order $\kappa \approx 2\pi 100\,\text{kHz}$, therefore several orders of magnitude smaller than K, which shows that a transmon behaves as an effective two-level system. The restriction of the transmon Hamiltonian to its two lowest levels (the qubit levels) can be simply written $H_q/\hbar = -\frac{\omega_{01}}{2}\sigma_z$, σ_z being the Pauli operator in the $|0\rangle, |1\rangle$ subspace. The transmon qubit is arguably the simplest and the most studied superconducting qubit circuit nowadays. Note that to obtain a complete overview of the transmon circuit properties, it is necessary to go beyond the Kerr nonlinearity approximation and to use the full Josephson Hamiltonian; this can be found in several references.[1,6]

1.2.4. *Flux-tunable resonators*

Finally we would like to discuss another interesting types of circuits that can be built with Josephson junctions by inserting a SQUID in a resonator (lumped element

Fig. 6.   The transmon qubit is the simplest Josephson resonator, consisting of one Josephson junction in parallel with a capacitor C. Its lowest energy levels $|n\rangle$ are schematically shown in the right panel.

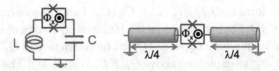

Fig. 7. Resonators including a SQUID have their resonance frequency $\omega_r(\Phi_x)$ tunable by the magnetic flux Φ_x threading the SQUID loop.

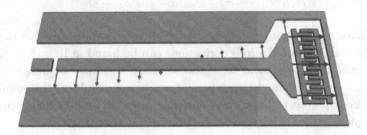

Fig. 8. In circuit QED experiments, a qubit (here a transmon qubit) is coupled (here capacitively) to a linear resonator (here a CPW distributed resonator). The resonator is coupled to waveguides that serve to send microwave pulses either to drive the qubit or to readout its state.

or distributed), as shown in Fig. 7. As explained above, the SQUID is a flux-tunable inductance $L_J(\Phi_x)$; therefore the resonator frequency itself becomes dependent on Φ_x. For a lumped element resonator (Fig. 7 left) the frequency is simply $\omega_0(\Phi_x) = 1/\sqrt{(L + L_J(\Phi_x))C}$; for a distributed resonator the formula involves additional numerical factors that can be found in Ref. 7. Often one would like this tunable resonator to be also linear in the incoming field; in that case, arrays of SQUIDs with large critical current Josephson junctions are of interest.[7]

1.3. *Transmon coupled to a resonator: Circuit QED*

Superconducting quantum circuits such as described in the previous paragraph can not only be easily designed and fabricated, owing to the flexibility of Josephson junctions, but they can also straightforwardly coupled to each other, with coupling constants that exceed by far the dissipation rates of these circuits. In particular, the simplest circuit QED experiment consists in coupling a superconducting qubit (for instance of the transmon type) to a linear resonator, as shown in Fig. 8. The resonator is used first to readout the qubit state, as explained in the following paragraphs, and can also be used as a quantum bus to couple several qubits together.

1.3.1. *Coupling hamiltonian*

Consider the circuit shown in Fig. 8, where a transmon qubit is capacitively coupled to a resonator through a coupling capacitor C_c. From simple electrostatic arguments it can be seen that this adds to the Hamiltonian of each individual system a coupling term which has the form $C_c\hat{Q}_r\hat{Q}_q/(C_rC_q)$, C_r (C_q) being the resonator (transmon) capacitance and $\hat{Q}_r$ $(\hat{Q}_q)$ being the resonator (transmon) charge operator. (This formula is valid in the limit where $C_c \ll C_q, C_r$). The restriction of $\hat{Q}_q$ to the qubit levels is $\hat{Q}_q = \langle 0|\hat{Q}_q|1\rangle\sigma_x$; as explained above $\hat{Q}_r = C_r\delta V_0(a+a^\dagger)$. The coupling Hamiltonian can therefore be recast as $\hbar g\sigma_x(a+a^\dagger)$, with $\hbar g = \beta\delta V_0|\langle 0|\hat{Q}_q|1\rangle|$,[8] and $\beta = C_c/C_q$. One thus obtains a total system Hamiltonian of the Jaynes-Cummings type

$$H/\hbar = -\frac{\omega_{01}}{2}\sigma_z + \omega_r a^\dagger a + g(a^\dagger\sigma^- + a\sigma^+), \qquad (8)$$

where we have in addition performed the rotating wave approximation to suppress off-resonant terms. More rigorous derivations can be found in Refs. 8, 1, 6.

To see whether the system is in the so-called "'strong coupling regime"' as defined in cavity quantum electrodynamics, we have to evaluate the coupling constant g and compare it to the various damping rates. As already mentioned, $\delta V_0 = 2$–$3\,\mu$V; typical values of the gate capacitor are $C_c \approx 10\,$fF yielding $\beta = 0.2$, and $|\langle 0|\hat{Q}_q|1\rangle| \approx 2e$. One therefore obtains $g \approx 2\pi250\,$MHz. As already mentioned, all typical damping rates (at least the damping caused by uncontrolled internal losses) are on the order of $100\,$kHz. The qubit — resonator coupling constant is thus several orders of magnitude large enough that both systems are in the strong coupling limit. All the well established physics of cavity quantum electrodynamics[9] therefore also directly applies to the case of a superconducting qubit in a resonator, hence the name "circuit quantum electrodynamics" or cQED.

It is however worth mentioning that coupling Hamiltonians much richer than Jaynes-Cummings are also well within reach. For instance, strong coupling of a two-level system to a nonlinear resonator is straightfowardly obtained by replacing the linear resonator in Fig. 8 by a Kerr nonlinear resonator such as described in the previous paragraph. This "nonlinear Jaynes-Cummings model" can to our knowledge only be realized with superconducting circuits and has no comparable implementation in atomic physics.[3, 10]

Finally we would like to mention that the Jaynes-Cummings Hamiltonian is only an approximate description of a transmon coupled to a distributed resonator such as shown in Fig. 8, in several respects. (1) We have so far neglected higher transmon energy levels, although these energy levels also are coupled to the resonator mode. Whenever quantitative results are requested, these higher energy levels should be in fact taken into account, since they are relatively close in frequency to ω_{01}. Indeed as already mentioned $|\omega_{12} - \omega_{01}| \approx 2\pi300\,$MHz, whereas $g \approx 2\pi200\,$MHz.[6] (2) We have also neglected higher order modes of the resonator, although the transmon is also coupled to these modes. With real atoms this approximation is usually valid since the coupling constant is many orders of magnitude smaller than the frequency difference between the resonator modes, but this is no longer true in cQED. Again, quantitative account of the experiments usually necessitate to consider the higher order modes.[11] (3) We have finally also performed the rotating wave approximation. With $g/\omega_c \approx 0.03$, this starts being questionable. Even larger ratios have been obtained with flux-qubits coupled to resonators, in which case one can reach $g/\omega_c \approx 0.1$ or more. Deviations with respect to the rotating wave approximation have been observed in that case.[12]

1.3.2. *Dispersive regime*

One interesting limit situation of the Jaynes-Cummings Hamiltonian is when the qubit is far detuned from the resonator, namely $|\Delta| \gg g$, with $\Delta = \omega_r - \omega_{01}$. In that case the Jaynes-Cummings Hamiltonian can be approximated by the so-called dispersive Hamiltonian $H_d = \chi a^\dagger a \sigma_z$, with $\chi = g^2/\Delta$. This dispersive Hamiltonian is the basis for modern qubit state readout methods.[8] Indeed, it describes how the frequency of the resonator is shifted by $\chi \sigma_z$, indicating that the resonator frequency is $\omega_{01} + \chi$ if the qubit is in $|g\rangle$, and $\omega_{01} - \chi$ if the qubit is in $|e\rangle$. This change of resonance frequency can be probed by measuring the phase or amplitude of a microwave pulse, resonant with the cavity, reflected on it or transmitted through it. How this detection is performed in practice is briefly explained at the end of these notes.

2. Experimental methods in circuit QED

After this theoretical introduction to circuit QED, we now discuss in more details how to perform circuit QED experiments. One of the key challenges is to isolate the qubit sufficiently well from its environment to thermalize it to the cryostat base temperature, while still being able to readout efficiently its state.

2.1. *Temperatures*

To appreciate the experimental challenges in circuit QED, it is worth starting with a short qualitative discussion of the various temperatures involved in the experiments.

The temperature of the cryostat determines the phonon temperature T_{ph} which is typically around 20 mK. Another relevant temperature is the temperature of the superconducting condensate T_{sc}, which usually approaches T_{ph} but doesn't actually reach it in most experiments, because at very low temperatures quasiparticle excitations may become very long-lived, so that even a tiny excess energy deposited in the superconducting material is sufficient to significantly raise its effective temperature above the fridge base temperature. These "excess quasiparticles" (compared to the quasiparticle density expected from the cryostat temperature T_{ph}) have been observed in all superconducting circuit experiments, and are believed to contribute to shorten the qubit coherence times. Typical observations yield $T_{sc} \approx 150\,\text{mK}$,[13] probably originating from stray microwave radiation reaching the sample, or radioactive material, or even cosmic rays.

Finally, the most important temperature is the temperature of the electromagnetic field $T_{em}(\omega)$ in the circuit, at a frequency ω close to the circuit resonance frequency, say between 5 and 7 GHz. It is essential to understand that since superconducting electrodes have nearly no dissipation at microwave frequency, the field temperature T_{em} is nearly entirely decoupled from the two other temperatures T_{ph} and even T_{sc}. Instead, the bath that determines nearly completely T_{em} is electromagnetic radiation from the continuum of modes to which the particular circuit is coupled. For instance, if one connects with a coaxial cable with little attenuation a circuit to a room-temperature amplifier or apparatus, T_{em} may easily reach 300 K, even though the sample itself is cooled at a phonon temperature $T_{ph} = 20$ mK.

2.2. *Measuring a linear resonator*

Linear resonators are usually characterized by simply cooling them either to 1.5 K in a pumped ^4He cryostat if the resonator is in niobium ($T_c = 9.2$ K), or to 20 mK in a dilution cryostat if it is in aluminum ($T_c = 1.2$ K). The resonator is directly connected to a vector network analyzer at room-temperature, allowing to measure transmission or reflextion complex coefficients. Typical data are shown in Fig. 9 for a Niobium resonator measured in transmission at 1.5 K.

In these measurements, according to the discussion above, $T_{em} \approx 300$ K since the coaxial cable brings no attenuation, and one might wonder whether this is an issue for measuring the resonator intrinsic properties. The reason why this does not matter is a specific property of *linear* resonators: their transmission coefficient is entirely independent of the energy stored in the resonator, and therefore does not depend on T_{em}. As seen in Fig. 9, the resonance frequency and quality factor of the resonator are not affected by the $T_{em} \approx 300$ K temperature of the intra-resonator

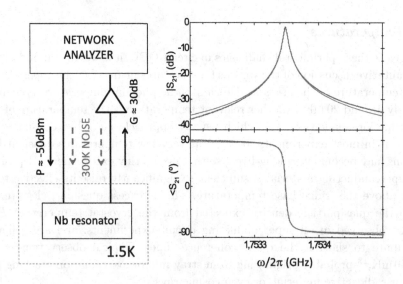

Fig. 9.   The resonance frequency and quality factor of a linear resonator are measured in a simple setup. The resonator is cooled in a cryostat, connected to a network analyzer that makes possible to measure its complex transmission coefficient. (Right panel) Measurement of a linear CPW resonator made in niobium and cooled at 1.5 K.

field which causes noise currents with rms value $\delta I_0 \sqrt{\frac{kT_{em}}{\hbar\omega_0}} \approx 1\,\mu A$ to flow through the resonator.

2.3.   *Measuring a Josephson resonator*

The situation is very different when the resonator includes a Josephson junction that makes it nonlinear. If the Josephson junction has a critical current on the order of $1\,\mu A$ as is typically the case, the currents generated by thermal noise may even overcome the junction critical current if $T_{em} = 300\,K$, in which case the junction becomes resistive, suppressing any resonance. To measure intrinsic properties of the Josephson resonator, it is therefore essential to lower T_{em} to a level at which the resonator nonlinearity does not manifest itself. These precautions are even more crucial for superconducting qubits which as explained earlier are nonlinear at the single photon level. Measuring superconducting qubits requires as unavoidable step to *thermalize* the electromagnetic field at the cryostat temperature, since even the slightest electromagnetic thermal excitation is detrimental to the quality of the quantum state prepared.

Cooling down the electromagnetic field can be achieved by inserting attenuators in the meauring lines thermalized at lower temperatures. Consider a cable connected to an apparatus at $T_{300} = 300\,K$ which generates $kT_{300}/(\hbar\omega_0) \approx 10^3$ noise photons per one Hertz bandwidth. If an attenuator A thermalized at temperature T_0 is inserted in the cable, the number of propagating noise photons becomes equal to $kT_{300}/(A\hbar\omega_0)$ plus the number of thermal photons generated by the attenuator itself $n_{th}(T_0) \equiv 1/(e^{\frac{\hbar\omega_0}{kT_0}} - 1) \approx kT_0/(\hbar\omega_0)$ if $kT_0/(\hbar\omega_0) \gg 1$. The resulting electromagnetic temperature is given by the relation $n_{th}(T_{em}) = kT_{300}/(A\hbar\omega_0) + n_{th}(T_0)$. We see that for sufficiently large attenuation, this may thermalize the field to $T_{em} \approx T_0$.

This strategy is adequate for the input line, where attenuation can be used without degrading the signal-to-noise ratio of the measurement. It is however not well suited for output measuring lines. Indeed, because of the Josephson resonator nonlinearity, it is also necessary to perform the measurements at extremely low power levels (in the femtoWatts or below, corresponding to one or a few photons in the resonator considered). Output signals should thus be carefully amplified and not attenuated; but at the same time, output lines also need to be protected from thermal noise, which as explained above seems to require low-temperature attenuation. The solution to this essential dilemma is provided by microwave circulators.[15] Circulators are three-port non-reciprocal microwave devices. Used as shown in Fig. 10, they make it possible to let all the output signal propagate towards the amplifier, whereas the noise coming from the opposite direction is channeled to a well thermalized 50Ω load that absorbs the noise without seeing its temperature raise significantly. Owing to imperfections of the device, one circulator can typically only provide 20 dB isolation from the noise; two circulators

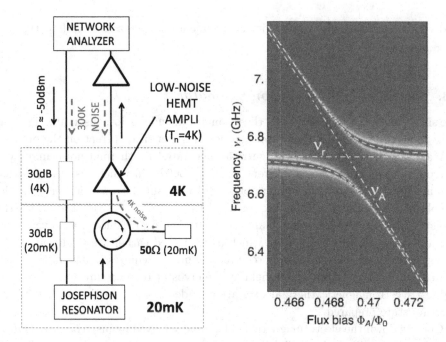

Fig. 10. (Left panel) Standard circuit QED setup. The Josephson resonator is protected from 300 K thermal noise by attenuation on the input measurement line (30dB at 4 K and 20 mK in this example), and by a microwave circulator in the output line. Complex transmission coefficients are measured at fW input power with a Vector Network Analyzer. (Right panel) Measurement of a vacuum Rabi splitting between a transmon qubit and a superconducting CPW resonator (reproduced with permission from[14]), with a setup similar to the one shown in the left panel.

in series are thus commonly used to obtain a 40 dB attenuation for the 4 K noise coming from the next stage amplifier, while letting nearly all the signal go through.

Having thermalized the electromagnetic field inside the resonator, it is also necessary to take care of maximizing the signal-to-noise ratio of the measurements. Since, as already mentioned, the measurement signals themselves need to be extremely low to avoid perturbing the Josephson resonoators behavior, it is crucial to amplify them after interaction with the sample while adding the minimal amount of noise possible. The noise power added by an amplifier at frequency ω and per unit bandwidth $E(\omega)$ is characterized by its noise temperature T_N, defined by the temperature at which a 50Ω load placed at the amplifier input would produce the same amount of thermal noise at the output $Gn_{th}(T_N)\hbar\omega = E(\omega)$, G being the amplifier power gain. The best commercial amplifiers for microwave signals have a gain of ≈ 35 dB and $T_N \approx 4$ K once cooled at low temperatures. Those amplifiers are cooled down to 4 K in standard setups for measuring Josephson resonators such as schematized in Fig. 10.

For the reasons discussed above, setups similar to the one shown schematically in Fig. 10 are used for all experiments in which Josephson resonators are to be measured at microwave frequencies, including circuit QED experiments; this is in a sense the "standard circuit QED setup". The first circuit QED experiment consists in performing the spectroscopy of the coupled "qubit-cavity" system, varying the qubit frequency in such a way that it goes through the cavity resonance. In this case, as is well known from cavity QED, an avoided level crossing is observed between the qubit and the cavity, with separation between the two lines reaching $2g$ at resonance. An example of such "vacuum Rabi splitting" is shown in Fig. 10, reprinted with permission from Ref. 14. The first observation of vacuum Rabi splitting in a qubit-cavity system[16] marks the starting point of circuit QED.

To emphasize the importance of a proper thermalization of the electromagnetic field at the sample level, we show in Fig. 11 data taken from Ref. 17 where spectroscopy of the qubit-resonator system, taken at resonance $\omega_{01} = \omega_r$, is measured for various intraresonator thermal photon numbers. The data show how strongly nonlinear a qubit-resonator system is at resonance, since even 0.1 thermal photons have measurable consequences on the spectrum.

Finally, spectroscopic measurements of a SQUID-based tunable resonator such as discussed in the first paragraph are shown in Fig. 12. They were performed using a setup similar to the one of Fig. 11.

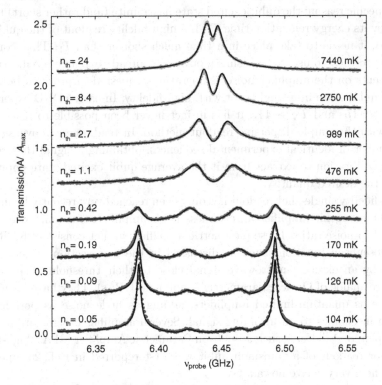

Fig. 11. Spectroscopy of a coupled qubit-resonator system at resonance, for various electromagnetic temperature T_{em} (reproduced with permission from Ref. 17).

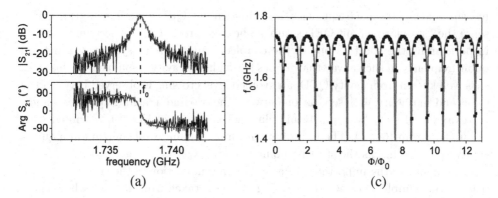

Fig. 12. (Left panel) Complex transmission coefficient of a SQUID-based resonator biased with a magnetic flux $\Phi = 0$. (Right panel) Dependence on the resonator frequency on Φ, showing a periodic dependence with period Φ_0. Reproduced with permission from Ref. 7.

2.4.   Qubit readout in circuit QED

In addition to the experimental challenges discussed above, reading out the state of a superconducting qubit using the dispersive readout method is even more difficult for one specific reason: the qubit excited state has a finite (and rather short) lifetime defined by its energy relaxation time T_1. If a high-fidelity readout of the qubit state is desired, it needs to take place in a time much shorter than T_1. This, combined with the necessity to use low measuring powers as already explained, puts stringent requirements on the amplifier noise temperature requested in order to be able to read the qubit state in a single shot with high fidelity. In early cQED experiments where $T_1 \approx 1\mu s$ and $T_N \approx 4\,K$, it has in fact never been possible to reach a high fidelity with the simple dispersive readout method. Instead, one can average out a large number of identical experimental sequences, obtaining a signal that requires a proper calibration to extract from it the average qubit excited state population, which is the desired quantity.

High-fidelity single-shot readout has only been reached very recently by inserting before the cryogenic amplifier an even less noisy amplifier, thus improving highly the signal-to-noise ratio. These parametric amplifiers in fact consist of KNRs built out of Josephson junctions such as discussed in the first section of these notes, pumped by an external microwave signal close to their threshold. They are the microwave analog of Optical Parametric Amplifiers (OPAs). They have been shown to behave as quantum-limited amplifiers, adding as little noise as permitted by quantum mechanics to an incoming signal. Several amplifier designs are presently being explored.[18,19] Using such Josephson Parametric Amplifier (JPA), high-fidelity single-shot readout of a transmon qubit was first reported in Ref. 20, opening a research field very active nowadays.

3.   Conclusion

The coupling of a superconducting qubit to a resonator offers a close to ideal implementation of the Jaynes-Cummings Hamiltonian, allowing to probe the physics of
strong coupling on a chip. Besides a simple theoretical introduction to this field, we
have tried to expose a few experimental important points that need to be addressed,
in particular the challenge of correctly thermalizing the electromagnetic field.

References

1. M. Devoret, B. Huard, R. Schoelkopf, and L. Cugliandolo, Eds., *Quantum Machines —
   Measurement and Control of Engineered Quantum Systems: Lecture Notes of the Les
   Houches Summer Schools*, Vol. 96 (Oxford University Press, 2013).
2. A. Megrant, C. Neill, R. Barends, B. Chiaro, Y. Chen, L. Feigl, J. Kelly, E. Lucero,
   M. Mariantoni, P. J. J. OMalley, D. Sank, A. Vainsencher, J. Wenner, T. C. White,
   Y. Yin, J. Zhao, C. J. Palmstrm, J. M. Martinis, and A. N. Cleland, Planar superconducting resonators with internal quality factors above one million, *Appl. Phys. Lett.*
   100, 113510, (2012).
3. F. R. Ong, M. Boissonneault, F. Mallet, A. Palacios-Laloy, A. Dewes, A. C.
   Doherty, A. Blais, P. Bertet, D. Vion, and D. Esteve, Circuit QED with a non-
   linear resonator: ac-Stark shift and dephasing, *Physical Review Letters* **106**(16),
   167002, (Apr., 2011). doi: 10.1103/PhysRevLett.106.167002. URL http://link.aps.org/
   doi/10.1103/PhysRevLett.106.167002.
4. I. Siddiqi, R. Vijay, F. Pierre, C. M. Wilson, M. Metcalfe, C. Rigetti, L. Frunzio, and
   M. H. Devoret, Rf-driven josephson bifurcation amplifier for quantum measurement,
   Phys. Rev. Lett. **93**, 207002, (2004). doi: 10.1103/PhysRevLett.93.207002.
5. I. Siddiqi, R. Vijay, F. Pierre, C. M. Wilson, L. Frunzio, M. Metcalfe, C. Rigetti, R. J.
   Schoelkopf, M. H. Devoret, D. Vion, and D. Esteve, Direct observation of dynamical
   bifurcation between two driven oscillation states of a josephson junction, *Physical
   Review Letters* **94**(2), 027005, (Jan., 2005). doi: 10.1103/PhysRevLett.94.027005. URL
   http://link.aps.org/doi/10.1103/PhysRevLett.94.027005.
6. J. Koch, T. M. Yu, J. Gambetta, A. A. Houck, D. I. Schuster, J. Majer, A. Blais,
   M. H. Devoret, S. M. Girvin, and R. J. Schoelkopf, Charge-insensitive qubit design
   derived from the cooper pair box, *Phys. Rev. A.* **76**, 042319, (2007). doi: 10.1103/Phys-
   RevA.76.042319.
7. A. Palacios-Laloy, F. Nguyen, F. Mallet, P. Bertet, D. Vion, and D. Esteve, Tunable
   resonators for quantum circuits, *J. Low Temp. Phys.* **151**, 1034, (2008).
8. A. Blais, R. Huang, A. Wallraff, S. M. Girvin, and R. J. Schoelkopf, Cavity quantum
   electrodynamics for superconducting electrical circuits: An architecture for quantum
   computation, *Phys. Rev. A.* **69**, 062320, (2004). doi: 10.1103/PhysRevA.69.062320.
9. J.-M. R. S. Haroche, *Exploring the Quantum.* (Oxford University Press, Oxford, 2007).
10. C. Laflamme and A. A. Clerk, Quantum-limited amplification with a nonlinear cav-
    ity detector, *Physical Review A.* **83**(3), 033803, (Mar., 2011). doi: 10.1103/Phys-
    RevA.83.033803. URL http://link.aps.org/doi/10.1103/PhysRevA.83.033803.
11. A. A. Houck, J. A. Schreier, B. R. Johnson, J. M. Chow, J. Koch, J. M. Gam-
    betta, D. I. Schuster, L. Frunzio, M. H. Devoret, S. M. Girvin, and R. J. Schoelkopf,

Controlling the spontaneous emission of a superconducting transmon qubit, *Phys. Rev. Lett.* **101**, 080502, (Aug, 2008). doi: 10.1103/PhysRevLett.101.080502. URL http://link.aps.org/doi/10.1103/PhysRevLett.101.080502.

12. P. Forn-Diaz, J. Lisenfeld, D. Marcos, J. J. Garcia-Ripoll, E. Solano, C. J. P. M. Harmans, and J. E. Mooij, Observation of the bloch-siegert shift in a qubit-oscillator system in the ultrastrong coupling regime, *Phys. Rev. Lett.* **105**, 237001, (Nov, 2010). doi: 10.1103/PhysRevLett.105.237001. URL http://link.aps.org/doi/10.1103/PhysRevLett.105.237001.

13. P. J. de Visser, J. J. A. Baselmans, S. J. C. Yates, P. Diener, A. Endo, and T. M. Klapwijk, Microwave-induced excess quasiparticles in superconducting resonators measured through correlated conductivity fluctuations, *Appl. Phys. Lett.* **100**, 162601, (2012). doi: 10.1126/science.1192739.

14. J. M. Fink, R. Bianchetti, M. Baur, M. Göppl, L. Steffen, S. Filipp, P. J. Leek, A. Blais, and A. Wallraff, Dressed collective qubit states and the tavis-cummings model in circuit qed, *Phys. Rev. Lett.* **103**, 083601, (Aug, 2009). doi: 10.1103/PhysRevLett.103.083601. URL http://link.aps.org/doi/10.1103/PhysRevLett.103.083601.

15. D. Pozar, *Microwave Engineering.* (John Wiley and Sons Ltd, 2011).

16. A. Wallraff, D. I. Schuster, A. Blais, L. Frunzio, R. Huang, J. Majer, S. Kumar, S. M. Girvin, and R. J. Schoelkopf, Strong coupling of a single photon to a superconducting qubit using circuit quantum electrodynamics, *Nature.* **431**, 162, (2004). doi: 10.1038/nature02851.

17. J. M. Fink, L. Steffen, P. Studer, L. S. Bishop, M. Baur, R. Bianchetti, D. Bozyigit, C. Lang, S. Filipp, P. J. Leek, and A. Wallraff, Quantum-to-classical transition in cavity quantum electrodynamics, *Phys. Rev. Lett.* **105**, 163601, (Oct, 2010). doi: 10.1103/PhysRevLett.105.163601. URL http://link.aps.org/doi/10.1103/PhysRevLett.105.163601.

18. M. A. Castellanos-Beltran and K. W. Lehnert, Widely tunable parametric amplifier based on a superconducting quantum interference device array resonator, *Appl. Phys. Lett.* **91**, 083509, (2007). doi: 10.1063/1.2773988.

19. N. Roch, E. Flurin, F. Nguyen, P. Morfin, P. Campagne-Ibarcq, M. H. Devoret, and B. Huard, Widely tunable, nondegenerate three-wave mixing microwave device operating near the quantum limit, *Phys. Rev. Lett.* **108**, 147701, (Apr, 2012). doi: 10.1103/PhysRevLett.108.147701. URL http://link.aps.org/doi/10.1103/PhysRevLett.108.147701.

20. R. Vijay, D. H. Slichter, and I. Siddiqi, Observation of quantum jumps in a superconducting artificial atom, *Phys. Rev. Lett.* **106**, 110502, (Mar, 2011). doi: 10.1103/PhysRevLett.106.110502. URL http://link.aps.org/doi/10.1103/PhysRevLett.106.110502.

Chapter 4

Quantum Open Systems

H. J. Carmichael

Department of Physics, University of Auckland
Private Bag 92019, Auckland, New Zealand
h.carmichael@auckland.ac.nz

An overview of the theory of Markov quantum open systems as commonly met in quantum optics is presented. Following a review of background material, the quantum trajectory formalism is developed from a consideration of photoelectron counting for output fields.

1. Introduction

The following lectures were first presented as a short course in quantum trajectory theory at the Joint Quantum Institute, University of Maryland at College Park in 2010. For this second presentation to the Singapore School of Physics, the lectures are expanded and brought into more systematic order. They cover the background material required for an understanding of quantum trajectory theory. The focus is therefore on the Markov treatment of quantum open systems as commonly met in quantum optics. In the quantum optics view the open system is approached as a generalized scattering center, i.e., the mediator of a transformation — "scattering" — from input fields to output fields, with inputs and outputs treated as excitations of a reservoir or bath. Ultimately, in the Markov case, outputs may be measured with no disturbance to the unconditional open system dynamics. Measuring (whether actual or virtual) then provides a foundation upon which a stochastic conditional dynamics may be built. Our aim in these lectures is to construct that conditional dynamics, while learning about Markov quantum open systems more broadly as we follow along our way.

We begin in Secs. 2 and 3 with a review of some standard background material: the Lindblad master equation, the quantum Langevin equation and input-output theory, correlation functions and quantum regression, and the Glauber-Sudarshan P representation, together with its associated quantum-classical correspondence. Aside from being an introduction to methods and ideas, these sections aim to bring us to a clear understanding of the distinction in quantum optics between so-called classical and nonclassical output fields. To this end, while perhaps not an obvious topic for lectures on quantum open systems, a brief introduction to the theory of photoelectron counting is included in Section 3.

Section 4 introduces the idea behind quantum trajectory theory at the level of a sketch. Our goal here is to familiarize ourselves with something unfamiliar: with the art of making an expansion of the density operator of an open system, not in terms of an *a priori* chosen set of basis states, but states conditioned upon a developing measurement record, with the measurement made continuously in time on the system outputs. The simplest possible example — spontaneous emission of a single photon — is used as an illustration, and to motivate a formal extension of the Bohr-Einstein quantum jump, from one stationary state to another, to jumps in the presence coherence, i.e., jumps for superpositions.

The sketch of Section 4 is put into solid form in Section 5, where a close connection between the quantum trajectory unraveling of a master equation and the theory of photoelectron counting (continuous record making) is developed. We see how the photoelectron count record probability density enters naturally into the expansion of the open system density operator, and explore the possibility of realizing sample records and their associated conditional states through Monte-Carlo simulation. To conclude Section 4, some examples are discussed: resonance fluorescence, electron shelving, and the damping of coherent states and coherent state superpositions.

Finally, in Section 6, cascaded systems are treated as a special class of quantum open systems. In the cascaded setup, the output of a first open system is fed as input to a second; in this way the range of inputs to the second system that can be treated is vastly expanded. A derivation of the cascaded system master equation is presented, along with its quantum trajectory unraveling. The formulation within quantum trajectory theory is then illustrated by solving the case of a damped coherent state fed from one cavity/resonator to another.

2.   Damped cavity/resonator

Possibly the simplest and most versatile example of a quantum open system is the harmonic oscillator coupled to a thermal reservoir or bath. Very often, in quantum optics, the reservoir is taken at zero temperature, since thermal photon numbers are extremely small for optical frequencies at room temperature. In this context, the oscillator represents a mode of an optical cavity or resonator with a vacuum field input and output field carrying whatever photons are lost to the environment from some prepared initial state; this might, for example, be a Fock state or a coherent state. The loss to the environment amounts to a damping of the intracavity field amplitude and energy, accompanied by the appearance of an exponential pulse of light at the cavity/resonator output. We will begin by treating this simplest situation, modeled, as sketched in Fig. 1, in one dimension. We first meet up with the basic elements from which the model is constructed, and then sketch how to treat

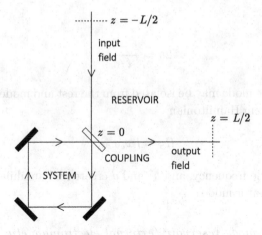

Fig. 1.   Schematic of a ring cavity/resonator modeled in 1D. Input and output fields couple to the cavity mode through the partially transmitting mirror at $z = 0$.

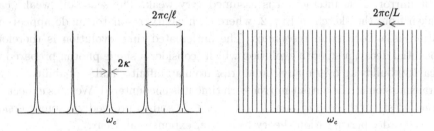

Fig. 2.   Sketch of the mode spectrum of the cavity/resonator (left) and the reservoir (right); reservoir modes have zero width.

the damped evolution in the Schrödinger picture with a Lindblad master equation, and in the Heisenberg picture via a quantum Langevin equation; the Heisenberg picture also introduces us to input-output theory. We follow with the extension to finite temperature, where thermal photons are exchanged between the reservoir and cavity/resonator, and finally look in more detail at the dynamics, first for incoherent and then coherent excitation.

2.1.   Basic elements

2.1.1.   The cavity/resonator mode and damping rate

The cavity/resonator is assumed to have modes that are well-separated in frequency space, as indicated in Fig. 2. For a ring resonator of length ℓ, longitudinal modes are separated by the free spectral range, $2\pi c/\ell$, which is assumed much larger than the mode width, given for an output mirror reflection coefficient (for light intensity) R by

$$2\kappa = \frac{1-R}{\ell/c}. \tag{1}$$

In this situation, one mode may be isolated from the rest and modeled as a harmonic oscillator, with system Hamiltonian

$$H_S = \hbar\omega_c a^\dagger a, \tag{2}$$

where ω_c is the mode frequency, and $a^\dagger$ and a create and annihilate photons in the chosen cavity/resonator mode.

2.1.2.   The multi-mode reservoir (external electromagnetic field)

Since the chosen mode has a width — the reflection coefficient is not $R = 1$ — there is an interaction between it and the electromagnetic environment seen through the output mirror. This interaction is assumed very weak. The sense of "weak" can be taken from the sketch in Fig. 2, where each cavity/resonator mode appears in isolation as a Lorentzian resonance. The anticipated time evolution is therefore exponential decay, a smooth evolution which (consider a single photon prepared in the cavity) builds up over many round trips from an infinitesimal probability, $1 - R$, for transmission at the output mirror each time it is encountered. We thus expect an interaction strength proportional to $\sqrt{1-R} \ll 1$, and to proceed via some version of second-order perturbation theory to recover exponential decay.

A photon transmitted by the output mirror excites a continuum of final states; it appears as a one-photon excitation of the external electromagnetic field. This field is expanded with periodic boundary conditions, period L, and has a similar spectrum of traveling-wave modes to the cavity/resonator, except that, with L very much larger than ℓ, the mode frequencies are very much closer together (Fig. 2). In anticipation of the continuum limit, $L \to \infty$, it is convenient to introduce the 1-D mode density (number of modes per unit frequency)

$$g(\omega) = \frac{L}{2\pi c}. \tag{3}$$

The reservoir (external electromagnetic field) Hamiltonian then sums — integrates in the limit — over harmonic oscillators:

$$H_R = \sum_j \hbar\omega_j r_j^\dagger r_j, \tag{4}$$

with ω_j the mode frequencies, and $r_j^\dagger$ and r_j creation and annihilation operators for photons in the mode of the external electromagnetic field of frequency ω_j. A similar sum over modes defines the positive frequency part of the external electric field operator in the Heisenberg picture:

$$E^{(+)}(z,t) = i \sum_j \sqrt{\frac{\hbar\omega_j}{2\epsilon_0 AL}} \, r_j(t) e^{i(\omega_j/c)z}, \tag{5}$$

with A an effective transverse cross-section (AL a quantization volume).

2.1.3. *Free field and photon flux units*

The sketch in Fig. 2 assumes a resonance with frequency ω_c much larger than the width 2κ (a resonance whose Q is very high). Considering then that frequencies lying within a few widths of the resonance are the only ones that matter, we may replace ω_j by ω_c in Eq. (5), multiply by $\sqrt{2\epsilon cA/\hbar\omega_c}$, and thus introduce the free external field in photon flux units:

$$\mathcal{E}_{\text{free}}^{(+)}(z,t) = \xi(t - z/c), \quad \xi(t) = i\sqrt{\frac{c}{L}} \sum_j r_j e^{-i\omega_j t}. \tag{6}$$

The sense of "photon flux units" can be appreciated by noting that $(c/L)r_j^\dagger r_j$ has units, not of photon number, but of photons per second through cross-section A. The designation "free" says that the free Heisenberg evolution, $r_j(t) = e^{-i\omega_j t} r_j$, has been introduced, i.e., this is the electromagnetic field in the absence of any interaction with the cavity/resonator. Note also that the assumed high Q allows us to extend the sum over j to both positive and negative infinity, in which case $\xi(t)$ satisfies the free-field commutation relation

$$[\xi(t), \xi^\dagger(t')] = \frac{c}{L} \sum_j e^{-i\omega_j(t-t')}$$

$$= \frac{c}{L} g(\omega) 2\pi \delta(t - t')$$

$$= \delta(t - t'), \tag{7}$$

where the sum is changed to an integral by introducing the (constant) 1-D mode density, Eq. (3).

2.1.4. *Interaction and coupling strength*

We must now introduce an interaction between the chosen cavity/resonator mode and modes of the external electromagnetic field reservoir to take the mode width, Eq. (1), into account. It will suffice to propose an interaction Hamiltonian,

$$H_{SR} = \sum_j \hbar(\kappa_j r_j a^\dagger + \kappa_j^* r_j^\dagger a), \tag{8}$$

written in the rotating-wave approximation, and with phenomenological coupling constants κ_j. Fermi's golden rule then relates the width, Eq. (1), to the 1-D mode density, Eq. (3), and the coupling constants κ_j. It gives a rate of transition (rate for the loss of one cavity photon) as the product of a final state density (reservoir mode density) and the interaction strength squared:

$$2\kappa = 2\pi g(\omega_c)|\kappa_{\omega_j=\omega_c}|^2. \tag{9}$$

The 1-D mode density is independent of frequency. If we make the approximation that the coupling strength is also frequency independent, we arrive at the coupling constant

$$\kappa_j = \sqrt{2\kappa}\sqrt{\frac{L}{c}}. \tag{10}$$

The phase is taken to be zero for convenience. In general there might be a phase shift associated with transmission through the output mirror. A phase introduced here would track through into the input term in the quantum Langevin equation (Section 2.3) and the source term in the output field (Section 2.4). It can always be absorbed into the definitions of the scaled input and output fields to preserve all expressions as they are given.

2.2.   The Lindblad master equation

We may work from this model either in the Schrödinger picture or in the Heisenberg picture. Very often a calculation uses both, e.g., using the Heisenberg picture to arrive at a formal expression for an output field correlation function and the Schrödinger picture for its evaluation.

We begin in the Schrödinger picture. A master equation is an equation of motion for the reduced density operator of the system, with the multi-mode reservoir traced out. Its derivation amounts to an elementary application of time-dependent perturbation theory. We begin with the Liouville equation satisfied by the density operator, χ, of the cavity/resonator mode and reservoir (external electromagnetic field) in the presence of their interaction:

$$\frac{d\chi}{dt} = \frac{1}{i\hbar}[H_S + H_R + H_{SR}, \chi]. \tag{11}$$

The steps then followed are pretty much the standard. We move to the interaction picture (quantities in the interaction picture are indicated by a tilde),

$$\frac{d\tilde{\chi}}{dt} = \frac{1}{i\hbar}[\tilde{H}_{SR}, \tilde{\chi}], \tag{12}$$

integrate the resulting Liouville equation formally,

$$\tilde{\chi}(t) = \chi(0) + \frac{1}{i\hbar}\int_0^t dt' [\tilde{H}_{SR}(t'), \tilde{\chi}(t')], \tag{13}$$

and substitute the solution back on the right-hand side of the Liouville equation:

$$\frac{d\tilde{\chi}}{dt} = \frac{1}{i\hbar}[\tilde{H}_{SR}(t), \chi(0)] - \frac{1}{\hbar^2}\int_0^t dt'[\tilde{H}_{SR}(t), [\tilde{H}_{SR}(t'), \tilde{\chi}(t')]]. \tag{14}$$

This begins a standard Dyson expansion. Truncation at second order would replace $\tilde{\chi}(t')$ by the initial density operator $\chi(0)$. Instead, we make a Born and Markov (no memory) approximation,

$$\tilde{\chi}(t') \rightarrow R_0\rho(t') \rightarrow R_0\rho(t),$$

where $\rho = \text{tr}_R(\chi)$ is the reduced density operator of the cavity/resonator and R_0 the (initial and unaltered) reservoir density operator. After taking the trace over Eq. (14), we then arrive at

$$\frac{d\tilde{\rho}}{dt} = -\frac{1}{\hbar^2}\int_0^t dt' \text{tr}_R[\tilde{H}_{SR}(t), [\tilde{H}_{SR}(t'), R_0\tilde{\rho}(t)]]. \tag{15}$$

The trace of the first term on the right-hand side of Eq. (14) has been assumed to be zero, though there is no loss of generality here. The assumption is equivalent to the statement that the mean of the reservoir field (mean external electromagnetic field) is zero; if it is not so — as, for example, when the cavity/resonator is driven by a coherent input (Section 2.6.3) — the first term is retained; it simply adds an interaction with a classical field.

Evaluating the trace calls for the evaluation of certain correlation functions, given a prescribed density operator R_0. With the reservoir (external electromagnetic field) in the vacuum state, most of the needed correlations vanish, i.e.,

$$\sum_{j,k}\kappa_j\kappa_k\langle r_jr_k\rangle_{R_0}e^{-iw_jt}e^{-iw_kt'} = \sum_{j,k}\kappa_j^*\kappa_k\langle r_j^{\dagger}r_k\rangle_{R_0}e^{iw_jt}e^{-iw_kt'} = 0. \tag{16}$$

The one non-vanishing correlation function is

$$\sum_{j,k}\kappa_j\kappa_k^*\langle r_jr_k^{\dagger}\rangle_{R_0}e^{-iw_jt}e^{iw_kt'} = \sum_j|\kappa_j|^2e^{-iw_j(t-t')}. \tag{17}$$

It appears inside the time integral on the right-hand side of Eq. (15) multiplied by $\exp[iw_c(t-t')]$, the exponential coming from the product $\tilde{a}^{\dagger}(t)\tilde{a}(t') = \exp[iw_c(t-t')]a^{\dagger}a$. Assuming t sufficiently long, we treat the integral and sum over frequencies as in a derivation of Fermi's golden rule: effectively, the reservoir is taken to be δ-correlated, with Eq. (17) — multiplied by $\exp[iw_c(t-t')]$ — evaluated as

$$\sum_j|\kappa_j|^2e^{-i(w_j-w_c)(t-t')} \rightarrow 2\pi g(w_c)|\kappa_{w_j=w_c}|^2\delta(t-t')$$

$$\rightarrow 2\kappa\delta(t-t'), \tag{18}$$

where Eq. (9) is used.

Working from the interaction Hamiltonian, Eq. (8), and using Eqs. (16) and (18) in Eq. (15), we arrive at the master equation for a damped cavity/resonator mode

$$\frac{d\rho}{dt} = \mathcal{L}\rho, \qquad (19)$$

with generalized Liouvillian (the dot shows where the argument goes)

$$\mathcal{L} = -i\omega_c[a^\dagger a, \cdot] + \Lambda[a], \quad \Lambda[\eta] = 2\eta \cdot \eta^\dagger - \eta^\dagger \eta \cdot - \cdot \eta^\dagger \eta. \qquad (20)$$

The master equation accords with the required (for a physically acceptable reduced evolution) Lindblad form.[1] The superoperator $\Lambda[\eta]$ is generic and designated as the Lindblad superoperator.

2.3.   The quantum Langevin equation

Let us turn now to the Heisenberg picture,[2,3] where we have coupled equations of motion for the annihilation operator of the cavity/resonator mode,

$$\frac{da}{dt} = -i\omega_c a - i\sum_j \kappa_j r_j, \qquad (21)$$

and for the annihilation operator of the reservoir (external electromagnetic field) mode of frequency ω_j,

$$\frac{dr_j}{dt} = -i\omega_j r_j - i\kappa^* a. \qquad (22)$$

We may follow a similar procedure to reach an autonomous equation of motion for the cavity/resonator with only the initial state of the reservoir entering the description. We work in the reverse interaction picture, i.e., with slowly varying operators $\tilde{a} = \exp(i\omega_c t)a$ and $\tilde{r}_j = \exp(i\omega_j t)r_j$, and solve formally for

$$\tilde{r}_j(t) = r_j(0) - ie^{i(\omega_j - \omega_c)t}\kappa_j^* \int_0^t dt'\, \tilde{a}(t')e^{-i(\omega_j - \omega_c)(t-t')}. \qquad (23)$$

Substituting into the equation of motion for $\tilde{a}$, and treating the time integral and frequency sum using Eq. (18) yields

$$\frac{d\tilde{a}}{dt} = -\kappa\tilde{a} - i\sum_j \kappa_j r_j(0)e^{-i(\omega_j - \omega_c)t}, \qquad (24)$$

Finally, returning to the Heisenberg picture, using Eq. (10), and introducing $\xi(t)$ from Eq. (6), we arrive at the advertised autonomous equation,

$$\frac{da}{dt} = -i\omega_c a - \kappa a - \sqrt{2\kappa}\xi(t). \qquad (25)$$

The result is the quantum Langevin equation for the damped cavity/resonator. It explicitly displays the expected damping (second term) and includes a δ-correlated operator noise $\xi(t)$ as the input or drive (third term). The noise operator obeys the free-field commutations relation, Eq. (7), from which the correlation function $\langle \xi(t)\xi^{\dagger}(t') \rangle = \delta(t - t')$ readily follows.

2.4.   *Input and output fields*

Whether or not the Schrödinger picture (master equation) or Heisenberg picture (quantum Langevin equation) is adopted for the damped cavity/resonator itself, the Heisenberg picture is natural, if not essential, for setting up a relationship between input and output fields.[4] We simply substitute the formal solution for $\tilde{r}_j(t)$, Eq. (23), into Eq. (5) to obtain the external electric field operator in the Heisenberg picture:

$$E^{(+)}(z,t) = i\sum_j \sqrt{\frac{\hbar\omega_j}{2\epsilon_0 AL}} r_j(0)e^{-i\omega_j(t-z/c)}$$

$$+ e^{-i\omega_c(t-z/c)} \sum_j \sqrt{\frac{\hbar\omega_j}{2\epsilon_0 AL}} \kappa_j^* \int_0^t dt'\, \tilde{a}(t)e^{i(\omega_j-\omega_c)(t'-t+z/c)}$$

$$= i\sum_j \sqrt{\frac{\hbar\omega_j}{2\epsilon_0 AL}} r_j(0)e^{-i\omega_j(t-z/c)} + \sqrt{\frac{\hbar\omega_c}{2\epsilon_0 Ac}} \left\{ \begin{array}{c} 0 \\ \frac{1}{2}\sqrt{2\kappa}a(t) \\ \sqrt{2\kappa}a(t-z/c) \end{array} \right\}. \qquad (26)$$

The time integral and frequency sum are treated, once again, using Eq. (18) — with $\sqrt{\omega_j}\kappa_j^*$ replacing $|\kappa_j|^2$ inside the sum — and the three options which appear in the curly bracket apply for $z < 0$, $z = 0$, and $z > 0$ (top to bottom). Adopting photon flux units, as in Section 2.1.3, from this expression we have an input field operator ($z < 0$),

$$\mathcal{E}_{\text{in}}^{(+)}(z,t) = \mathcal{E}_{\text{free}}^{(+)}(z,t), \qquad (27)$$

the noise operator, or drive, in the quantum Langevin equation, and an output field operator ($z > 0$)

$$\mathcal{E}_{\text{out}}^{(+)}(z,t) = \mathcal{E}_{\text{in}}^{(+)}(z,t) + \sqrt{2\kappa}a(t - z/c), \qquad (28)$$

which is a sum of fields reflected and transmitted at the cavity output mirror. There is clearly a discontinuity at the output mirror ($z = 0$); it is spanned with

$$\mathcal{E}^{(+)}(0,t) = \mathcal{E}_{\text{in}}^{(+)}(0,t) + \frac{1}{2}\sqrt{2\kappa}a(t). \qquad (29)$$

2.5.   Other examples

2.5.1.   Damped cavity/resonator with a thermal (chaotic) input

The extension to a reservoir (external electromagnetic field) in a thermal state at temperature T, or more generally one which carries a beam of broadband chaotic light, changes the reservoir correlation functions, Eqs. (16) and (17). There are now two nonvanishing correlation functions:

$$\sum_{j,k} \kappa_j^* \kappa_k \langle r_j^\dagger r_k \rangle_{R_0} e^{i\omega_j t} e^{-i\omega_k t'} = \sum_j |\kappa_j|^2 \bar{n}_j e^{i\omega_j (t-t')}, \tag{30}$$

and

$$\sum_{j,k} \kappa_j \kappa_k^* \langle r_j r_k^\dagger \rangle_{R_0} e^{-i\omega_j t} e^{i\omega_k t'} = \sum_j |\kappa_j|^2 (\bar{n}_j + 1) e^{-i\omega_j (t-t')}, \tag{31}$$

with $\bar{n}_j$ a mean photon number. Treating them again in the manner of Eq. (18), they define nonvanishing correlations of the input field,

$$\langle \mathcal{E}_{\text{in}}^{(-)}(0,t) \mathcal{E}_{\text{in}}^{(+)}(0,t') \rangle = 2\kappa \bar{n} \delta(t - t'), \tag{32}$$

and

$$\langle \mathcal{E}_{\text{in}}^{(+)}(0,t) \mathcal{E}_{\text{in}}^{(-)}(0,t') \rangle = 2\kappa(\bar{n} + 1)\delta(t - t'). \tag{33}$$

These are the new correlations defining the properties of the operator noise $\xi(t) = \mathcal{E}_{\text{in}}^{(+)}(0,t)$ in the quantum Langevin equation. They trace through in the Schrödinger picture to give a master equation with generalized Liouvillian

$$\mathcal{L} = -i\omega_c[a^\dagger a, \cdot] + \kappa(\bar{n} + 1)\Lambda[a] + \kappa\bar{n}\Lambda[a^\dagger]. \tag{34}$$

Comparing Eq. (20), we see that there are now two Lindblad superoperators. Their physical significance will be made clear in Section 2.6.1.

2.5.2.   Two-state atom

The Lindblad form is extremely versatile and appears in the modeling of many open systems with experimentally addressable inputs and outputs. Considering historical significant, spontaneous emission is perhaps even more of a fundamental example than the damped cavity/resonator. Master equations for a radiatively damped two-state atom are similar to those we have met, defined by generalized Liouvillians, Eqs. (20) and (34), but with creation and annihilation operators, $a^\dagger$ and a replaced by raising and lowering operators $\sigma_+ = |e\rangle\langle g|$ and $\sigma_- = |g\rangle\langle e|$. The energy damping rate is introduced not from a phenomenological interaction, but from the dipole coupling of an atom to the radiation field. It works out just as in Wigner-Weisskopf theory[5] to be the Einstein A coefficient,

$$\gamma = \frac{1}{4\pi\epsilon_0} \frac{4\omega_a^3 \mu_{eg}^2}{3\hbar c^3}, \tag{35}$$

with ω_a the atomic resonance frequency and μ_{eg} the dipole moment. The output or scattered field (dipole field in the far-field limit) can again be written as the sum of an input, or free field, and a source term:[6]

$$\mathbf{E}_{\text{out}}^{(+)}(\mathbf{r},t) = \mathbf{E}_{\text{in}}^{(+)}(\mathbf{r},t) + \frac{\omega_a^2}{4\pi\epsilon_0 c^2 r}[(\hat{\boldsymbol{\mu}}_{eg} \times \hat{\mathbf{r}}) \times \hat{\mathbf{r}}]\sigma_-(t - r/c). \tag{36}$$

This vector expression accounts for scattering in all directions. There is, from one point of view, a continuum of output channels, considering that the field may be looked at from any direction and the photon flux depends on the polar angle with the dipole moment. The angular distribution belongs nonetheless to one and the same dipole and a single mode (single spherical harmonic). A simple formula in photon flux units, comparable to Eq. (28), may be obtained by integrating the output photon flux over all directions. Multiplying the squared output field amplitude by $2\epsilon_0 c/\hbar\omega_a$ and integrating over a sphere of radius r gives back the Einstein A coefficient. Thus the output field in units of the integrated photon flux is

$$\mathcal{E}_{\text{out}}^{(+)}(\mathbf{r},t) = \mathcal{E}_{\text{in}}^{(+)}(z,t) + \sqrt{\gamma}\sigma_-(t - r/c). \tag{37}$$

2.6. What's in a master equation?

2.6.1. Diagonal matrix elements

We noted that deriving the Lindblad master equation is, essentially, an exercise in second-order perturbation theory, closely aligned with the derivation of Fermi's golden rule. We expect, then, to see transition rates between initial and final states. The view is recovered by taking diagonal matrix elements of the master equation for a cavity/resonator in contact with a thermal reservoir (external electromagnetic field) at temperature T. The generalized Liouvillian is Eq. (34). Taking diagonal matrix elements yields a familiar rate equation (Pauli master equation[7]) for the photon occupation probability $p_n = \rho_{n,n}$:

$$\frac{dp_n}{dt} = \kappa(\bar{n} + 1)[2(n + 1)p_{n+1} - 2np_n] + \kappa\bar{n}[2np_{n-1} - 2(n + 1)p_n]. \tag{38}$$

This equation describes a birth-death process, balancing the rate of change of an occupation probability with incoherent transitions — spontaneous and stimulated emission, and absorption — into and out of the considered state (see Fig. 3); Lindblad term $\kappa(\bar{n} + 1)\Lambda[a]$ accounts for spontaneous and stimulated emission, while $\kappa\bar{n}\Lambda[a^\dagger]$ accounts for absorption. The rate equation is solved in steady state using detailed balance, i.e., by equating the rates of transition between neighboring states. This yields the recurrence relation and solution,

$$2\kappa(\bar{n} + 1)(n + 1)p_{n+1} = 2\kappa\bar{n}(n + 1)p_n \implies p_n = \frac{\bar{n}^n}{(\bar{n} + 1)^{n+1}}. \tag{39}$$

The result is the thermal distribution, with mean $\bar{n} = [\exp(\hbar\omega_c/kT) - 1]^{-1}$ the Bose-Einstein distribution (in photon energy $\hbar\omega_c$).

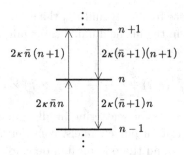

Fig. 3.   Transitions for a cavity/resonator due to absorption, and spontaneous and stimulated emission at temperature T.

2.6.2.   Initial coherent state

Master equations of the Lindblad type are not limited, however, to rate equations and incoherent transitions. A density operator may have non-vanishing off-diagonal as well as diagonal elements, and the master equation describes their evolution as well. Consider Eqs. (19) and (20) written in the interaction picture, i.e., the master equation

$$\frac{d\tilde{\rho}}{dt} = \kappa(2a\tilde{\rho}a^\dagger - a^\dagger a\tilde{\rho} - \tilde{\rho}a^\dagger a), \tag{40}$$

and decay of the cavity/resonator from an initial coherent state $|\alpha(0)\rangle$. Classically, we would expect a cavity filled with coherent light to decay with the release of a similarly coherent exponential pulse. There is no apparent source of incoherence, so perhaps we should try an *ansatz* that preserves coherence throughout the decay. Substituting $\tilde{\rho}(t) = |\tilde{\alpha}(t)\rangle\langle\tilde{\alpha}(t)|$, with ket $|\tilde{\alpha}(t)\rangle$ a coherent state, we deduce from Eq. (40) that it must obey the equation of motion

$$\frac{d|\tilde{\alpha}\rangle}{dt} = \kappa(\tilde{\alpha}^*\tilde{\alpha} - \tilde{\alpha}a^\dagger)|\tilde{\alpha}\rangle. \tag{41}$$

Note now that any coherent state may be generated from the vacuum, i.e.,

$$|\tilde{\alpha}\rangle = \exp(-|\tilde{\alpha}|^2/2)\exp(\tilde{\alpha}a^\dagger)|0\rangle. \tag{42}$$

It follows that $|\tilde{\alpha}(t)\rangle$ must also satisfy the equation of motion

$$\frac{d|\tilde{\alpha}\rangle}{dt} = \left[-\frac{1}{2}\left(\frac{d\tilde{\alpha}^*}{dt}\tilde{\alpha} + \tilde{\alpha}^*\frac{d\tilde{\alpha}}{dt}\right) + \frac{d\tilde{\alpha}}{dt}a^\dagger\right]|\tilde{\alpha}\rangle. \tag{43}$$

The two equations are consistent if $d\tilde{\alpha}/dt = -\kappa\tilde{\alpha}$. Thus the *ansatz* works, with coherent state amplitude $\tilde{\alpha}(t) = \alpha(0)\exp(-\kappa t)$, or transforming back from the interaction to the Schrödinger picture,

$$\alpha(t) = \alpha(0)\exp[-(\kappa + i\omega_c)t]. \tag{44}$$

2.6.3. *Coherently driven cavity*

The response of a cavity/resnator to a coherent drive can be approached as a straightforward extension of the solution just given. To begin, though, we must consider the form of the master equation. The drive is accounted for by taking one mode of the reservoir (external electromagnetic field) to be in a coherent state. The mean external field is then nonzero, and the first term on the right-hand side of Eq. (14) must be retained. For drive frequency ω and coherent state amplitude α, there is the interaction at the additional first order,

$$\text{tr}_R \left\{ \frac{1}{i\hbar} [\tilde{H}_{SR}(t), R_0\tilde{\rho}] \right\} = -i[\mathcal{E}e^{-i\omega t}\tilde{a}^\dagger(t) + \mathcal{E}^* e^{i\omega t}\tilde{a}(t), \tilde{\rho}], \tag{45}$$

where, using Eq. (10),

$$\mathcal{E} = \sqrt{2\kappa}\sqrt{\frac{c}{L}}\alpha. \tag{46}$$

The master equation has generalized Liouvillian

$$\mathcal{L} = -i\omega_c[a^\dagger a, \cdot] - i[\mathcal{E}e^{-i\omega t}a^\dagger + \mathcal{E}^* e^{i\omega t}a, \cdot] + \kappa\Lambda[a]. \tag{47}$$

Considering resonant driving for simplicity, the master equation in the interaction picture is

$$\frac{d\tilde{\rho}}{dt} = \kappa(2b\tilde{\rho}b^\dagger - b^\dagger b\tilde{\rho} - \tilde{\rho}b^\dagger b), \tag{48}$$

with $b = a + i\mathcal{E}/\kappa$ a displaced annihilation operator.

The approach to steady state may now be deduced from the solution to Eq. (40). Taking mode a initially in the vacuum state, mode b starts out in a coherent state of amplitude $i\mathcal{E}/\kappa$. It decays with amplitude $\tilde{\beta}(t) = i(\mathcal{E}/\kappa)\exp(-\kappa t)$. Thus, making the displacement and transforming back to the Schrödinger picture, the response of the cavity/resonator to an on-resonance ($\omega = \omega_c$) coherent drive finds the cavity mode in a coherent state of amplitude

$$\alpha(t) = -i(\mathcal{E}/\kappa)[1 - \exp(-\kappa t)]\exp(-i\omega_c t). \tag{49}$$

The off-resonance case is readily treated by the method of Section 2.6.2.

3. Correlation functions and quantum-classical correspondence

We have now met with the basic ideas; but the master equation, viewed on its own, is far from an end to our story. It gives us the ability to calculate one-time averages, but not directly two-time or multi-time averages. A commonly encountered example of a two-time average is the intensity correlation function (in photon flux units)

$$G^{(2)}(t, t+\tau) = \langle \mathcal{E}_{\text{out}}^{(-)}(t) \mathcal{E}_{\text{out}}^{(-)}(t+\tau) \mathcal{E}_{\text{out}}^{(+)}(t+\tau) \mathcal{E}_{\text{out}}^{(+)}(t) \rangle$$

$$= (2\kappa)^2 \langle a^\dagger(t) a^\dagger(t+\tau) a(t+\tau) a(t) \rangle. \tag{50}$$

The second line drops the input field from Eq. (28). This step is permitted because $\mathcal{E}_{\text{in}}^{(+)}(t+\tau)$ commutes with $a(t)$ and can be passed to the right, where it acts, along with $\mathcal{E}_{\text{in}}^{(+)}(t)$, on the reservoir vacuum state when the average is taken. Note that while we begin with an average over the cavity/resonator and the reservoir (external electromagnetic field), the second line averages over the cavity/resonator only.

By considering arbitrary correlation functions, we expand our view to approach the cavity/resonator, along with its input and output, as an example of a quantum stochastic process. This point of view lies relatively close to the surface when we look at the quantum Langevin equation, Eq. (25), where it is apparent from the explicit operator noise; a formal solution to this equation — given correlation functions for the noise — implicitly defines output field correlation functions to all orders. It is less apparent from the master equation. The extension that brings it to the surface is the quantum regression procedure, which provides a scheme for calculating multi-time averages from the master equation, or, more specifically, from repeated evolution under the generalized Liouvillian $\mathcal{L}$. We first meet the procedure as it applies to the intensity correlation function, Eq. (50), and illustrate it by considering the famous Hanbury Brown and Twiss photon bunching effect. We then delve into the mysteries of nonclassical versus classical light. A clear definition of this distinction, as it is usually made in quantum optics, is draw from the so-called quantum-classical correspondence, which for some output fields (classical) casts photoelectron theory in semiclassical form, denying the recasting to others (nonclassical output fields). We meet up along the way with the Glauber-Sudarshan P representation, Fokker-Planck and Langevin equations (classical Langevin equations), and the rudiments of photoelectron counting theory.

3.1.   Quantum regression

3.1.1.   By solving the master equation twice

We might make a summary of what we have learned, at the formal level, about the master equation approach. The reduced density operator at time t is constructed through the sequence of relations:

$$\rho(t) = \text{tr}_R[\chi(t)] = \text{tr}_R[e^{L(t-t_0)}\chi(t_0)] \rightarrow e^{\mathcal{L}(t-t_0)}\rho(t_0). \tag{51}$$

Formally, the end result of the development that removes (evaluates) the trace is a substitution: $\chi \rightarrow \rho$ and $L \rightarrow \mathcal{L}$. When calculating a two-time, or, more generally, a multi-time average, we make the same substitution in the Heisenberg picture. Let the unitary time evolution operator be

$$U(t) = \exp[-i(H/\hbar)t], \tag{52}$$

with H the Hamiltonian in the Liouville equation, Eq. (11). Consider the two-time average, Eq. (50), as an example, working in the Heisenberg picture in the standard way yields:

$$
\begin{aligned}
\langle a^\dagger(t)a^\dagger(t+\tau)a(t+\tau)a(t)\rangle &= \mathrm{tr}_{S\otimes R}[\chi(0)U^\dagger(t)a^\dagger U(t)U^\dagger(t+\tau)a^\dagger U(t+\tau) \\
&\qquad \times U^\dagger(t+\tau)aU(t+\tau)U^\dagger(t)aU(t)] \\
&= \mathrm{tr}_{S\otimes R}[\chi(0)U^\dagger(t)a^\dagger U^\dagger(\tau)a^\dagger aU(\tau)aU(t)],
\end{aligned}
$$

where we use

$$U(t)U^\dagger(t+\tau) = U^\dagger(\tau), \quad U(t+\tau)U(t+\tau) = 1, \quad U(t+\tau)U^\dagger(t) = U(\tau).$$

Now employing the cyclic property of the trace,

$$
\begin{aligned}
\langle a^\dagger(t)a^\dagger(t+\tau)a(t+\tau)a(t)\rangle &= \mathrm{tr}_{S\otimes R}[a^\dagger aU(\tau)aU(t)\chi(0)U^\dagger(t)a^\dagger U^\dagger(\tau)] \\
&= \mathrm{tr}_{S\otimes R}\{a^\dagger a e^{\mathcal{L}\tau}[a\chi(t)a^\dagger]\} \\
&\to \mathrm{tr}_S\{a^\dagger a e^{\mathcal{L}\tau}[a\rho(t)a^\dagger]\}, \tag{53}
\end{aligned}
$$

where, finally, the trace over the reservoir is removed, making the noted substitution: $\chi \to \rho$, $L \to \mathcal{L}$.

A straightforward rewriting of Eq. (53) introduces the idea of a conditional state. If we multiply and divide by the mean photon number at time t,

$$\langle a^\dagger(t)a^\dagger(t+\tau)a(t+\tau)a(t)\rangle = \mathrm{tr}_S\{a^\dagger a e^{\mathcal{L}\tau}[\rho_{\mathrm{cond}}(t)]\}\mathrm{tr}_S[a\rho(t)a^\dagger], \tag{54}$$

with conditional density operator at time t

$$\rho_{\mathrm{cond}}(t) = \frac{a\rho(t)a^\dagger}{\mathrm{tr}_S[a\rho(t)a^\dagger]}. \tag{55}$$

Equation (54) is read as the product of the mean photon number at time t, and the mean photon number at time $t + \tau$, *subject to the condition* that a photon was deleted from the field at time t. It is clear that Eq. (54) requires the master equation to be solved twice, first for $\rho(t)$, and then for $\rho_{\mathrm{cond}}(t+\tau) = \exp(\mathcal{L}\tau)\rho_{\mathrm{cond}}(t)$.

3.1.2. *In the manner of Lax*

The quantum regression procedure is attributed to Lax,[8–10] who introduced it in a different, though equivalent, form. Imagine that from the master equation we are able to derive a coupled hierarchy of linear equations for operator expectation values,

$$\frac{d\langle \boldsymbol{O}(t)\rangle}{dt} = \boldsymbol{M}\langle \boldsymbol{O}(t)\rangle, \tag{56}$$

where $\langle \boldsymbol{O}(t) \rangle$ is a column vector of those expectation values — perhaps the first element is $\langle a(t) \rangle$, the second $\langle a^\dagger(t) \rangle$, the third $\langle (a^\dagger a)(t) \rangle$, the fourth $\langle (aa^\dagger a)(t) \rangle$, and so on — and $\boldsymbol{M}$ is a matrix, possibly of infinite dimension, though usually either finite or truncated to finite dimension. It is always possible to set up a hierarchy like this. Lax's procedure then amounts to something rather simple. We first make the change $t \to t + \tau$, $dt \to d\tau$. Having done this, we may remove the angular brackets, multiply on the left and right, respectively, by any two operators $A(t)$ and $B(t)$ (one might be the unit operator), and then replace the angular brackets. The procedure is summarized by saying that if Eq. (56) holds, then correlation functions obey exactly the same set of equations:

$$\frac{d\langle A(t)\boldsymbol{O}(t+\tau)B(t) \rangle}{d\tau} = \boldsymbol{M}\langle A(t)\boldsymbol{O}(t+\tau)B(t) \rangle. \tag{57}$$

We again see the need to solve twice, since Eq. (56) must be solved for the initial conditions required for the solution of Eq. (57).

3.2. Example: Hanbury Brown and Twiss

The photon bunching of Hanbury Brown and Twiss[11–14] stimulated much of the early work on photoelectron counting. Both Mandel[15] and Glauber[16,17] open their seminar papers with references to Hanbury Brown and Twiss. Davies[18] picks up the same theme as the reason for developing a quantum mechanical theory of stochastic processes. The proposal of Hanbury Brown and Twiss amounts to the measurement of an intensity autocorrelation, or, in the time domain (they were concerned with spatial correlations), a delayed photon coincidence rate:

$$\begin{pmatrix} \text{probability for ``start'' in } [t, t + \Delta t) \\ \text{and ``stop'' in } [t + \tau, t + \tau + \Delta t) \end{pmatrix} = (2\kappa\Delta t)^2 \langle a^\dagger(t)a^\dagger(t+\tau)a(t+\tau)a(t) \rangle. \tag{58}$$

Evaluation of the right-hand side provides a simple illustration of the quantum regression procedure applied in the manner of Lax.

Figure 4 shows a cavity/resonator set up to model a source of filtered thermal light. The two partially transmitting mirrors are unbalanced. Photons are absorbed and emitted predominantly through the mirror to the left; a small flux is emitted through the mirror (with vacuum-field input) to the right, where the correlation function is measured. The master equation is taken with generalized Liouvillian, Eq. (34), which gives the equation of motion for the mean photon number,

$$\frac{d\langle (a^\dagger a)(t) \rangle}{dt} = -2\kappa[\langle (a^\dagger a)(t) \rangle - \bar{n}], \tag{59}$$

with solution

$$\langle (a^\dagger a)(t) \rangle = \bar{n}[1 - \exp(-2\kappa t)] \xrightarrow[t\to\infty]{} \bar{n}. \tag{60}$$

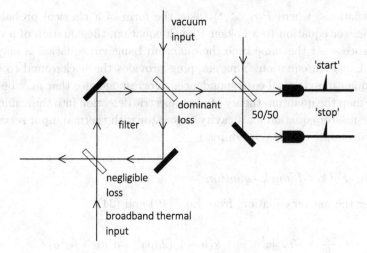

Fig. 4. Hanubury Brown and Twiss measurement of the temporal intensity autocorrelation of filtered thermal light.

The required two-time average satisfies the same equation of motion, though now viewed as a function of delay:

$$\frac{d\langle a^\dagger(t)[(a^\dagger a)(t+\tau)]a(t)\rangle}{d\tau} = -2\kappa\{\langle a^\dagger(t)[(a^\dagger a)(t+\tau)]a(t)\rangle - \bar{n}\langle(a^\dagger a)(t)\rangle\}$$
$$= -2\kappa\{\langle a^\dagger(t)[(a^\dagger a)(t+\tau)]a(t)\rangle - \bar{n}^2\}. \tag{61}$$

In the second line we adopt the limit $t \to \infty$. For a thermal field, from the photon number distribution, Eq. (39), the photon number variance is

$$\langle(a^\dagger a)^2\rangle - \langle a^\dagger a\rangle^2 = \bar{n}(\bar{n}+1) \implies \langle a^{\dagger 2}a^2\rangle + \bar{n} - \bar{n}^2 = \bar{n}(\bar{n}+1)$$
$$\implies \langle a^{\dagger 2}a^2\rangle = 2\bar{n}^2. \tag{62}$$

Thus we obtain the normalized correlation function,

$$g^{(2)}(\tau) = \lim_{t\to\infty}\frac{\langle a^\dagger(t)[(a^\dagger a)(t+\tau)]a(t)\rangle}{\bar{n}^2} = 1 + \exp(-2\kappa\tau). \tag{63}$$

It shows the photon bunching effect, starting out at $g^{(2)}(0) = 2$ and decaying to $g^{(2)}(\infty) = 1$.

3.3. *The Glauber-Sudarshan P representation*

The Glauber-Sudarshan P representation[16, 19] is built around a diagonal expansion of the density operator in coherent states:

$$\rho(t) = \int d^2\alpha P(\alpha, \alpha^*, t)|\alpha\rangle\langle\alpha|. \tag{64}$$

In some situations, where $P(\alpha, \alpha^*, t)$ takes the form of a classical probability, it maps the master equation to a Fokker-Planck equation, the equation of a classical diffusion process. At the same time the quantum Langevin equation is mapped to a classical Langevin equation. This mapping provides the background to what is called the quantum-classical correspondence, a correspondence that may be carried through to map the quantum theory of photoelectric detection into the semiclassical theory. The master equation for a cavity/resonator with thermal input serves as an elementary example of this procedure.

3.3.1.   The Fokker-Planck equation

We consider the master equation, from Eqs. (19) and (34),

$$\frac{d\rho}{dt} = -i\omega_c[a^\dagger a, \rho] + \kappa(\bar{n} + 1)(2a\rho a^\dagger - a^\dagger a\rho - \rho a^\dagger a)$$
$$+ \kappa\bar{n}(2a^\dagger \rho a - aa^\dagger \rho - \rho aa^\dagger), \qquad (65)$$

and substitute the diagonal coherent state expansion, Eq. (64), for ρ. We then aim to replace all operator actions on $|\alpha\rangle\langle\alpha|$ by operations (multiplication/differentiation) involving α and α^*. To this end, note that

$$|\alpha\rangle\langle\alpha| = \exp(-\alpha^*\alpha)\exp(\alpha a^\dagger)|0\rangle\langle 0|\exp(-\alpha^* a), \qquad (66)$$

from which we may write

$$a|\alpha\rangle\langle\alpha| = \alpha|\alpha\rangle\langle\alpha|, \qquad\qquad |\alpha\rangle\langle\alpha|a^\dagger = \alpha^*|\alpha\rangle\langle\alpha|$$
$$a^\dagger|\alpha\rangle\langle\alpha| = \left(\alpha^* + \frac{\partial}{\partial\alpha}\right)|\alpha\rangle\langle\alpha|, \quad |\alpha\rangle\langle\alpha|a = \left(\alpha + \frac{\partial}{\partial\alpha^*}\right)|\alpha\rangle\langle\alpha|. \qquad (67)$$

All of this sits inside an integration with respect to α and α^*. Integration by parts can then be used to move the partial derivatives from $|\alpha\rangle\langle\alpha|$ to $P(\alpha, \alpha^*, t)$, with boundary terms assumed to vanish. These few steps take us to the Fokker-Planck equation

$$\frac{\partial P}{\partial t} = \left[(\kappa + i\omega_c)\frac{\partial}{\partial\alpha}\alpha + (\kappa - i\omega_c)\frac{\partial}{\partial\alpha^*}\alpha^* + \kappa\bar{n}\frac{\partial^2}{\partial\alpha\partial\alpha^*}\right]P. \qquad (68)$$

A partial differential equation like this is well known from the theory of classical diffusion processes. First- and second-order derivatives bring about a drift and diffusion (spreading), respectively, of the probability distribution P.

With a solution for $P(\alpha, \alpha^*, t)$, it is clear, from Eq. (64), that the expectation of any normal-ordered product of $a^\dagger$'s and a's can be calculated as a corresponding moment of α^* and α:

$$\langle (a^\dagger)^p a^q \rangle = \mathrm{tr}[(a^\dagger)^p a^q \rho] = \int d^2\alpha P(\alpha, \alpha^*) \mathrm{tr}[(a^\dagger)^p a^q |\alpha\rangle\langle\alpha|]$$

$$= \int d^2\alpha (\alpha^*)^p \alpha^q P(\alpha, \alpha^*). \tag{69}$$

3.3.2. The Langevin equation

For every Fokker-Planck equation there exists an equivalent stochastic differential equation satisfied by a continuous family of random variables labeled by t. The stochastic differential equation equivalent to Eq. (68) is the classical Langevin equation,

$$d\alpha_t = -(\kappa + i\omega_c)\alpha_t dt + \sqrt{2\kappa\bar{n}}dZ_t, \tag{70}$$

where α_t denotes the continuous family of random variables, whose realizations map out a path in the complex plane, and dZ_t is a complex-valued Weiner increment with covariance matrix

$$\langle dZ_t dZ_t \rangle_{\mathrm{path}} = \langle dZ_t^* dZ_t^* \rangle_{\mathrm{path}} = 0, \quad \langle dZ_t^* dZ_t \rangle_{\mathrm{path}} = dt. \tag{71}$$

The label indicates an average over paths realized by the complex-valued Wiener (diffusion) process Z_t. In this formulation, the quantum-classical correspondence allows time- and normal-ordered products of creation and annihilation operators to be computed as averages over paths in the complex plane (2-D harmonic oscillator phase space) realized by α_t, e.g.,

$$\langle (a^\dagger a)(t) \rangle = \langle |\alpha_t|^2 \rangle_{\mathrm{path}}, \tag{72}$$

and

$$\langle a^\dagger(t)[(a^\dagger a)(t+\tau)]a(t) \rangle = \langle |\alpha_{t+\tau}|^2 |\alpha_t|^2 \rangle_{\mathrm{path}}. \tag{73}$$

Having introduced the notion of paths, it is useful to introduce a new notation for the P function with the time variable no longer a separate argument but a parametric label on α_t: we make the substitution $d^2\alpha P(\alpha, \alpha^*, t) \rightarrow d\alpha_t P(\alpha_t)\delta^{(2)}(\alpha - \alpha_t)$ and replace Eq. (64) by

$$\rho(t) = \int d\alpha_t P(\alpha_t)|\alpha_t\rangle\langle\alpha_t|. \tag{74}$$

The integral is now viewed as an average over paths of the location of the phase-space variable at time t.

3.4.   *Quantum-classical correspondence*

3.4.1.   *Backaction: conditional states and 'a priori' paths*

The word "backaction" is normally associated with quantum physics, where the connotation is the "action" of a measurement "back" onto the state of the system measured — what is sometimes referred to as the "collapse of the wavefunction" or the "reduction" (post measurement) of the quantum state. The word might also be applied to a system governed by classical statistics. The sense in this case is the conditioning of the predictor (probability) of subsequent observations on the result of current observations. We may illustrate this sense by evaluating $\rho_{\text{cond}}(t)$, Eq. (55), in the P representation:

$$\rho_{\text{cond}}(t) = \frac{\int d\alpha_t P(\alpha_t)|\alpha_t|^2|\alpha_t\rangle\langle\alpha_t|}{\int d\alpha_t P(\alpha_t)|\alpha_t|^2} = \int d\alpha_t P_{\text{cond}}(\alpha_t)|\alpha_t\rangle\langle\alpha_t|, \tag{75}$$

with conditional P function

$$P_{\text{cond}}(\alpha_t) = \frac{P(\alpha_t)|\alpha_t|^2}{\int d\alpha_t P(\alpha_t)|\alpha_t|^2}. \tag{76}$$

Note that while the weight function changes from $P(\alpha_t)$ to $P_{\text{cond}}(\alpha_t)$ — this is the "backaction" — there is no change made to the path at time t, which remains, in the diagonal expansion of the density operator, as $|\alpha_t\rangle$. The central point is that on the annihilation (photoelectric detection) of a photon at time t, the action of the annihilation operator brings no change to the coherent state: $a|\alpha_t\rangle\langle\alpha_t|a^\dagger \to |\alpha_t|^2|\alpha_t\rangle\langle\alpha_t|$, with the factor $|\alpha_t|^2$ absorbed into updated path weights. In this sense, an ensemble of paths is set *a priori* — through an equation like Eq. (70) — when the P representation maps the master equation to a classical diffusion process. The probabilities for photoelectric detection and the updating of post detection states (probability distributions) is built on top of the *a priori* paths in an entirely classical sense. A full elaboration of the idea requires us to look into the elements of photoelectron counting theory.

3.4.2.   *Elements of photoelectron counting theory*

Constant classical intensity: Consider an intensity I (in photon flux units), time step dt, and counting interval $[0, T)$, with $T = N dt$. The probability for the photoelectric detector to "click" in time step dt is $I dt$ and the probability for "no click" is $1 - I dt$; we assume dt sufficiently short that the probability for two or more clicks in a time step is negligible. It follows that

$$\begin{pmatrix} \text{probability for } n \\ \text{``clicks'' in } [0, T) \end{pmatrix} = \frac{N(N-1)\cdots(N-n+1)}{n!}(I dt)^n(1 - I dt)^{N-n}. \tag{77}$$

The limit $N \to \infty$, $dt \to 0$, with $Ndt = T$ constant yields:

$$\left(\begin{array}{c}\text{probability for } n \\ \text{``clicks'' in } [0, T)\end{array}\right) = \frac{N(N-1)\cdots(N-n+1)}{n!}(Idt)^n$$

$$\times \sum_{k=1}^{N-n}(-Idt)^k \frac{(N-n)(N-n-1)\cdots(N-n-k+1)}{k!}$$

$$= \frac{(INdt)^n}{n!}\left(1 - \frac{1}{N}\right)\cdots\left(1 - \frac{(n-1)}{N}\right)$$

$$\times \sum_{k=1}^{N-n}\frac{(-INdt)^k}{k!}\left(1 - \frac{n}{N}\right)\cdots\left(1 - \frac{(n-k+1)}{N}\right)$$

$$\to \frac{(IT)^n}{n!}\exp(-IT). \tag{78}$$

It is hardly a surprise to arrive at the well-known Poisson distribution. A number of generalizations build on this result in pretty much self-evident ways:

Time-varying classical intensity: When the light intensity is not constant in time, something more than a combinatorial factor — first line of Eq. (78) — is required to account for the sum over different placements of n counts in N available boxes. The unsurprising result has

$$IT \to \int_0^T I(t)dt. \tag{79}$$

Stochastic classical intensity: When the classical intensity is stochastic, the integral over time is a stochastic integration,

$$\int_0^T I(t)dt \to \int_0^T I_t dt, \tag{80}$$

and the Poisson distribution must be averaged over realizations of I_t:

$$\left(\begin{array}{c}\text{probability for } n \\ \text{``clicks'' in } [0, T)\end{array}\right) \to \left\langle \frac{\left(\int_0^T I_t dt\right)^n}{n!}\exp\left(-\int_0^T I_t dt\right)\right\rangle_{\text{path}}. \tag{81}$$

This is the semiclassical photoelectron counting formula of Mandel.[15, 20]

Quantum photon flux: The quantum version of this formula, usually attributed to Kelly and Kleiner,[21] is superficially a straightforward generalization, with

$$I_t dt \to \mathcal{E}_{\text{out}}^{(-)}(t)\mathcal{E}_{\text{out}}^{(+)}(t)dt, \tag{82}$$

and the average over realizations of I_t replaced by a time- and normal-ordered quantum average:

$$\left(\begin{array}{c} \text{probability for } n \\ \text{"clicks" in } [0, T) \end{array}\right) \rightarrow \left\langle : \frac{\left[\int_0^T \mathcal{E}_{\text{out}}^{(-)}(t)\mathcal{E}_{\text{out}}^{(+)}(t)dt\right]^n}{n!} \exp\left[-\int_0^T \mathcal{E}_{\text{out}}^{(-)}(t)\mathcal{E}_{\text{out}}^{(+)}(t)dt\right] : \right\rangle. \tag{83}$$

The time- and normal- ordered correlation functions of Glauber coherence theory[17, 22] underly the quantum photoelectron counting formula. The ordering is key to the distinction drawn between classical and nonclassical light. The Glauber-Sudarshan P representation makes the connection.

3.4.3.   Classical versus nonclassical light

The distinction is based upon the fact that the quantum counting formula, Eq. (83), can be mapped into the semiclassical formula, Eq. (81), if the quantum field possess a well-behaved positive definite P function. We sketch the mapping here. The aim is to show that an arbitrary correlation function written in time and normal order can be evaluated as an average over *a priori* stochastic paths, realizations of the intensity (in photon flux units) $2\kappa|\alpha_t|^2$. The demonstration is made in four steps.

1. Drop input (free) fields: Expanding the exponential in Eq. (83) yields a sum of integrals over time- and normal-ordered averages, e.g.,

$$G^{(k)}(t_1, \ldots, t_k) = \langle \mathcal{E}_{\text{out}}^{(-)}(t_1), \ldots \mathcal{E}_{\text{out}}^{(-)}(t_k)\mathcal{E}_{\text{out}}^{(+)}(t_k) \ldots \mathcal{E}_{\text{out}}^{(+)}(t_1)\rangle, \tag{84}$$

with $t_1 < t_2 \ldots < t_k$. A generalization of the comment below Eq. (50) allows us to move all $\mathcal{E}_{\text{in}}^{(-)}(t_j)$ to the left and all $\mathcal{E}_{\text{in}}^{(+)}(t_j)$ to the right in the sequence of operators, act with them on the vacuum and, thus, in effect, drop all input field operators. In effect, we have the substitution

$$\mathcal{E}_{\text{out}}^{(-)}(t_j) \rightarrow \sqrt{2\kappa}a^\dagger(t_j), \quad \mathcal{E}_{\text{out}}^{(+)}(t_j) \rightarrow \sqrt{2\kappa}a(t_j). \tag{85}$$

2. Recast as multi-time quantum regression: We may now recast the correlation function in superoperator form and, in a generalization of the quantum regression procedure of Section 3.1.1, evaluate the trace over the reservoir (external electromagnetic field) in the Born-Markov approximation:

$$\begin{aligned} G^{(k)}(t_1, \ldots, t_k) &= (2\kappa)^k \text{tr}_{S \otimes R}[\chi(0)a^\dagger(t_1) \ldots a^\dagger(t_k)a(t_k) \ldots a(t_1)] \\ &= (2\kappa)^k \text{tr}_{S \otimes R}[a(t_k) \ldots a(t_1)\chi(0)a^\dagger(t_1) \ldots a^\dagger(t_k)] \\ &= \text{tr}_{S \otimes R}[\mathcal{S}e^{L(t_k - t_{k-1})}\mathcal{S} \ldots \mathcal{S}e^{L(t_2 - t_1)}\mathcal{L}e^{Lt_1}\chi(0)], \end{aligned} \tag{86}$$

where we introduce the source superoperator

$$S = 2\kappa(a \cdot a^\dagger). \tag{87}$$

Then, as in Section 3.1.1, the trace over the reservoir is evaluated with the substitution: $L \to \mathcal{L}$, $\chi \to \rho$. This yields the multi-time quantum regression formula

$$G^{(k)}(t_1, \ldots, t_k) = \text{tr}_S[Se^{\mathcal{L}(t_k - t_{k-1})}S \ldots Se^{\mathcal{L}(t_2 - t_1)}Se^{\mathcal{L}t_1}\rho(0)]. \tag{88}$$

3. Evaluate in the P representation: If a well-defined P function exists, e.g., one satisfying a Fokker-Planck equation like Eq. (68), the following correspondences hold, as we work our way from right to the left in Eq. (88):

$$e^{\mathcal{L}t_1}\rho(0) \to \int d^2\alpha_1 P(\alpha_1, \alpha_1^*, t_1)|\alpha_1\rangle\langle\alpha_1|, \tag{89}$$

$$Se^{\mathcal{L}t_1}\rho(0) \to 2\kappa \int d^2\alpha_1 |\alpha_1|^2 P(\alpha_1, \alpha_1^*, t_1)|\alpha_1\rangle\langle\alpha_1|, \tag{90}$$

$$e^{\mathcal{L}(t_2 - t_1)}Se^{\mathcal{L}t_1}\rho(0) \to 2\kappa \int d^2\alpha_1 |\alpha_1|^2 P(\alpha_1, \alpha_1^*, t_1)e^{\mathcal{L}(t_2 - t_1)}|\alpha_1\rangle\langle\alpha_1|, \tag{91}$$

and

$$e^{\mathcal{L}(t_2 - t_1)}|\alpha_1\rangle\langle\alpha_1| \int d^2\alpha_2 P(\alpha_2, \alpha_2^*, t_2|\alpha_1, \alpha_1^*, t_1)|\alpha_2\rangle\langle\alpha_2|, \tag{92}$$

where in the last expression $P(\alpha_2, \alpha_2^*, t_2|\alpha_1, \alpha_1^*, t_1)$ is a conditional distribution, i.e., the P function representing the state at time t_2 given that the state is the coherent state $|\alpha_1\rangle$ at time t_1. Contracting the conditional distribution with $P(\alpha_1, \alpha_1^*, t_1)$ to get a joint distribution,

$$P(\alpha_2, \alpha_2^*, t_2|\alpha_1, \alpha_1^*, t_1)P(\alpha_1, \alpha_1^*, t_1) = P(\alpha_2, \alpha_2^*, t_2; \alpha_1, \alpha_1^*, t_1), \tag{93}$$

we complete the first step of an iterative procedure that ultimately reexpresses the correlation function as the phase-space average

$$G^{(k)}(t_1, \ldots, t_k) = (2\kappa)^k \int d^2\alpha_1 \ldots d^2\alpha_k \left(\prod_{j=1}^k |\alpha_j|^2\right) P(\alpha_k, \alpha_k^*, t_k; \ldots; \alpha_1, \alpha_1^*, t_1). \tag{94}$$

4. Rewrite as a sum over paths: Finally, we generalize the rewriting of Eq. (64) as Eq. (74). We introduce the change of notation,

$$d^2\alpha_1 \ldots d^2\alpha_k P(\alpha_k, \alpha_k^*, t_k; \ldots; \alpha_1, \alpha_1^*, t_1) \to d\alpha_t P(\alpha_t) \prod_{j=1}^k \delta^{(2)}(\alpha_{t_j} - \alpha_j), \tag{95}$$

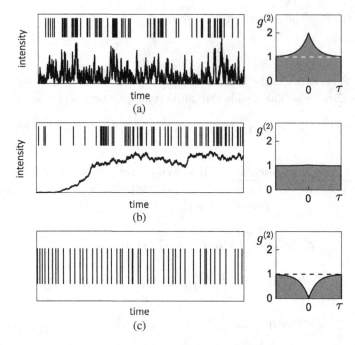

Fig. 5. Sample photoelectron counting sequences (vertical lines) with corresponding *a priori* count rates: (a) filtered thermal or chaotic light, (b) turn-on of a laser (near coherent light), and (c) resonance fluorescence (antibunched light), where there is no *a priori* count rate.

and obtain

$$G^{(k)}(t_1, \ldots, t_k) = (2\kappa)^k \int d\alpha_t \left(\prod_{j=1}^{k} |\alpha_{t_j}|^2 \right) P(\alpha_t). \tag{96}$$

The correlation function is now evaluated as an average over *a priori* realizations of the intensity (in photon flux units) $2\kappa|\alpha_t|^2$.

By treating all multi-time correlations in this manner, Eq. (83) is mapped into Eq. (81), with *a priori* classical intensity $I_t = 2\kappa|\alpha_t|^2$. Figure 5 illustrates these ideas with two photoelectron counting sequences generated from an *a priori* count rate, the first modeling filtered thermal light, with α_t a realization of the (classical) Langevin equation, Eq. (70), and the second the turn-on of a laser close to threshold, with α_t a realization of[23]

$$d\alpha_t = -\kappa\alpha_t(1 - \wp + \wp|\alpha_t|^2)dt + \sqrt{2\kappa}n_{\text{sat}}^{-1/2}dZ_t, \tag{97}$$

where $\wp$ is the pump parameter and n_{sat} is the saturation photon number. The third sequence is antibunched (resonance fluorescence). For it there is no *a priori* count rate; it is generated as a quantum trajectory, by following the prescription set out in Section 5.4.1.

4. Quantum trajectories: Don't trace, disentangle

4.1. *Sketch #1: Labeling kets with "records"*

When deriving a master equation we trace over the reservoir, removing from the description all information about the external electromagnetic field — information, for example, like the number of photons it contains. The novelty of the quantum trajectory approach is to disentangle rather than trace. In this way, correlations between what is (or might be) observed in the environment and the state of the cavity/resonator are retained. In effect, an ideal measurement is made on the environment (external electromagnetic field), which thus refines the state of the cavity/resonator to be in accord with the measurement result. The conditions that justify the Born-Markov treatment guarantee that measuring the environment does not interfere with the dynamics of the cavity/resonator, in the sense that putting a measuring instrument (photodetector) physically in place does not change anything at the level of the master equation description; the same master equation holds whether or not a photodetector intercepts the output field as shown in Fig. 6.

We sketch these ideas in a little more detail noting that the density operator, χ, of the cavity/resonator and its external electromagnetic field describes an entangled state:

$$\chi(t) = |\psi_{S \otimes R}(t)\rangle\langle\psi_{S \otimes R}(t)|, \tag{98}$$

$$|\psi_{S \otimes R}(t)\rangle = |0\rangle|\phi_S^{(0)}(t)\rangle + |1\rangle|\phi_S^{(0)}(t)\rangle + |2\rangle|\phi_S^{(0)}(t)\rangle + \cdots, \tag{99}$$

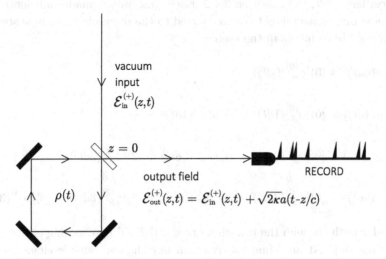

Fig. 6. Schematic of a ring cavity/resonator modeled in 1D, with input and output fields and photoelectron counting record.

where the kets $|0\rangle$, $|1\rangle$, $|2\rangle$, etc. denote states of the external (output) field populated by 0, 1, 2, etc. photons, and $|\phi_S^{(0)}(t)\rangle$, $|\phi_S^{(1)}(t)\rangle$, $|\phi_S^{(2)}(t)\rangle$, etc., are cavity/resonator states reached after emitting 0, 1, 2, etc. photons. We of course assume that the cavity input field is in the vacuum state. Structure of this kind is expected because the interaction between the system and reservoir has been taken in a form, Eq. (8), that preserves photon number (because of the rotating-wave approximation). At the simplest level, disentangling the cavity/resonator from the reservoir (external electromagnetic field) amounts to setting up a decomposition of the density operator in the form

$$\rho(t) = \sum_{n=0}^{\infty} |\phi_S^{(n)}(t)\rangle\langle\phi_S^{(n)}(t)|, \tag{100}$$

or

$$\rho(t) = \sum_{n=0}^{\infty} \langle\phi_S^{(n)}(t)|\phi_S^{(n)}(t)\rangle \frac{|\phi_S^{(n)}(t)\rangle\langle\phi_S^{(n)}(t)|}{\langle\phi_S^{(n)}(t)|\phi_S^{(n)}(t)\rangle}. \tag{101}$$

It is clear that what we have here does correspond to the reduced density operator $\text{tr}_R[\chi(t)]$, since the reservoir kets $|0\rangle$, $|1\rangle$, $|2\rangle$, etc. are orthogonal and normalized. Having access to the pieces of the decomposition is more informative than $\rho(t)$ on its own, though: $\langle\phi_S^{(n)}(t)|\phi_S^{(n)}(t)\rangle$ is the probability that n photons have been emitted into the reservoir up to time t, and $|\phi_S^{(n)}(t)\rangle$ — in Eq. (101) the square root of the denominator provides normalization — is the cavity/resonator state conditioned upon the fact that, indeed, n photons were emitted up to time t.

It is possible to take the idea a step further by disentangling, not at time t, but continuously in time. Imagine the time evolution divided up into a sequence of time steps, length dt, as in Section 3.4.2. Each time step is made sufficiently short that at most one photon might be transferred to the reservoir in a single step. The evolution then branches as in the sketch:

$$|\psi_{S\otimes R}(0dt)\rangle = |0\rangle|\phi_S^{(0)}(0dt)\rangle$$
$$\Downarrow \qquad \searrow$$
$$|\psi_{S\otimes R}(1dt)\rangle = |0\rangle|\phi_S^{(0)}(1dt)\rangle + |1\rangle|\phi_S^{(1)}(1dt)\rangle$$
$$\Downarrow \qquad \searrow \quad \Downarrow \qquad \searrow$$
$$|\psi_{S\otimes R}(2dt)\rangle = |0\rangle|\phi_S^{(0)}(2dt)\rangle + |1\rangle|\phi_S^{(1)}(2dt)\rangle + |2\rangle|\phi_S^{(2)}(2dt)\rangle$$
$$\Downarrow \qquad \searrow \quad \Downarrow \quad \searrow \quad \Downarrow \qquad \searrow$$
$$|\psi_{S\otimes R}(3dt)\rangle = |0\rangle|\phi_S^{(0)}(3dt)\rangle + |1\rangle|\phi_S^{(1)}(3dt)\rangle + |2\rangle|\phi_S^{(2)}(3dt)\rangle + |3\rangle|\phi_S^{(3)}(3dt)\rangle$$

A particular path through the branching tree is defined by choosing either $\Downarrow$ or $\searrow$ at each time step. At any time $t = Ndt$, all but the two outside elements in the expansion of the state can be reached by more than one path, so a decomposition of the density operator over paths, or records, is not quite the same thing as a decomposition over the number of photons emitted up to time t. It is written

$$\rho(t) = \sum_{\text{REC}} P_{\text{REC}}(t)|\Phi_{\text{REC}}(t)\rangle\langle\Phi_{\text{REC}}(t)|, \tag{102}$$

with P_{REC} the probability for a particular record (path) finishing at time t, and $|\Phi_{\text{REC}}(t)\rangle$ the state of the cavity/resonator conditioned upon that record. Clearly, the evolution of $|\Phi_{\text{REC}}\rangle$ is conditional and stochastic. The questions to answer are: what are the branching ratios and how does $|\Phi_{\text{REC}}\rangle$ evolve along a particular branch?

4.2. Sketch #2: Familiar and unfamiliar expansions of ρ

4.2.1. Bohr-Einstein quantum jumps: Unconditional expansion of ρ

The rate equations of Section 2.6.1 provide a description that might be based on the quantum jumps introduced by Bohr and Einstein in the time of the old quantum theory; the picture is elaborated most fully in Einstein A and B theory,[24, 25] with its three processes of spontaneous emission, stimulated emission, and absorption. We aim for the moment to simply look at this type of dynamic from a different point of view, adding nothing but an unfamiliar (though not fundamentally new) conditional probability and some notation to go along with it. Since tracing everything through gets a little complicated even if we consider the thermal excitation of a two-state system rather than a cavity/resonator mode, we will carry the exercise through for the near trivial case of a single downwards jump, i.e., for spontaneous emission, at rate γ, from an excited state $|e\rangle$ to a ground state $|g\rangle$. The trivial rate equation is

$$\frac{dp_e}{dt} = -\gamma p_e, \tag{103}$$

with solution $p_e(t) = p_e(0)\exp(-\gamma t)$, $p_g(t) = 1 - p_e(0)\exp(-\gamma t)$. The standard expansion of the density operator is made on the basis of these probabilities to occupy the two orthogonal basis states:

$$\rho(t) = p_e(t)|e\rangle\langle e| + p_e(t)|g\rangle\langle g|. \tag{104}$$

4.2.2. Record probabilities and expansion in conditional states

Compare this familiar form with the alternative expansion written as a sum over records,

$$\rho(t) = P_{\text{NULL}}\left[\frac{p_e(0)e^{-\gamma t}}{p_g(0) + p_e(0)e^{-\gamma t}}|e\rangle\langle e| + \frac{p_g(0)}{p_g(0) + p_e(0)e^{-\gamma t}}|g\rangle\langle g|\right]$$

$$+ \int_0^t P_{\text{JUMP}}(t')dt'\,|g\rangle\langle g|. \tag{105}$$

Even for this trivial example there are infinitely many records finishing at time t; although they fall into just two classes so far as the number of photons emitted goes: either there has been no photon emitted up to time t, or there has been one photon emitted. In the latter case, the time of the emission distinguishes one record from another. We label the no emission case as the NULL event. The record probabilities are, for no photon emitted,

$$P_{\text{NULL}} = p_g(0) + p_e(0) \exp(-\gamma t), \tag{106}$$

and for the emission of a photon at time t',

$$P_{\text{JUMP}}(t')dt' = \gamma dt' p_e(0) \exp(-\gamma t'). \tag{107}$$

Note that the NULL record corresponds to two possibilities that are unresolved in the conditional density operator appearing inside the square bracket in Eq. (105). It might be that the initial state was $|g\rangle$, in which case there has definitely been no photon emitted up to time t; the probability for this possibility is $p_g(0)$. On the other hand, it may be that the initial state was $|e\rangle$ but no photon is yet emitted; the probability for this possibility is $p_e(0) \exp(-\gamma t)$. Over a long enough time the ambiguity is resolved, as the density operator conditioned on the NULL record (in the square bracket) is asymptotically $|g\rangle\langle g|$ — the first possibility is definitely the case if no photon is ever emitted. Note how the resolution develops continuously over time, along with the changing weights in front of $|e\rangle\langle e|$ and $|g\rangle\langle g|$ inside the square bracket: the longer the wait without an emitted photon, the more likely the first possibility becomes over the second.

The probability for the record with a photon emitted at time t' is the product of the probability, $p_e(0) \exp(-\gamma t')$, that the initial state is $|e\rangle$ and no photon is emitted prior to time t', and the probability, $\gamma dt'$, for a photon to be emitted in the interval dt' at time t'. The integral in Eq. (101) sums over possible emission times.

Equation (101) may appear as a somewhat perverse way to expand the density operator. It nevertheless arises quite naturally from a formal solution to the master equation.

4.2.3. *From the quantum master equation*

The master equation for spontaneous emission from a two-state system (Section 2.5.2) has generalized Liouvillian

$$\mathcal{L} = -i\omega_a[\sigma_+\sigma_-, \cdot] + (\gamma/2)\Lambda[\sigma_-]. \tag{108}$$

We may decompose $\mathcal{L}$ as the sum of a jump term and a no-jump term by splitting the Lindblad operator $(\gamma/2)\Lambda[\sigma_-]$ into two pieces:

$$\mathcal{L} = \mathcal{S} + (\mathcal{L} - \mathcal{S}), \tag{109}$$

with

$$S = \gamma \sigma_- \cdot \sigma_+, \tag{110}$$

and

$$\mathcal{L} - S = -i\omega_a[\sigma_+\sigma_-, \cdot] - \frac{\gamma}{2}[\sigma_+\sigma_-, \cdot]_+$$

$$= \frac{1}{i\hbar}(H \cdot - \cdot H^\dagger), \tag{111}$$

where

$$H = \hbar\left(\omega_a - i\frac{\gamma}{2}\right)\sigma_+\sigma_- \tag{112}$$

is a non-Hermitian Hamiltonian. We now develop a Dyson expansion for $\rho(t)$, with $\mathcal{L} - S$ and S taken in the respective roles of free propagator and interaction. The steps run parallel to those of time-dependent perturbation theory. We begin with the master equation, Eq. (19), written in the "interaction" picture:

$$\frac{d\tilde{\rho}}{dt} = e^{-(\mathcal{L}-S)t}Se^{(\mathcal{L}-S)t}\tilde{\rho}, \tag{113}$$

with $\tilde{\rho} = \exp[-(\mathcal{L} - S)t]\rho(t)$. The formal solution is

$$\tilde{\rho}(t) = \rho(0) + \int_0^t dt' e^{-(\mathcal{L}-S)t'}Se^{(\mathcal{L}-S)t'}\tilde{\rho}(t'). \tag{114}$$

Continuing then with the standard procedure, this formal solution is substituted in the integral on the right-hand side of Eq. (113), noting that, for the spontaneous emission example [i.e., from Eqs. (123) and (111)] $S\exp[(\mathcal{L} - S)(t' - t'')]S = 0$, because $\mathcal{L} - S$ preserves the number of excitations and the two appearances of S attempt to destroy two — the initial state offers only one. Thus the Dyson expansion truncates at first order in S,

$$\tilde{\rho}(t) = \rho(0) + \int_0^t dt' e^{-(\mathcal{L}-S)t'}Se^{(\mathcal{L}-S)t'}\rho(0), \tag{115}$$

and the formal solution for the density operator is

$$\rho(t) = e^{(\mathcal{L}-S)t}\rho(0) + \int_0^t dt' e^{(\mathcal{L}-S)(t-t')}Se^{(\mathcal{L}-S)t'}\tilde{\rho}(0). \tag{116}$$

The connection to the expansion over records, Eq. (105), and record probabilities, Eq. (106) and Eq. (107), is completed after normalizing to obtain conditional density operators:

$$\rho(t) = P_{\text{NULL}}(t)\frac{e^{(\mathcal{L}-S)t}\rho(0)}{\text{tr}[e^{(\mathcal{L}-S)t}\rho(0)]}$$

$$+ \int_0^t P_{\text{JUMP}}(t')dt'\frac{e^{(\mathcal{L}-S)(t-t')}Se^{(\mathcal{L}-St')}\rho(0)}{\text{tr}[e^{(\mathcal{L}-S)(t-t')}Se^{(\mathcal{L}-S)t'}\rho(0)]}, \tag{117}$$

with record probability,

$$P_{\text{NULL}}(t) = \text{tr}[e^{(\mathcal{L}-\mathcal{S})t}\rho(0)], \tag{118}$$

for no photon emitted up to time t, and

$$P_{\text{JUMP}}(t')dt' = \text{tr}[e^{(\mathcal{L}-\mathcal{S})(t-t')}\mathcal{S}e^{(\mathcal{L}-\mathcal{S})t'}\rho(0)]dt', \tag{119}$$

for a photon emitted at time t'. From these last expressions it is easy enough to recover Eq. (105), and Eq. (106) and Eq. (107), from the definitions of $\mathcal{S}$ and $\mathcal{L} - \mathcal{S}$, and initial density operator

$$\rho(0) = p_e(0)|e\rangle\langle e| + p_g(0)|g\rangle\langle g|. \tag{120}$$

4.2.4.   Extension to pure states

The formal expansion, Eq. (117), gains something over the example based on Bohr-Einstein quantum jumps, since it may also be applied to an initial superposition state,

$$|\psi(0)\rangle = c_e(0)|e\rangle + c_g(0)|g\rangle. \tag{121}$$

In this case all the pieces of the expansion look the same, except for the state conditioned on the NULL record, which, replaces the mixed state in the square bracket in Eq. (105) by the pure state

$$|\psi_{\text{NULL}}(t)\rangle = \frac{c_e(0)e^{-(\gamma/2+i\omega_a)t}|e\rangle + c_g(0)|g\rangle}{\sqrt{|c_g(0)|^2 + e^{-\gamma t}|c_e(0)|^2}}. \tag{122}$$

Thus, most generally, we have a decomposition based on the record of quantum jumps that includes superpositions and the evolution of coherence.

The ket $|\psi_{\text{NULL}}(t)\rangle$ evolves over time to the ground state. In a fraction $|c_g(0)|^2$ of realizations (see Section 5.3) there will be no jump and the ground state will be reached through this continuous evolution. Conversely, in a fraction $|c_e(0)|^2$ of realizations, a jump does occur at some point during the evolution and the ground state is reached through this jump. We may interpret the observation of a "jump" or "no jump" in a single realization as a measurement of what the initial state must have been; albeit a measurement that takes several decay lifetimes to be completed. In this sense the situation is the same as with the initial incoherent ensemble. Viewed more broadly, though, the whole set up is quite different and a significant extension of the original Bohr-Einstein quantum jump: an appropriately designed measurement can distinguish a prepared superposition from a prepared incoherent ensemble; the physical situations are different; transient coherence can be detected.

5. Quantum trajectories and photoelectron counting

What we have seen so far about quantum trajectories is only a sketch. The sketch suggests that we decompose the density operator of the cavity/resonator, not based on any *a priori* choice of basis states, but in terms of states labeled by (conditioned upon) a record over time of photon emissions. We have seen how this can be done for a simple example involving just one emission. Building upon quantum jumps in the sense of Einstein A and B theory, such a decomposition is, perhaps, unfamiliar, but ultimately only a reorientation of our statistical ideas — the use of conditional rather than unconditional probabilities. Moreover, for a single emission, the decomposition is seen to fall quite naturally out of a Dyson expansion for the solution to the Lindblad master equation.

Moving forward in this section, we would like to make the sense of conditioning on a record more concrete. The original notion of a quantum jump — as in the old quantum theory — might be thought of in concrete terms, i.e., in terms of jumps that are real events; but the idea does not carry through to modern quantum theory in this way: quantum evolution, even for an open system, is not simply a case of stochastic jumping between stationary states; there is coherent evolution and there are superpositions. Action of the superoperator $\mathcal{S}$, e.g., in the sequence $e^{(\mathcal{L}-\mathcal{S})(t-t')}\mathcal{S}e^{(\mathcal{L}-\mathcal{S})t'}$, might appear to execute some sort of jump. It is, however, at bottom, only a jump in some formal sense. If we are to make these ideas more concrete, "conditional" should derive its meaning from the counting of photons in the output field, with the counting performed by a — possibly only imagined, i.e., virtual — photoelectric detector, as shown in Fig. 6. The records, then, are to be seen as continuous measurement, or photoelectron counting records.

In this section we outline the connection between photoelectron counting theory, met in Section 3.4.2, and the Dyson expansion for the density operator. We then see how to set up a Monte Carlo scheme to realize simulated photoelectron counting records, along with the decomposition (unraveling[26]) of the density operator conditioned upon those records.

5.1. *Photoelectron counting records and conditional states*

5.1.1. *Master equation unraveling*

In order to set the scene, let us first extend the Dyson expansion to all orders and cast it as an unraveling of the density operator along the lines indicated by Eq. (117). Returning to the cavity/resonator example avoids the truncation at first-order; splitting of generalized Liouvillian is carried through with

$$\mathcal{S} = 2\kappa a \cdot a^\dagger, \qquad (123)$$

and

$$\mathcal{L} - \mathcal{S} = \frac{1}{i\hbar}(H \cdot - \cdot H^\dagger), \quad H = \hbar(\omega_c - i\kappa)a^\dagger a. \tag{124}$$

The full Dyson expansion follows in the usual way from the repeated iteration of Eq. (114), with the resulting expansion,

$$\rho(t) = \sum_{n=0}^{\infty} \int_0^t dt_n \int_0^{t_n} dt_{n-1} \cdots \int_0^{t_2} dt_1 e^{(\mathcal{L}-\mathcal{S})(t-t_n)} \mathcal{S} e^{(\mathcal{L}-\mathcal{S})(t_n-t_{n-1})} \mathcal{S} \cdots$$
$$\cdots \mathcal{S} e^{(\mathcal{L}-\mathcal{S})t_1} \rho(0). \tag{125}$$

The first step towards a connection with photoelectron counting is made through a rewriting of this formula in the style of Eq. (102). The form taken by $\mathcal{S}$ and $\mathcal{L} - \mathcal{S}$, and the assumption of a pure initial state, allows for the rewriting,

$$\rho(t) = \sum_{n=0}^{\infty} \int_0^t dt_n \int_0^{t_n} dt_{n-1} \cdots \int_0^{t_2} dt_1 P_{\text{REC}}(t) |\psi_{\text{REC}}\rangle \langle \psi_{\text{REC}}(t)|, \tag{126}$$

with record probabilities and conditional states,

$$P_{\text{REC}}(t)dt_1 \ldots dt_n = \langle \bar{\psi}_{\text{REC}}(t) | \bar{\psi}_{\text{REC}}(t)\rangle dt_1 \ldots dt_n, \tag{127}$$

and

$$|\psi_{\text{REC}}(t)\rangle = \frac{|\bar{\psi}_{\text{REC}}(t)\rangle}{\sqrt{\langle \bar{\psi}_{\text{REC}}(t) | \bar{\psi}_{\text{REC}}(t)\rangle}}, \tag{128}$$

where

$$|\bar{\psi}_{\text{REC}}(t)\rangle = \exp\left[-\frac{i}{\hbar}H(t - t_n)\right] J \exp\left[-\frac{i}{\hbar}H(t_n - t_{n-1})\right] J \cdots$$
$$\cdots J \exp\left[-\frac{i}{\hbar}Ht_1\right] |\psi(0)\rangle, \tag{129}$$

with

$$J = \sqrt{2\kappa}a \tag{130}$$

the jump operator ($\mathcal{S} = J \cdot J^\dagger$). In this it is suggested that $P_{\text{REC}}(t)$, so defined, is the probability density for counting n photons in the output field at times $t_1, t_2, \ldots, t_n$, and none at any other times in the interval 0 to t. The second step which completes the connection with photoelectron counting is the demonstration that $P_{\text{REC}}(t)$ is, indeed, the continuous measurement record probability density.

5.1.2. *Record probability density*

We met some of the basic ideas and formulas of photoelectron counting in Section 3.4.2, focusing on the probability for a specified number of counts over a specified interval of time. More generally, counting sequences are defined, as a stochastic process, by assigning one of two different hierarchies of probabilities. Considering a sequence of n counts over a given period of time, we might assign either a

$$
\begin{array}{c}
\text{nonexclusive} \\
\text{probability}
\end{array}
= \text{Prob}
\left(
\begin{array}{cccc}
\text{"click"} & \text{"click"} & & \text{"click"} \\
dt_1 & dt_2 & & dt_n \\
\hline
\| \quad \text{anything} \quad \| & \| \quad \cdots\cdots\cdots & & \| \\
t_1 & t_2 & & t_n
\end{array}
\right)
$$

$$
= \langle \mathcal{E}_{\text{out}}^{(-)}(t_1) \ldots \mathcal{E}_{\text{out}}^{(-)}(t_n) \mathcal{E}_{\text{out}}^{(+)}(t_n) \ldots \mathcal{E}_{\text{out}}^{(+)}(t_1) \rangle \, dt_1 \ldots dt_n, \tag{131}
$$

or an

$$
\begin{array}{c}
\text{exclusive} \\
\text{probability}
\end{array}
= \text{Prob}
\left(
\begin{array}{cccc}
\text{"click"} & \text{"click"} & & \text{"click"} \\
dt_1 & dt_2 & & dt_n \\
\hline
\| \quad \text{no "click"} \quad \| & \| \quad \cdots\cdots\cdots & & \| \\
t_1 & t_2 & & t_n
\end{array}
\right)
$$

$$
= \Big\langle \, : \exp\left[-\int_{t_n}^{t} dt'\, \mathcal{E}_{\text{out}}^{(-)}(t')\mathcal{E}_{\text{out}}^{(+)}(t')\right] \mathcal{E}_{\text{out}}^{(-)}(t_n)\mathcal{E}_{\text{out}}^{(+)}(t_n) \cdots
$$

$$
\cdots \mathcal{E}_{\text{out}}^{(-)}(t_1)\mathcal{E}_{\text{out}}^{(+)}(t_1) \exp\left[-\int_{0}^{t_1} dt'\, \mathcal{E}_{\text{out}}^{(-)}(t')\mathcal{E}_{\text{out}}^{(+)}(t')\right] : \, \Big\rangle \, dt_1 \ldots dt_n. \tag{132}
$$

A full hierarchy of either type completely defines the counting process. The first type, the nonexclusive probabilities, are simply the multi-coincidence rates entering into Glauber's theory of optical coherence.[16, 17, 22] The exclusive probabilities include additional exponential factors at the start and end of the sequence — shown in Eq. (132) — and also in between each pair of "clicks" — not shown in Eq. (132). The significance of these factors can be appreciated from Eq. (83), where the same exponential determines the probability for no counts during the assigned counting interval. The exponentials add the condition that "clicks" are to be excluded in between the n count times. The exclusive probabilities are record probabilities, as they apply to fully assigned records, with either a "click" or "no click" assigned at each time step. Thus we aim now to show that the exclusive photoelectron counting probabilities are precisely the probabilities $P_{\text{REC}}(t)dt_1 \ldots dt_n$ appearing inside the Dyson expansion, Eq. (126).

The demonstration starts out running in parallel with the initial steps sketched in Section 3.4.3:

1. Drop input (free) fields: Expanding the exponentials yields a sum of integrals over time- and normal-ordered averages, just as before, though in this case an n-fold sum. By the argument previously given we may, in effect, drop all input field operators and make the substitution

$$\mathcal{E}_{\text{out}}^{(-)}(t) \to \sqrt{2\kappa} a^\dagger(t), \quad \mathcal{E}_{\text{out}}^{(+)}(t) \to \sqrt{2\kappa} a(t). \tag{133}$$

2. Recast in superoperator form: As in step 2 of Section 3.4.3, the normal ordering and time evolution are written out explicitly and the trace reexpressed in superoperator form:

$$\begin{pmatrix} \text{exclusive} \\ \text{probability} \\ \text{density} \end{pmatrix} = \text{tr}_{S \otimes R} \left\{ \exp_T \left[-\int_{t_n}^t dt' e^{-Lt'} \mathcal{S} e^{Lt'} \right] e^{-Lt_n} \mathcal{S} e^{Lt_n} \cdots \right.$$

$$\left. \cdots e^{-Lt_1} \mathcal{S} e^{Lt_1} \exp_T \left[-\int_0^{t_1} dt' e^{-Lt'} \mathcal{S} e^{Lt'} \right] \chi(0) \right\}$$

$$= \text{tr}_{S \otimes R} [e^{(L-\mathcal{S})(t-t_n)} \mathcal{S} e^{(L-\mathcal{S})(t_n-t_{n-1})} \cdots \mathcal{S} e^{(L-\mathcal{S})t_1} \chi(0)], \tag{134}$$

where $\exp_T$ denotes a time-ordered operator expansion, and we have made use of the identity,

$$e^{Lt_k} \exp_T \left[-\int_{t_{k-1}}^{t_k} dt' e^{-Lt'} \mathcal{S} e^{Lt'} \right] e^{-Lt_{k-1}} = e^{(L-\mathcal{S})(t_k-t_{k-1})}, \tag{135}$$

which follows, for example, from perturbation theory, by considering L as a free propagator and $-\mathcal{S}$ as an interaction. This is a very important step because it absorbs the condition that there is to be "no click" in between the specified "click" times into a change of the effective Liouvillian from L to $L - S$; this change is the precursor of the $\mathcal{L} - \mathcal{S}$ which appears in the Dyson expansion of $\rho(t)$ through the splitting up of the generalized Liouvillian $\mathcal{L}$.

3. Evaluate the trace over the reservoir: In the Born-Markov approximation, tr_R is removed with the substitution: $L \to \mathcal{L}$, $\chi \to \rho$. This yields

$$\begin{pmatrix} \text{exclusive} \\ \text{probability} \\ \text{density} \end{pmatrix} = \text{tr}_S [e^{(\mathcal{L}-\mathcal{S})(t-t_n)} \mathcal{S} e^{(\mathcal{L}-\mathcal{S})(t_n-t_{n-1})} \cdots \mathcal{S} e^{(\mathcal{L}-\mathcal{S})t_1} \rho(0)]$$

$$= \langle \bar{\psi}_{\text{REC}}(t) | \bar{\psi}_{\text{REC}}(t) \rangle, \tag{136}$$

the advertised result.

As an aside it is worth noting how the probability for n photoelectron counts in the interval 0 to t is recovered from the Dyson expansion. Taking the trace of Eq. (126) yields

$$1 = \sum_{n=0}^{\infty} \int_0^t dt_n \int_0^{t_n} dt_{n-1} \cdots \int_0^{t_2} dt_1 P_{\text{REC}}(t)$$

$$= \sum_{n=0}^{\infty} \left(\begin{array}{c} \text{probability for } n \\ \text{"clicks" in } [0,t) \end{array} \right),$$

with

$$\left(\begin{array}{c} \text{probability for } n \\ \text{"clicks" in } [0,t) \end{array} \right) = \int_0^t dt_n \int_0^{t_n} dt_{n-1} \cdots \int_0^{t_2} dt_1 P_{\text{REC}}(t). \qquad (137)$$

The result is nothing more than a summation of n-"click" record probabilities over "click" times; with Eq. (127) and Eq. (128), it is a counting formula in the style of Davies.[18, 27] It has been claimed[28, 29] that the Davies formula is more correct than Kelly-Kleiner, Eq. (83). Here, however, we have effectively seen one derived from the other. The essential point in making the connection is the open-system (photo-emissive) character of the source of the photons counted. The Kelly-Kleiner formula presumes this, i.e., that the detector simply records emitted photons; it does not "suck" photons from a system that would not otherwise emit them. Mandel makes this point[30] in reply to the criticism of the Kelly-Kleiner formula.

5.2. *Sum over records in S and in $S \otimes R$*

Before moving on to the Monte-Carlo simulation of records, let us return to the previously introduced sketch of the branching tree,

$$|\psi_{S \otimes R}(0dt)\rangle = |0\rangle|\phi_S^{(0)}(0dt)\rangle$$
$$\Downarrow \qquad \searrow$$
$$|\psi_{S \otimes R}(1dt)\rangle = |0\rangle|\phi_S^{(0)}(1dt)\rangle + |1\rangle|\phi_S^{(1)}(1dt)\rangle$$
$$\Downarrow \qquad \searrow \qquad \Downarrow \qquad \searrow$$
$$|\psi_{S \otimes R}(2dt)\rangle = |0\rangle|\phi_S^{(0)}(2dt)\rangle + |1\rangle|\phi_S^{(1)}(2dt)\rangle + |2\rangle|\phi_S^{(2)}(2dt)\rangle$$
$$\Downarrow \qquad \searrow \qquad \Downarrow \qquad \searrow \qquad \Downarrow \qquad \searrow$$
$$|\psi_{S \otimes R}(3dt)\rangle = |0\rangle|\phi_S^{(0)}(3dt)\rangle + |1\rangle|\phi_S^{(1)}(3dt)\rangle + |2\rangle|\phi_S^{(2)}(3dt)\rangle + |3\rangle|\phi_S^{(3)}(3dt)\rangle.$$

Consider the expansion of the density operator at time $t = 3dt$. There are eight records, corresponding to zero, one, two, or at most three "clicks" ($\emptyset$ and ! denote "no click" and "click"):

$$R1 = \{\emptyset\,\emptyset\,\emptyset\};\qquad\qquad\qquad\qquad\qquad 0 \text{ "clicks"}$$

$$R2 = \{!\,\emptyset\,\emptyset\},\quad R3 = \{\emptyset\,!\,\emptyset\},\quad R4 = \{\emptyset\,\emptyset\,!\};\quad 1 \text{ "click"}$$

$$R5 = \{\emptyset\,!\,!\},\quad R6 = \{!\,\emptyset\,!\},\quad R7 = \{\emptyset\,!\,!\};\quad 2 \text{ "clicks"}$$

$$R8 = \{!\,!\,!\};\qquad\qquad\qquad\qquad\qquad 3 \text{ "clicks"}.$$

The density operator is expanded over records as

$$\rho(3dt) = P_{R1}|\psi_{R1}\rangle\langle\psi_{R1}| + (P_{R2}dt)|\psi_{R2}\rangle\langle\psi_{R2}| + (P_{R3}dt)|\psi_{R3}\rangle\langle\psi_{R3}|$$
$$+ (P_{R4}dt)|\psi_{R4}\rangle\langle\psi_{R4}| + (P_{R5}dt^2)|\psi_{R5}\rangle\langle\psi_{R5}| + (P_{R6}dt^2)|\psi_{R6}\rangle\langle\psi_{R6}|$$
$$+ (P_{R7}dt^2)|\psi_{R7}\rangle\langle\psi_{R7}| + (P_{R8}dt^3)|\psi_{R8}\rangle\langle\psi_{R8}|, \tag{138}$$

where the sum over rows sums the number of "clicks", n, while the sum across each row carries out the integration over the "click" times. The record probabilities and conditional states are ($k = 1, 2, \ldots 8$; $n = 0, 1, 2,$ or 3)

$$P_{Rk}dt^n = \langle\bar\psi_{Rk}|\bar\psi_{Rk}\rangle dt^n, \quad |\bar\psi_{Rk}\rangle = \frac{|\bar\psi_{Rk}\rangle}{\sqrt{\langle\bar\psi_{Rk}|\bar\psi_{Rk}\rangle}}, \tag{139}$$

with unnormalized kets (e.g., for 1 "click"):

$$|\bar\psi_{R2}\rangle = J \exp\left(-\frac{i}{\hbar}H2dt\right)|\psi(0)\rangle, \tag{140}$$

$$|\bar\psi_{R3}\rangle = \exp\left(-\frac{i}{\hbar}Hdt\right)J \exp\left(-\frac{i}{\hbar}Hdt\right)|\psi(0)\rangle, \tag{141}$$

$$|\bar\psi_{R4}\rangle = \exp\left(-\frac{i}{\hbar}H2dt\right)J|\psi(0)\rangle. \tag{142}$$

Everything is contained in the unnormalized ket $|\bar\psi_{Rk}\rangle$. It is interesting to sketch how it connects with the evolution in $S \otimes R$ shown above. Reservoir states are now viewed as detector states, which are labeled by records that decohere, so that at each branching a stochastic selection may be made (of course, the Schrödinger equation will not do this for us); for example, for the second entry on the last line of the branching tree,

$$|1\rangle|\psi_S^{(1)}(3dt)\rangle \to |1\rangle_{R2}(|\bar\psi_{R2}\rangle\sqrt{dt}) + |1\rangle_{R3}(|\bar\psi_{R3}\rangle\sqrt{dt}) + |1\rangle_{R4}(|\bar\psi_{R4}\rangle\sqrt{dt}), \tag{143}$$

where the $\sqrt{dt}$ arises from the stochastic branching — the units of J are the square root of inverse time — and with

$$_{R2}\langle 1|1\rangle_{R3} = {}_{R2}\langle 1|1\rangle_{R4} = {}_{R3}\langle 1|1\rangle_{R4} = 0. \tag{144}$$

The probability for 1 "click" after $3dt$ is then the sum of three record probabilities:

$$\langle\psi_S^{(1)}(3dt)|\psi_S^{(1)}(3dt)\rangle = \langle\bar{\psi}_{R2}|\bar{\psi}_{R2}\rangle dt + \langle\bar{\psi}_{R3}|\bar{\psi}_{R3}\rangle dt + \langle\bar{\psi}_{R4}|\bar{\psi}_{R4}\rangle dt. \qquad (145)$$

5.3. *Monte Carlo algorithm for jump trajectories*

Although the beginnings of quantum trajectory theory were tied to photoelectron counting,[31, 32] and to work on resonance fluorescence that effectively — though without knowledge — treated photon counting in a Davies-type approach,[33, 34] attention in quantum optics soon came to focus on Monte-Carlo simulation[35–38] as a tool for solving a master equation, i.e., constructing the density operator as an ensemble average over of realizations of pure conditional states. Such a tool is particularly useful when the dimension of a problem becomes so large that direct numerical solution of the density matrix equations is intractable, though a pure-state expansion can still be handled. We now return to the simple example of free photon decay from a cavity/resonator with an eye on Monte-Carlo simulations.

5.3.1. *Following paths: Bayesian inference*

Decomposing the density operator as a sum over records is perhaps interesting and useful in simple cases, where the outlined formal expansion can be put in place analytically. On the other hand, exclusive probability densities are more generally not so easy to calculate — hence the power of the normal-ordered approach based on Glauber's multi-coincidence rates — so it is desirable to devise a broadly applicable numerical scheme. Such a thing is provided by a Monte Carlo scheme that is capable of tacking individual paths through the branching tree, deciding from numerically generated random numbers which direction to take at each branch point. A simulation like this realizes a particular record, a sequence of decisions — "click" or "no click" — at each branch point, and the conditional state $|\psi_{\text{REC}}\rangle$ that goes along with it. The challenge is to put together an algorithm that realizes records and associated conditional states at the correct frequency, i.e., in such a way that a histogram constructed from an ensemble of realizations shows each particular record realized at the correct probability density, $P_{\text{REC}}(t) = \langle\bar{\psi}_{\text{REC}}(t)|\bar{\psi}_{\text{REC}}(t)\rangle$. Let us imagine, for example, that the after three time steps the realized record is $R3 = \{\emptyset!\emptyset\}$. At the next time step there are two options for updating the record:

$$R3\emptyset = \{\emptyset!\emptyset\emptyset\} \quad \text{and} \quad R3! = \{\emptyset!\emptyset!\}. \qquad (146)$$

The crucial question in making a choice between them is: what is the branching ratio? with what relative probability should $R3\emptyset$ be chosen over $R3!$ or *visa versa*? The answer follows from Bayesian inference. The following sketch gives explicit expressions each of the (unnormalized) conditional states involved at the considered branch point:

$$\exp\left(-\frac{i}{\hbar}Hdt\right)J\exp\left(-\frac{i}{\hbar}Hdt\right)|\psi(0)\rangle$$

$$\downarrow$$

$$|1\rangle_{R3}|\bar\psi_{R3}\rangle$$

$$\Downarrow \qquad\qquad \searrow$$

$$|1\rangle_{R3\emptyset}|\bar\psi_{R3\emptyset}\rangle \qquad |2\rangle_{R3!}|\bar\psi_{R3!}\rangle$$

$$\uparrow \qquad\qquad\qquad \uparrow$$

$$\exp\left(-\frac{i}{\hbar}H2dt\right)J\exp\left(-\frac{i}{\hbar}Hdt\right)|\psi(0)\rangle \quad J\exp\left(-\frac{i}{\hbar}Hdt\right)J\exp\left(-\frac{i}{\hbar}Hdt\right)|\psi(0)\rangle.$$

We wish to calculated the branching probabilities and hopefully express them in terms of quantities that are known prior to the branching. The required probabilities are conditioned on the past record, i.e, the record up to and including the third time step. Reading the notation $\Downarrow$ and $\searrow$ off the branching tree, from a simple application of Bayes formula they are the conditional probabilities

$$\mathrm{Prob}\,(\Downarrow|R3) = \frac{\mathrm{Prob}\,(R3\wedge\Downarrow)}{\mathrm{Prob}\,(R3)} = \frac{\mathrm{Prob}\,(R3\emptyset)}{\mathrm{Prob}\,(R3)} = \frac{\langle\bar\psi_{R3\emptyset}|\bar\psi_{R3\emptyset}\rangle dt}{\langle\bar\psi_{R3}|\bar\psi_{R3}\rangle dt}, \tag{147}$$

and

$$\mathrm{Prob}\,(\searrow|R3) = \frac{\mathrm{Prob}\,(R3\wedge\searrow)}{\mathrm{Prob}\,(R3)} = \frac{\mathrm{Prob}\,(R3!)}{\mathrm{Prob}\,(R3)} = \frac{\langle\bar\psi_{R3!}|\bar\psi_{R3!}\rangle dt^2}{\langle\bar\psi_{R3}|\bar\psi_{R3}\rangle dt}. \tag{148}$$

Then reading the explicit expressions for the conditional states off the sketch above, the required probabilities are calculated from

$$\langle\bar\psi_{R3\emptyset}|\bar\psi_{R3\emptyset}\rangle dt = \langle\bar\psi_{R3}|\exp\left[\frac{i}{\hbar}(H^\dagger - H)dt\right]|\bar\psi_{R3}\rangle dt$$

$$= \langle\bar\psi_{R3}|\exp(-2\kappa a^\dagger a dt)|\bar\psi_{R3}\rangle dt, \tag{149}$$

and

$$\langle\bar\psi_{R3!}|\bar\psi_{R3!}\rangle dt^2 = \langle\bar\psi_{R3}|J^\dagger J|\bar\psi_{R3}\rangle dt^2$$

$$= 2\kappa\langle\bar\psi_{R3}|(a^\dagger a)|\bar\psi_{R3}\rangle dt^2. \tag{150}$$

The branching probabilities are

$$\mathrm{Prob}\,(\Downarrow|R3) = \frac{\langle\bar\psi_{R3}|\exp(-2\kappa a^\dagger a dt)|\bar\psi_{R3}\rangle}{\langle\bar\psi_{R3}|\bar\psi_{R3}\rangle}$$

$$= \langle\psi_{R3}|\exp(-2\kappa a^\dagger a dt)|\psi_{R3}\rangle$$

$$= 1 - 2\kappa\langle\psi_{R3}|(a^\dagger a)|\psi_{R3}\rangle dt, \tag{151}$$

and

$$\text{Prob}\left(\diagdown \,|R3\right) = \frac{2\kappa\langle\bar\psi_{R3}|a^\dagger a|\bar\psi_{R3}\rangle dt}{\langle\bar\psi_{R3}|\bar\psi_{R3}\rangle}$$

$$= 2\kappa\langle\psi_{R3}|(a^\dagger a)|\psi_{R3}\rangle dt. \tag{152}$$

One or two comments are in order. There is nothing particularly Bayesian, in the distinct interpretational sense, in the derivation of these probabilities, as we are merely manipulating frequencies of events in photoelectron counting sequences. The interpretation nevertheless becomes distinctly Bayesian once conditioning on the record is passed to an updated conditional quantum state.

Secondly, it is an important feature that these probabilities are determined by the *normalized* state prior to the branching. It is not necessary to know the record probability (density) itself, i.e., the norm $\langle\bar\psi_{R3}|\bar\psi_{R3}\rangle$. Although this norm is easy enough to keep track of over a small number of time steps, it becomes smaller and smaller as the length of the record increases since the number of alternate records is growing exponentially with time; thus, the ket $|\bar\psi_{\text{REC}}\rangle$ decays towards zero in a long numerical simulation.

5.3.2. *A simple Monte Carlo procedure*

Let $|\bar{\bar\psi}_{\text{REC}}\rangle$ denote a ket that is occasionally rescaled to keep the norm within practical bounds — possibly but not necessarily renormalized at each time step. At time t this ket is known. From what we see above, it and the record are advanced one more time step by the following simple Monte Carlo procedure:

1. Compute the conditional mean photon number,

$$\langle(a^\dagger a)(t)\rangle_{\text{REC}} = \frac{\langle\bar{\bar\psi}_{\text{REC}}(t)|(a^\dagger a)|\bar{\bar\psi}_{\text{REC}}(t)\rangle}{\langle\bar{\bar\psi}_{\text{REC}}(t)|\bar{\bar\psi}_{\text{REC}}(t)\rangle}. \tag{153}$$

2. Generate a random number R uniformly distributed on the unit line.

3. If $R < 2\kappa\langle(a^\dagger a)(t)\rangle_{\text{REC}}dt$, advance the conditional ket by executing a quantum jump:

$$|\bar{\bar\psi}_{\text{REC}}(t)\rangle \rightarrow |\bar{\bar\psi}_{\text{REC}}(t+dt)\rangle = J|\bar{\bar\psi}_{\text{REC}}(t)\rangle. \tag{154}$$

4. If $R > 2\kappa\langle(a^\dagger a)(t)\rangle_{\text{REC}}dt$, advance the conditional ket under "free" propagation by the non-Hermitian Hamiltonian:

$$|\bar{\bar\psi}_{\text{REC}}(t)\rangle \rightarrow |\bar{\bar\psi}_{\text{REC}}(t+dt)\rangle = \exp\left(-\frac{i}{\hbar}Hdt\right)|\bar{\bar\psi}_{\text{REC}}(t)\rangle. \tag{155}$$

Let $|\psi_{\text{REC}}^{(k)}(t)\rangle$, $k = 1, 2, \ldots, N$, denote an ensemble of kets generated in this way. The expansion of $\rho(t)$ as a sum over records, Eq. (126), is approximated by

$$\varrho(t) = \frac{1}{N} \sum_{k=1}^{N} |\psi_{\text{REC}}^{(k)}(t)\rangle\langle\psi_{\text{REC}}^{(k)}(t)|. \tag{156}$$

5.4.   *Examples*

5.4.1.   *Resonance fluorescence*

Resonance fluorescence extends the spontaneous emission example (Section 2.5.2) by adding a coherent drive (Section 2.6.3). The master equation, Eq. (19), has generalized Liouvillian

$$\mathcal{L} = -i\omega_a[\sigma_+\sigma_-, \cdot] - i[\mathcal{E}e^{-i\omega t}\sigma_+ + \mathcal{E}^*e^{i\omega t}\sigma_-, \cdot] + (\gamma/2)\Lambda[\sigma_-], \tag{157}$$

with ω_a the atomic resonance frequency, ω the drive frequency, $\mathcal{E}$ the amplitude of the drive, and γ the Einstein A coefficient. After transforming to a frame rotating at drive frequency ω, the quantum trajectory unraveling of the master equation has jump operator

$$J = \sqrt{\gamma}\sigma_-, \tag{158}$$

and non-Hermitian Hamiltonian

$$H = \hbar\left(\Delta\omega - i\frac{\gamma}{2}\right)\sigma_+\sigma_- + \hbar(\mathcal{E}\sigma_+ + \mathcal{E}^*\sigma_-), \tag{159}$$

where $\Delta\omega = \omega_a - \omega$ is a detuning.

As an example to illustrate quantum trajectory ideas, resonance fluorescence is a particularly simple case, since every quantum jump returns the atom to precisely the same state, the ground state $|g\rangle$. As a result, the evolution in between the jumps always begins from the ground state and the entire jump record is determined by a repeated application of the so-called waiting time distribution, i.e., the distribution of times separating one jump and the next.[31] Considered as a record probability, the waiting time distribution is the probability density for the record that starts with the atom in the ground state and consists of a NULL, or "no click", interval of duration t followed by a "click" (jump) at t:

$$\langle\bar{\psi}_{\text{WAIT}}(t)|\bar{\psi}_{\text{WAIT}}(t)\rangle = \langle g|\exp\left(\frac{i}{\hbar}H^\dagger t\right)J^\dagger J\exp\left(-\frac{i}{\hbar}Ht\right)|g\rangle. \tag{160}$$

In the case of exact resonance ($\Delta\omega = 0$), the straightforward solution for the non-unitary evolution under H yields the compact formula

$$\langle\bar{\psi}_{\text{WAIT}}(t)|\bar{\psi}_{\text{WAIT}}(t)\rangle = \gamma\exp[-(\gamma/2)t]|\mathcal{E}|^2\frac{\sinh^2\left(\sqrt{(\gamma/4)^2 - |\mathcal{E}|^2}t\right)}{(\gamma/4)^2 - |\mathcal{E}|^2}. \tag{161}$$

This distribution vanishes at $t = 0$ and decays to zero at ∞; it is therefore, in a sense, peaked about its mean; although at weak drive it is more-or-less a slowly decaying exponential with a narrow slice (width γ^{-1}) chopped out at $t = 0$. Its peaked character is most evident for $|\mathcal{E}| = \gamma/4$. The vanishing at $t = 0$ and resulting "peaked" character are an indication of the antibunching of photons in resonance fluorescence [Fig. 5, frame (c)].

The way in which the conditional state evolves changes with the strength of the drive relative to the decay rate through the dependence of Eq. (161) on the factor $(\gamma/4)^2 - |\mathcal{E}|^2$. Figure 7 illustrates the three different types of trajectories that occur for weak, intermediate, and strong drive. At sufficiently weak drive — in the perturbative regime — a steady state is reached prior to each photon emission. At intermediate drive no steady state is reached; instead, a photon emission (jump) interrupts the evolution, before the atom reaches full excitation. Strong driving induces the Rabi oscillations for which resonance fluorescence is so well known. Quantum jumps interrupt and reinitiate the Rabi oscillation at random times, thus dephasing the oscillation in the ensemble average.

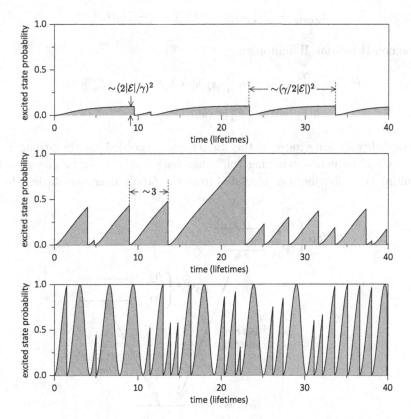

Fig. 7. Sample trajectories for resonance fluorescence: weak excitation, $2\mathcal{E}/\gamma = 0.3$ (top); intermediate excitation, $2\mathcal{E}/\gamma = 0.7$ (middle); strong excitation, $2\mathcal{E}/\gamma = 2.3$ (bottom).

5.4.2.   Electron shelving

The electron shelving scheme dramatically illustrates the role of the NULL, or "no-click", measurement response in quantum trajectory theory. The scheme was conceived by Dehmelt[39] as an amplifier for monitoring the state — ground or excited — of a weak metastable transition: if a second, strong transition, sharing the same ground state is driven, seeing fluorescence indicates that the prior state must have been the shared ground state of the weak transition, while no fluorescence indicates it was the excited state. When both transitions are continuously and simultaneously driven, a switching "on" and "off" of fluorescence is predicted and provides a direct monitoring of quantum jumps on the weak transition. A series of experiments observed this switching "on" and "off" in 1986.[40–42]

An example of the level scheme employed is shown in Fig. 8. The master equation involved is a straightforward generalization from the master equation for resonance fluorescence, to include both a strong and a weak drive, and both strong and weak decay channels. With both drives on resonance, the quantum trajectory unraveling has jump operators,

$$J_{\text{strong}} = \sqrt{\gamma_s}\,|g\rangle\langle s|, \quad J_{\text{weak}} = \sqrt{\gamma_w}\,|g\rangle\langle w|, \tag{162}$$

and the non-Hermitian Hamiltonian

$$H = -i\hbar\frac{\gamma_s}{2}|s\rangle\langle s| - i\hbar\frac{\gamma_w}{2}|s\rangle\langle s| + \hbar(\mathcal{E}_s|s\rangle\langle g| + \mathcal{E}_s^*|g\rangle\langle s|)$$
$$+ \hbar(\mathcal{E}_w|w\rangle\langle g| + \mathcal{E}_w^*|g\rangle\langle w|). \tag{163}$$

The possibility of long gaps in the fluorescence recorded on the strong transition — "click" sequence switching "off" then back "on" — can be discerned from the waiting time distribution calculated from Eq. (160), after introducing the new

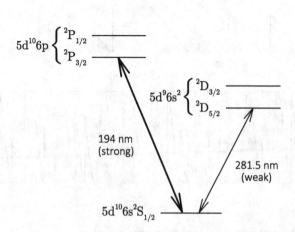

Fig. 8.   Typical level scheme (Hg II ion[42]) used in electron shelving.

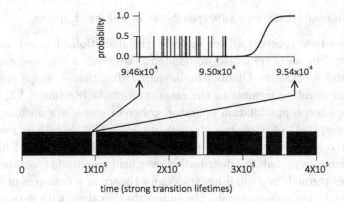

Fig. 9.   Sample record showing the spontaneous turning "on" and turning "off" of strong transition fluorescence in electron shelving.

H and with J_{strong} replacing J: the distribution has a long tail of small integrated area (weight), corresponding to a small probability for an extremely long wait in between photon emissions.[43] A quantum trajectory simulation reveals the underlying role of the NULL-measurement dynamic in bringing about the electron shelving.

A typical simulation shows fluorescence on the strong transition switching "on" and "off", as in Fig. 9. There are no actual *jumps* to the metastable state when the fluorescence switches "off", though. With the weak transition driven coherently, the evolution of the conditional state to the excited state, $|w\rangle$ (see Fig. 8), is continuous. It is a statistical inference draw from an unusually long NULL measurement interval. In between the scattering of any pair of consecutive photons on the strong transition, the conditional state is a superposition,

$$|\psi_{\text{REC}}(t)\rangle = c_g(t)|g\rangle + c_s(t)|s\rangle + c_w(t)|w\rangle. \tag{164}$$

For the most part $c_w(t)$ remains very close to zero, while $c_g(t)$ and $c_s(t)$ evolve more-or-less in the same way as in resonance fluorescence. Occasionally, however, there will be an unusually long interval with no "click" in the record recording the fluorescence (jump on the strong transition). When this happens, $c_w(t)$ begins to evolve continuously, from something very close to zero eventually to unity. An example of the behavior is shown in the expanded plot of the first turn "off" of the fluorescence in Fig. 9. The evolution marks our growing confidence that the long wait is not simply a statistically rare case of a long wait between emissions on the strong transition, but a result of the atom not being able to emit again on the strong transition because it is, in fact, "shelved" in the metastable state. In the quantum trajectory simulation it is driven there by NULL-measurement backaction. The importance of this NULL-measurement dynamic in electron shelving was first pointed out by Porrati and Putterman.[44]

5.4.3.   Coherent states and superpositions of coherent states

We have seen how coherent states underlie the definition of nonclassical versus classical light. For an output field designated "classical", a diagonal expansion in coherent states — positive Glauber-Sudarshan P function — maps the quantum photoelectron counting formula to the classical formula (Section 3.4.3). As eigenstates of the photon annihilation operator, coherent states are unchanged by the removal (absorption) of a photon through photoelectric detection; measurement backaction has an entirely classical interpretation, as the conditional updating of the P function as a probability distribution over field amplitude (Section 3.4.1). All of this carries through into quantum trajectory theory in a transparent way: if the conditional state is a coherent state, the jumps that go along with detector "clicks" make no change to the conditional state at all; there are "clicks" but there is no longer any "jumping"; there is no call for "jumping" to account for backaction.

Positive P functions exist for mixtures of coherent states. Superpositions are an entirely different story, and quantum trajectories for coherent state superpositions provide a nice illustration of the distinction, at the level of photoelectron counting, between classical and nonclassical light. We explore a couple of examples in what follows. They are particularly illuminating, as the calculations can for the most part be carried through analytically. In order to set the background, we return first to the damped coherent state, treated at the level of the master equation in Section 2.6.2.

1. Damped coherent state:
Working in the interaction picture, the damped cavity/resonator is described by the jump operator and non-Hermitian Hamiltonian,

$$J = \sqrt{2\kappa}a, \quad H = -i\hbar\kappa a^\dagger a. \tag{165}$$

The initial conditional state is $|\bar{\psi}_{\text{REC}}(0)\rangle = |\alpha\rangle$, and following the direction taken in Section 2.6.2 suggests we adopt the *ansatz*

$$|\bar{\psi}_{\text{REC}}(t)\rangle = A(t)|\alpha(t)\rangle, \tag{166}$$

with

$$|\alpha(t)\rangle = \exp[\alpha(t)a^\dagger - \alpha^*(t)a]|0\rangle. \tag{167}$$

The norm $\langle\bar{\psi}_{\text{REC}}(t)|\bar{\psi}_{\text{REC}}(t)\rangle = |A(t)|^2$ is the record probability density. If we are able to solve the trajectory equations, we solve not only a mater equation but also a photon counting problem.

Consider a record of n counts up to time t. The evolution of the conditional state consists of a sequence of jumps, at the ordered count times $t_1, t_2, \ldots, t_n$, with continuous evolution in between. A jump at time t_k preserves the *ansatz* while changing the norm, with

$$A(t_k) \rightarrow \sqrt{2\kappa}\alpha(t_k)A(t_k). \tag{168}$$

The evolution in between the jumps must satisfy the equation of motion

$$\frac{d|\bar{\psi}_{\text{REC}}\rangle}{dt} = -(\kappa a^\dagger a)|\bar{\psi}_{\text{REC}}\rangle. \tag{169}$$

It preserves the *ansatz* too, provided $\alpha(t)$ and $A(t)$ satisfy the equations

$$\frac{d\alpha}{dt} = -\kappa\alpha, \tag{170}$$

and

$$\frac{1}{A}\frac{dA}{dt} = -\frac{d\alpha^*}{dt}\alpha = -\kappa|\alpha|^2, \tag{171}$$

with solutions $\alpha(t) = \alpha\exp(-\kappa t)$ and $A(t) = A(t_k)\exp[-\frac{1}{2}|\alpha|^2(e^{-2\kappa t_k} - e^{-2\kappa t})]$, $t_k \le t < t_{k+1}$. From the pieces we can construct an analytical expression for the conditional state,

$$|\bar{\psi}_{\text{REC}}(t)\rangle = (\sqrt{2\kappa}\alpha e^{-\kappa t_n})\dots(\sqrt{2\kappa}\alpha e^{-\kappa t_1})\exp\left[-\frac{1}{2}|\alpha|^2(1 - e^{-2\kappa t})\right]|\alpha e^{-\kappa t}\rangle, \tag{172}$$

from which we obtain the record probability density

$$\langle\bar{\psi}_{\text{REC}}(t)|\bar{\psi}_{\text{REC}}(t)\rangle = (2\kappa|\alpha|^2 e^{-2\kappa t_n})\dots(2\kappa|\alpha|^2 e^{-2\kappa t_1})\exp[-|\alpha|^2(1 - e^{-2\kappa t})]. \tag{173}$$

The result is just what is expected from the semiclassical treatment of photo-electron counting for a classical field of decaying intensity (in photon flux units) $2\kappa|\alpha|^2\exp(-2\kappa t)$: the exponential factor is the probability for no counts over an interval of length t [Eqs. (78) and (79) with $n = 0$], i.e., the probability for the sequence of NULL measurement results, and $2\kappa|\alpha|^2 e^{-2\kappa t_k}$ is the probability density for a count at time t_k. From the record probability density we obtain the count distribution — the probability for n counts in time T — by summing (integrating) over all possible count times. Using

$$(2\kappa|\alpha|^2)^n\int_0^T dt_n\int_0^{t_n} dt_{n-1}\cdots\int_0^{t_2} dt_1 e^{-2\kappa t_n}\dots e^{-2\kappa t_1} = \frac{[2\kappa|\alpha|^2\int_0^t e^{-2\kappa t'}\,dt']^n}{n!}$$

to carry out the sum brings us to the expected Poisson distribution [Eqs. (78) and (79)]:

$$P(n, T) = \frac{[|\alpha|^2(1 - e^{-2\kappa t})]^n}{n!}\exp[-|\alpha|^2(1 - e^{-2\kappa t})]. \tag{174}$$

2. Damped even and odd coherent state superpositions:

Superpositions of coherent states provide the canonical illustration of the mysteries of Schrödinger cats and the role of decoherence in the rapid — much faster than the energy decay time — destruction of quantum interference.[45, 46] Figure 10 plots

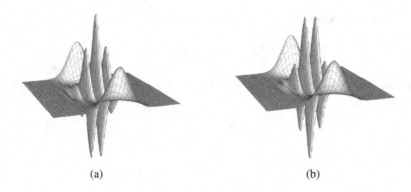

(a)   (b)

Fig. 10.   Wigner functions for even (a) and odd (b) coherent state superpositions.

the Wigner functions of even and odd coherent state superpositions, which are distinguished from a mixture of coherent states by the interference fringe that runs in quadrature to the displacement of the Gaussian peaks. Since the evolution of the unnormalized conditional ket, $|\bar{\psi}_{\text{REC}}(t)\rangle$ is linear, it is straightforward to construct trajectories for damped coherent state superpositions from what we have just seen. Considering the initial state

$$|\bar{\psi}_{\text{REC}}(0)\rangle = \frac{1}{\sqrt{2}} \frac{|\alpha\rangle + |-\alpha\rangle}{\sqrt{1 + \exp(-2|\alpha|^2)}}, \tag{175}$$

the two pieces of the superposition differ only in whether Eq. (168) is taken over as $A(t_k) \to +\sqrt{2\kappa}\alpha(t_k)A(t_k)$ or $A(t_k) \to -\sqrt{2\kappa}\alpha(t_k)A(t_k)$. It follows, using Eq. (172), that

$$|\bar{\psi}_{\text{REC}}(t)\rangle = \left(\sqrt{2\kappa}\alpha e^{-\kappa t_n}\right) \ldots \left(\sqrt{2\kappa}\alpha e^{-\kappa t_1}\right) \exp\left[-\frac{1}{2}|\alpha|^2(1 - e^{-2\kappa t})\right]$$

$$\times \frac{1}{\sqrt{2}} \frac{|\alpha e^{-\kappa t}\rangle + (-1)^n|-\alpha e^{-\kappa t}\rangle}{\sqrt{1 + \exp(-2|\alpha|^2)}}. \tag{176}$$

Considering an ensemble of such trajectories, the number of photon emissions (jumps) up to time t is unpredictable. After a time on the order of the time waited for the loss from the cavity/resonator of the very first photon — $t \sim (2\kappa|\alpha|^2)^{-1}$ — any ensemble will contain a similar number of sequences with n even as with n odd. Clearly, the interference fringe is then canceled in the ensemble average. When the mean number of photons $|\alpha|^2$ is large, this can be a very short time indeed. We see that preserving the coherence between the two pieces of the superposition is a matter of tracking every last photon transferred to the environment. It is not so much that the prepared cat "dies". It simply leaves the cavity/resonator to exist partly in the output field; it is necessary to know what part is on the inside and what on the outside (down to the level of one photon) if we are to continue to have access to the coherence of the prepared cat.

3. Damped coherent state superposition:

Even and odd coherent state superpositions are a special case, since only the relative phase of the superposition is changed by the evolution. More generally, both the relative phase and the relative amplitude are changed. Consider the initial state

$$|\bar{\psi}_{\text{REC}}(0)\rangle = \frac{1}{\sqrt{2}} \frac{|\alpha\rangle + |\beta\rangle}{\sqrt{1 + \text{Re}\langle\alpha|\beta\rangle}}, \tag{177}$$

where α and β are allowed to be any two complex numbers. Once again, the linearity of the(unnormalized) conditional state evolution means that $|\bar{\psi}_{\text{REC}}(t)\rangle$ is simply a sum of terms written down from Eq. (172):

$$|\bar{\psi}_{\text{REC}}(t)\rangle = |\bar{\psi}_{\text{REC}}^{(\alpha)}(t)\rangle + |\bar{\psi}_{\text{REC}}^{(\beta)}(t)\rangle, \tag{178}$$

with

$$|\bar{\psi}_{\text{REC}}^{(\lambda)}(t)\rangle = \left(\sqrt{2\kappa}\lambda e^{-\kappa t_n}\right) \dots \left(\sqrt{2\kappa}\lambda e^{-\kappa t_1}\right) \exp\left[-\frac{1}{2}|\lambda|^2\left(1 - e^{-2\kappa t}\right)\right] |\lambda e^{-\kappa t}\rangle. \tag{179}$$

In the general case, when $|\alpha|$ and $|\beta|$ are different, the components $|\bar{\psi}_{\text{REC}}^{(\alpha)}(t)\rangle$ and $|\bar{\psi}_{\text{REC}}^{(\beta)}(t)\rangle$ compete with one another to dominate the sum. One component grows in norm relative to the other. Which one grows depends on the sequence of count times. A low number of counts causes the component corresponding to the smaller of $|\alpha|$ and $|\beta|$ to dominate — the result of the default evolution (without jumps) $A_\lambda(t) = A_\lambda(t_k) \exp[-\frac{1}{2}|\lambda|^2(e^{-2\kappa t_k} - e^{-2\kappa t})]$, $t_k \leq t < t_{k+1}$, which favors a smaller $|\lambda|^2$. A large number of counts causes the component corresponding to the larger of $|\alpha|$ and $|\beta|$ to dominate — a result of the jump evolution $A_\lambda(t_k) \to \sqrt{2\kappa}\lambda(t_k)A_\lambda(t_k)$, which favors a larger $|\lambda|^2$. There is once again decoherence, with the initial superposition turning into a mixture under the average over trajectories; but in this case the state localizes on one or other component of the superposition in each realization, i.e., the stochastic evolution models decoherence in the sense of making a choice between alternative "meter" states in a quantum measurement.

6. Cascaded systems

Throughout these lectures we have for the most part kept our eyes on the damped cavity/resonator model, set out Section 2, and the similar damped two-state atom. We encountered master equations for a vacuum input — Eq. (20) and Eq. (108) — a thermal or chaotic light input — Eq. (34) — and a coherent input — Eq. (47) and Eq. (157). Broadband squeezed light is another commonly met input field.[4,47] The number of different input fields considered is small, measured against the huge range of possibilities. How, for example, might we treat a *filtered* thermal light input (modeled in Section 3.2) or an input of *filtered* squeezed light? Beyond this, how might we treat an antibunched input, or an input field with some other form of exotic photon statistics?

In a particular case various methods might be available (e.g., non-Markov master equations for filtered light). One approach that is broadly applicable is to model the source of the input field and the target at which it is directed as a single composite, or cascaded system; a schematic is shown in Fig. 11. As the final topic of these lectures we derive the master equation for a such a set up and explore one simple application of its quantum trajectory unraveling. The approach taken is that of Ref. 48, though other derivations of the master equation exist.[49, 50] Applications range from the obvious driving of a target system by an exotic source,[51, 52] to quantum state transfer between a source and target[53] and the study of entanglement between a source and target.[54–56]

6.1.   Model

6.1.1.   Hamiltonian

The model illustrated in Fig. 11 is a straightforward extension of the model outlined in Section 2. In the present set up, however, the system is composed of two parts, the source subsystem and the target subsystem:

$$H_S = H_{\text{source}} + H_{\text{target}}. \tag{180}$$

Each subsystem is a cavity/resonator mode. The two modes are designated a (source) and b (target), with free Hamiltonians (in the absence of any intracavity interactions) $H_{\text{source}} = \hbar \omega_a a^\dagger a$ and $H_{\text{target}} = \hbar \omega_b b^\dagger b$. The reservoir Hamiltonian is again given by Eq. (4), and there are two parts to the interaction between system

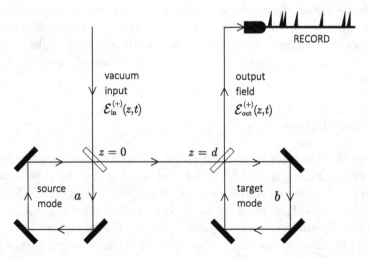

Fig. 11.   Schematic of cascaded ring cavity/resonators modeled in 1D, with photoelectron counting record at the target resonator output.

and reservoir:

$$H_{SR} = H_{SR}^{\text{source}} + H_{SR}^{\text{target}}. \tag{181}$$

The interactions are conveniently written in terms of the reservoir electric field operator expressed in photon flux units, which is achieved using Eq. (5) multiplied by $\sqrt{2\epsilon Ac/\hbar\omega_{a,b}}$, and Eq. (10):

$$H_{SR}^{\text{source}} = i\hbar\sqrt{2\kappa_a}[a\mathcal{E}^{(-)}(0) - a^\dagger\mathcal{E}^{(+)}(0)], \tag{182}$$

$$H_{SR}^{\text{target}} = i\hbar\sqrt{2\kappa_b}[b\mathcal{E}^{(-)}(d) - b^\dagger\mathcal{E}^{(+)}(d)], \tag{183}$$

with $\mathcal{E}^{(\pm)}(z) = \mathcal{E}^{(\pm)}(z,0)$. Any difference between the two photon energies, $\hbar\omega_a$ and $\hbar\omega_b$, is ignored when scaling the electric field.

6.1.2.   Inputs and outputs

In a straightforward extension of Section 2.4, the field carried by the reservoir is written in the Heisenberg picture as a sum of the input (free) field, Eq. (27), and fields originating from the source cavity/resonator and the target cavity/resonator:

$$\mathcal{E}^{(+)}(z,t) = \mathcal{E}_{\text{in}}^{(+)}(z,t) + \mathcal{E}_{\text{source}}^{(+)}(z,t) + \mathcal{E}_{\text{target}}^{(+)}(z,t), \tag{184}$$

with, following from Eqs. (28) and (29),

$$\mathcal{E}_{\text{source}}^{(+)}(z,t) = \begin{cases} \sqrt{2\kappa_a}a(t') & z > 0 \\ \frac{1}{2}\sqrt{2\kappa_a}a(t') & z = 0 \\ 0 & z < 0, \end{cases} \tag{185}$$

$t' = t - z/c$, and

$$\mathcal{E}_{\text{target}}^{(+)}(z,t) = \begin{cases} \sqrt{2\kappa_b}b(t'') & z > d \\ \frac{1}{2}\sqrt{2\kappa_b}b(t'') & z = d \\ 0 & z < d, \end{cases} \tag{186}$$

$t'' = t - (z - d)/c$.

6.2.   The Lindblad master equation

Beginning from the Liouville equation, Eq. (11), we first perform a unitary transformation to introduce the source-retarded density operator

$$\chi_{\text{ret}} = \exp\left[-i\frac{1}{\hbar}(H - H_{SR}^{\text{target}})\left(-\frac{d}{c}\right)\right]\chi\exp\left[i\frac{1}{\hbar}(H - H_{SR}^{\text{target}})\left(-\frac{d}{c}\right)\right], \tag{187}$$

which satisfies the transformed equation of motion,

$$\frac{d\chi_{\text{ret}}}{dt} = \frac{1}{i\hbar}[H_{\text{ret}}, \chi_{\text{ret}}], \tag{188}$$

where the transformed Hamiltonian is

$$\begin{aligned} H_{\text{ret}} &= H - H_{SR}^{\text{target}} + i\hbar\sqrt{2\kappa_b}\left\{b\left[\mathcal{E}^{(-)}(0) + \frac{1}{2}\sqrt{2\kappa_a}a^\dagger\right] - \text{H.c.}\right\} \\ &= H'_S + H_R + H_{SR}, \end{aligned} \tag{189}$$

and we introduce the modified system Hamiltonian,

$$H' = H_{\text{source}} + H_{\text{target}} + i\hbar\sqrt{\kappa_a\kappa_b}(a^\dagger b - ab^\dagger), \tag{190}$$

and modified interaction

$$H'_{SR} = i\hbar\left[(\sqrt{2\kappa_a}a + \sqrt{2\kappa_b}b)\mathcal{E}^{(-)}(0) - \text{H.c.}\right]. \tag{191}$$

It is now possible to proceed with the derivation of a Lindblad master equation in the standard way, as, in Eq. (191), both cavity/resonators now interact with the reservoir at the same spatial location. The result is an equation of motion for the source-retarded reduced density operator:

$$\frac{d\rho_{\text{ret}}}{dt} = \mathcal{L}\rho_{\text{ret}}, \tag{192}$$

with generalized Liouvillian,

$$\mathcal{L} = \frac{1}{i\hbar}\left[H_{\text{source}} + H_{\text{target}} + i\hbar\sqrt{\kappa_a\kappa_b}(a^\dagger b - ab^\dagger), \cdot\right] + \Lambda[J], \tag{193}$$

and jump operator

$$J = \sqrt{2\kappa_a}a + \sqrt{2\kappa_b}b. \tag{194}$$

6.3.   *Quantum trajectory unraveling*

6.3.1.   *Jump operator and non-Hermitian Hamiltonian*

The jump operator appears as a sum over fields because the detection of a photon at the output of cavity/resonator b is unable to tell whether its origin was the source or the target. A second very natural piece of physics appears from the non-Hermitian Hamiltonian, which is given, following the standard prescription for dividing up the generalized Liouvillian (see Section 4.2.3), by

$$\begin{aligned} H &= H_{\text{source}} + H_{\text{target}} + i\hbar\sqrt{\kappa_a\kappa_b}(a^\dagger b - ab^\dagger) - i\hbar\frac{1}{2}J^\dagger J \\ &= (H_{\text{source}} - i\hbar\kappa_a a^\dagger a) + (H_{\text{target}} - i\hbar\kappa_b b^\dagger b) - i\hbar 2\sqrt{\kappa_a\kappa_b}ab^\dagger. \end{aligned} \tag{195}$$

Note the cancelation of terms involving $a^\dagger b$ after substituting for $i\hbar J^\dagger J$. As a result, the photon exchange is unidirectional, from source to target, just as it should be in an open systems treatment.

6.3.2. *Example: Initial coherent state*

The decay of an initial coherent state is, perhaps, not such an exciting example, as it involves none of the interesting features mentioned at the beginning of the section. Nevertheless, the calculations can be carried through analytically and provide us with a straightforward check on the formulation, since the results for coherent states should be the same would be obtained using classical fields.

We generalize from Example 1. of Section 5.4.3, unraveling the master equation in quantum trajectories; we solve for a record probability density and from this obtain the photoelectron counting distribution of the output field. With initial state

$$|\psi_{\text{REC}}^{\text{ret}}(0)\rangle = |\alpha\rangle_a|0\rangle_b, \tag{196}$$

the *ansatz*, Eq. (166), is carried over in the form,

$$|\bar{\psi}_{\text{REC}}^{\text{ret}}(t)\rangle = A(t)|\alpha(t)\rangle_a|\beta(t)\rangle_b, \tag{197}$$

with $A(t)$ determining the conditional state norm (record probability density) and coherent states

$$|\alpha(t)\rangle_a = \exp[\alpha(t)a^\dagger - \alpha^*(t)a]|0\rangle_a, \tag{198}$$

$$|\beta(t)\rangle_b = \exp[\beta(t)b^\dagger - \beta^*(t)b]|0\rangle_b. \tag{199}$$

We consider a record comprising n counts at the ordered times $t_1, t_2, \ldots t_n < t$. The counting is performed at the output of cavity/resonator b, as shown in Fig. 11. The calculation of the record probability density and photoelectron counting distribution proceeds in four easy steps:

1. Jump at time t_k:
As coherent states are eigenstates of the annihilation operator, the jump maps the conditional state norm while keeping the *ansatz* intact; thus, for the jump we have the generalization of Eq. (168),

$$A(t_k) \to \left[\sqrt{2\kappa_a}\alpha(t_k) + \sqrt{2\kappa_b}\beta(t_k)\right]A(t_k). \tag{200}$$

2. Evolution between jumps:
Working in the interaction picture and with both cavity/resonators on resonance, $H_{\text{source}} = H_{\text{target}} = 0$, and in place of Eq. (169), the equation of motion to be solved in between the jumps is

$$\frac{d|\bar{\psi}_{\text{REC}}^{\text{ret}}\rangle}{dt} = -(\kappa_a a^\dagger a + \kappa_b b^\dagger b + 2\sqrt{\kappa_a\kappa_b}ab^\dagger)|\bar{\psi}_{\text{REC}}^{\text{ret}}\rangle. \tag{201}$$

The *ansatz* is preserved if $\alpha(t)$, $\beta(t)$, and $A(t)$ satisfy the equations:

$$\frac{d\alpha}{dt} = -\kappa_a \alpha, \tag{202}$$

$$\frac{d\beta}{dt} = -\kappa_b \beta - 2\sqrt{\kappa_a \kappa_b}\,\alpha, \tag{203}$$

and

$$\frac{1}{A}\frac{dA}{dt} = -\frac{d\alpha^*}{dt}\alpha - \frac{d\beta^*}{dt}\beta$$

$$= -\kappa_a|\alpha|^2 - \kappa_b|\beta|^2 - 2\sqrt{\kappa_a \kappa_b}\,\alpha\beta^*$$

$$\rightarrow -\frac{1}{2}\left|\sqrt{2\kappa_a}\,\alpha + \sqrt{2\kappa_b}\,\beta\right|^2, \tag{204}$$

where the last line neglects the term $-\sqrt{\kappa_a \kappa_b}(\alpha\beta^* - \alpha^*\beta)$, which is permitted as this simply amounts to omitting an overall phase on the amplitude A. The solutions are:

$$\alpha(t) = \alpha \exp(-\kappa_a t), \tag{205}$$

$$\beta(t) = \alpha \frac{2\sqrt{\kappa_a \kappa_b}}{\kappa_a - \kappa_b}[\exp(-\kappa_a t) - \exp(-\kappa_b t)], \tag{206}$$

and

$$A(t) = A(t_k)\exp\left[-\frac{1}{2}\int_{t_k}^{t}\left|\sqrt{2\kappa_a}\,\alpha(t') + \sqrt{2\kappa_b}\,\beta(t')\right|^2 dt'\right], \quad t_k \leq t < t_{k+1}. \tag{207}$$

3. Conditional state and record probability:

Bringing together the pieces, the jump, Eq. (200), and evolution between jumps, Eq. (207), allows us to write down the complete expression for the conditional state up to time t:

$$|\bar{\psi}_{\text{REC}}^{\text{ret}}(t)\rangle = \left[\sqrt{2\kappa_a}\,\alpha(t_n) + \sqrt{2\kappa_b}\,\beta(t_n)\right]\dots\left[\sqrt{2\kappa_a}\,\alpha(t_1) + \sqrt{2\kappa_b}\,\beta(t_1)\right]$$

$$\times \exp\left[-\frac{1}{2}\int_{0}^{t}\left|\sqrt{2\kappa_a}\,\alpha(t') + \sqrt{2\kappa_b}\,\beta(t')\right|^2 dt'\right], \tag{208}$$

with norm and record probability density

$$\langle\bar{\psi}_{\text{REC}}^{\text{ret}}(t)|\bar{\psi}_{\text{REC}}^{\text{ret}}(t)\rangle = \left|\sqrt{2\kappa_a}\,\alpha(t_n) + \sqrt{2\kappa_b}\,\beta(t_n)\right|^2 \dots \left|\sqrt{2\kappa_a}\,\alpha(t_1) + \sqrt{2\kappa_b}\,\beta(t_1)\right|^2$$

$$\times \exp\left[-\int_{0}^{t}\left|\sqrt{2\kappa_a}\,\alpha(t') + \sqrt{2\kappa_b}\,\beta(t')\right|^2 dt'\right]. \tag{209}$$

4. Photoelectron counting distribution:

The probability to count n photoelectrons in time T now follows by summing over ordered sequences of count times. We define the intensity in units of photon flux, $|\xi(t)|^2 = |\sqrt{2\kappa_a}\alpha(t) + \sqrt{2\kappa_b}\beta(t)|^2$, and, from Eq. (209), obtain

$$
\begin{aligned}
P(n,T) &= \int_0^T dt_n \int_0^{t_n} dt_{n-1} \cdots \int_0^{t_2} dt_1 |\xi(t_n)|^2 \cdots |\xi(t_1)|^2 \exp\left[- \int_0^T |\xi(t')|^2 dt' \right] \\
&= \frac{\left[\int_0^T |\xi(t')|^2 dt' \right]^n}{n!} \exp\left[- \int_0^T |\xi(t')|^2 dt' \right].
\end{aligned}
\tag{210}
$$

The result clearly agrees with Eq. (83), due to the explicit normal ordering of that expression and the fact that $|\psi_{\mathrm{REC}}(t)\rangle = |\alpha(t)\rangle_a |\beta(t)\rangle_b$ is a coherent state; the normalized state is entirely independent of the counting record. Since the output field is classical (coherent), the result also agrees with the semiclassical version of the photoelectron counting formula.

Acknowledgments

Support for travel and accommodation from the Singapore School of Physics is gratefully acknowledged.

References

1. G. Lindblad, On the generators of quantum dynamical semigroups, *Commun. Math. Phys.* **48**, 119–1304, (1976).
2. I. R. Senitzky, Dissipation in quantum mechanics: The harmonic oscillator, *Phys. Rev.* **119**, 670–679, (1960).
3. I. R. Senitzky, Dissipation in quantum mechanics: The harmonic oscillator. II, *Phys. Rev.* **124**, 642–648, (1961).
4. C. W. Gardiner and M. J. Collett, Input and output in damped quantum systems: Quantum stochastic differential equations and the master equation, *Phys. Rev. A* **31**, 3761–3774, (1985).
5. V. G. Weisskopf and E. Wigner, *Z. Phys.* **63**, 54, (1930).
6. B. R. Mollow, Power spectrum of light scattered by two-level atoms, *Phys. Rev.* **188**, 1969–1975, (1969).
7. W. Pauli, in *Probleme der Modernen Physik, Arnold Sommerfeld zum 60. Geburtstage gewidmet von seinen Schülern*, Vol. 1, ed. P. Debye (Hirzel Verlag, Leipzig, 1928).
8. M. Lax, Formal theory of quantum fluctuations from a driven system, *Phys. Rev.* **129**, 2342–2348, (1963).
9. M. Lax, Quantum noise. X. Density-matrix treatment of field and population-difference equations, *Phys. Rev.* **157**, 213–231, (1967).
10. M. Lax, The Lax-Onsager regression 'theorem' revisited, *Optics Commun.* **179**, 463–476, (2000).
11. R. Hanbury Brown and R. Q. Twiss, Correlation between photons in coherent beams of light, *Nature* **177**, 27–29, (1956).
12. R. Hanbury Brown and R. Q. Twiss, A test of a new type of stellar interferometer on Sirius, *Nature* **178**, 1046–1048, (1956).

152 *H. J. Carmichael*

13. R. Hanbury Brown and R. Q. Twiss, Interferometry of the intensity fluctuations in light I. Basic theory: The correlation between photons in coherent beams of radiation, *Proc. Roy. Soc. Lond. A* **242**, 300–324, (1957).
14. R. Hanbury Brown and R. Q. Twiss, Interferometry of the intensity fluctuations in light II. An experimental test of the theory for partially coherent light, *Proc. Roy. Soc. Lond. A* **243**, 291–319, (1958).
15. L. Mandel, Fluctuations of photon beams and their correlations, *Proc. Phys. Soc.* **72**, 1037–1048, (1958).
16. R. J. Glauber, Photon correlations, *Phys. Rev. Lett.* **10**, 84–86, (1963).
17. R. J. Glauber, The quantum theory of optical coherence, *Phys. Rev.* **130**, 2529–2539, (1963).
18. E. B. Davies, Quantum stochastic processes, *Commun. Math. Phys.* **15**, 277–304, (1969).
19. E. C. G. Sudarshan, Equivalence of semiclassical and quantum mechanical descriptions of statistical light beams, *Phys. Rev. Lett.* **10**, 277–279, (1963).
20. L. Mandel, Fluctuations of photon beams: The distribution of the photo-electrons, *Proc. Phys. Soc.* **74**, 233–243, (1959).
21. P. L. Kelley and W. H. Kleiner, Theory of electromagnetic field measurement and photoelectron counting, *Phys. Rev.* **136**, A316–A334, (1964).
22. R. J. Glauber, Coherent and incoherent states of the radiation field, *Phys. Rev.* **131**, 2766–2788, (1963).
23. H. J. Carmichael, *Statistical Methods in Quantum Optics 1: Master Equations and Fokker-Planck Equations*, Corrected Second Printing, p. 275 (Springer, Berlin, Heidelberg, 1999).
24. A. Einstein, Strahlungsemission und — absorption nach der Quantentheorie, *Verhandlungen der Deutschen Physikalischen Gesellschaft* **18**, 328–323, (1916).
25. A. Einstein, Quantentheorie der Strahlung, *Physikalische Zeitschrift* **18**, 121–128, (1917). An English translation appears in B. L. van der Waerden, *Sources of Quantum Mechanics*, Chap. 1 (North Holland, Amsterdam, 1967).
26. The term "unraveling" traces to p. 28 of A. Alsing and H. J. Carmichael, Spontaneous dressed-state polarization of a coupled atom and cavity mode, *Quantum Opt.* **3**, 13–32, (1991): "The formalism 'unravels' the dynamics of an operator master equation like equation (41), expressing the evolution of the density operator as a generalized sum over quantum state 'trajectories'."
27. E. B. Davies, *Quantum Theory of Open Systems* (Academic Press, London, 1976).
28. M. D. Srinivas and E. B. Davies, Photon counting probabilities in quantum optics, *Optica Acta* **28**, 981–996, (1981).
29. M. D. Srinivas and E. B. Davies, What are the photon counting probabilities for open systems — a reply to Mandel's comments, *Optica Acta* **29**, 235–238, (1982).
30. L. Mandel, Comment on 'Photon counting probabilities in quantum optics', *Optica Acta* **28**, 1447–1450, (1981).
31. H. J. Carmichael, S. Singh, R. Vyas, and P. R. Rice, Photoelectron waiting times and atomic state reduction in resonance fluorescence, *Phys. Rev. A* **39**, 1200–1218, (1989).
32. M. Wolinsky and H. J. Carmichael, Photoelectron counting statistics for the degenerate parametric oscillator, *Coherence and Quantum Optics VI*, eds. L. Mandel and E. Wolf, Proceedings of the Sixth Rochester Conference on Coherence and Quantum Optics, Rochester, N. Y., June 26–28, 1989 (Plenum, New York, 1989), pp. 1239–1242.
33. B. R. Mollow, Pure-state analysis of resonant light scattering: Radiative damping, saturation and multiphoton effects, *Phys. Rev. A* **12**, 1919–1943, (1975).

34. R. J. Cook, Photon statsistics in resonance fluorescence from laser deflection of an atomic beam, *Opt. Commun.* **35**, 347–350, (1980); Photon number statistics in resonance fluorescence, *Phys. Rev. A* **23**, 1243–1250, (1981).

35. H. J. Carmichael and L. Tan, Quantum measurement theory of photoelectric detection, *OSA Technical Digest Series*, Vol. 15 (Opt. Soc. Am., Washington, 1990), p. 3.

36. H. J. Carmichael, *An Open Systems Approach to Quantum Optics*, Lecture Notes in Physics, New Series m-Monographs, Vol. m18 (Springer, Berlin, Heidelberg, 1993).

37. J. Dalibard, Y. Castin, and K. Mølmer, Wave-function approach to dissipative processes in quantum optics, *Phys. Rev. Lett.* **68**, 580–583, (1992).

38. R. Dum, P. Zoller, and H. Ritsch, Monte-Carlo simulation of the atomic master equation for spontaneous emission, *Phys. Rev. A* **45**, 4879–4887, (1992).

39. H. G. Dehmelt, *Bulletin of the American Physical Society* **20**, 60, (1975).

40. W. Nagourney, J. Sandberg, and H. Dehmelt, Shelved optical electron amplifier: Observation of quantum jumps, *Phys. Rev. Lett.* **56**, 2797–2799, (1986).

41. Th. Sauter, W. Neuhauser, R. Blatt, and P. E. Toschek, Observation of quantum jumps, *Phys. Rev. Lett.* **57**, 1696–1698, (1986).

42. J. C. Berquist, R. G. Hulet, W. M. Itano, and D. J. Wineland, Observation of quantum jumps in a single atom, *Phys. Rev. Lett.* **57**, 1699–1702, (1986).

43. P. Zoller, M. Marte, and D. F. Walls, Quantum jumps in qatomic system, *Phys. Rev. A* **35**, 198–207, (1987).

44. M. Porrati and S. Putterman, Wave-function collapse due to null measurements: The origin of intermittent atomic fluorescence, *Phys. Rev. A* **36**, 929–932, (1987).

45. M. Brune, E. Hagley, J. Dreyer, X, Maître, A. Maali, C. Wunderlich, J. M. Raimond, and S. Haroche, Observing the progressive decoherence of the "meter" in a quantum measurement, *Phys. Rev. Lett.* **77**, 4887–4890, (1996).

46. S. Degélise, I. Dotsenko, C. Sayrin, J. Bernu, M. Brune, J. M. Raimond, and S. Haroche, *Nature* **455**, 510–514, (2008).

47. C. W. Gardiner, Inhibition of atomic phase decays by squeezed light: A direct effect of squeezing, *Phys. Rev. Lett.* **56**, 1917–1920, (1986).

48. H. J. Carmichael, Quantum trajectory theory for cascaded open systems, *Phys. Rev. Lett.* **70**, 2273–2276, (1993).

49. M. I. Kolobov and I. V. Sokolov, *Opt. Spektrosk.* **62**, 112, (1987).

50. C. W. Gardiner, Driving a quantum system with the output from another driven quantum system, *Phys. Rev. Lett.* **70**, 2269–2272, (1993).

51. P. Kochan and H. J. Carmichael, Photon statistics dependence of single-atom absorption, *Phys. Rev. A* **50**, 1700–1709, (1994).

52. C. W. Gardiner and A. S. Parkins, Driving atoms with light of arbitrary statistics, *Phys. Rev. A* **50**, 1792–1806, (1994).

53. J. I. Cirac, P. Zoller, H. J. Kimble, and H. Mabuchi, Quantum State transfer and entanglement distribution among distant nodes in a quantum network, *Phys. Rev. Lett.* **78**, 3221–3224, (1997).

54. S. Clark, A. Peng, M. Gu, and S. Parkins, Unconditional preparation of entanglement between atoms in optical cavities, *Phys. Rev. Lett.* **91**, 177901–4, (2003).

55. H. Nha and H. J. Carmichael, Decoherence of a two-state atoms driven by coherent light, *Phys. Rev. A* **71**, 013805-1-6, (2005).

56. C. Noh and H. J. Carmichael, Disentanglement of source and target and the laser quantum state, *Phys. Rev. Lett.* **100**, 120405-1-4, (2008).

Chapter 5

Basic Concepts in Quantum Information

Steven M. Girvin

Sloane Physics Laboratory, Yale University
New Haven, CT 06520-8120, USA
steven.girvin@yale.edu

In the last 25 years a new understanding has evolved of the role of information in quantum mechanics. At the same time there has been tremendous progress in atomic/optical physics and condensed matter physics, and particularly at the interface between these two formerly distinct fields, in developing experimental systems whose quantum states are long-lived and which can be engineered to perform quantum information processing tasks. These lecture notes will present a brief introduction to the basic theoretical concepts behind this 'second quantum revolution.' These notes will also provide an introduction to 'circuit QED.' In addition to being a novel test bed for non-linear quantum optics of microwave electrical circuits which use superconducting qubits as artificial atoms, circuit QED offers an architecture for constructing quantum information processors.

1. Introduction and overview

Quantum mechanics is now more than 85 years old and is one of the most successful theories in all of physics. With this theory we are able to make remarkably precise predictions of the outcome of experiments in the microscopic world of atoms, molecules, photons, and elementary particles. Yet despite this success, the subject remains shrouded in mystery because quantum phenomena are highly counter-intuitive to human beings whose intuition is built upon experience in the macroscopic world. Indeed, the founders of the field struggled mightily to comprehend it. Albert Einstein, who made truly fundamental contributions to the theory, never fully believed it. Ironically this was because he had the deepest understanding of the counter-intuitive predictions of the theory. In his remarkable 'EPR' paper with Podolsky and Rosen,[1] he famously proposed a thought experiment showing that quantum entanglement produced experimental outcomes that seemed to be obviously impossible. In an even greater irony, modern experiments based on John Bell's analysis of the EPR 'paradox' now provide the best proof not only that quantum mechanics is correct, but that reality itself is simply not what we naively think it is. In the last 25 years we have suddenly and quite unexpectedly come to a deeper understanding of the role of information in the quantum world. These ideas could very easily have been understood in 1930 because of the laws of quantum mechanics were known at the time. Yet, because of the counter-intuitive nature of these concepts, it would be nearly another 60 years before these further implications of the quantum laws came to be understood. We begin with a high-level 'executive summary' overview of these concepts and ideas which will then be developed in more detail in later sections.

From the earliest days of the quantum theory, the Heisenberg uncertainty principle taught us that certain physical quantities could never be known to arbitrary precision. This, combined with the genuine randomness of the outcome of certain measurements, seemed to indicate that somehow we were 'worse off' than we had been in the old world of classical physics. Recently however, we have come to understand that quantum uncertainty, far from being a 'bug', is actually a 'feature' in the program of the universe. Quantum uncertainty and its kin, entanglement, are actually powerful resources which we can use to great advantage to, encode, decode, transmit and process information in highly efficient ways that are impossible in the classical world. In addition to being a profound advance in our understanding of quantum mechanics, these revolutionary ideas have spurred development of a new kind of quantum engineering whose goal is to develop working quantum 'machines'.[2] The new ideas that have been developed also have profound applications for precision measurements and atomic clocks. For example, ordinary atomic clocks keep track of time by following the evolution of product states of N spins, each precessing at frequency Ω_0:

$$|\Psi(t)\rangle = [|0\rangle + e^{-i\Omega_0 t}|1\rangle] \otimes [|0\rangle + e^{-i\Omega_0 t}|1\rangle] \otimes \ldots \otimes [|0\rangle + e^{-i\Omega_0 t}|1\rangle]. \qquad (1)$$

By using maximally entangled states of the form

$$|\Psi(t)\rangle = [|0000\ldots0\rangle + e^{-iN\Omega_0 t}|1111\ldots1\rangle] \qquad (2)$$

one gets a clock which runs N times faster and therefore gains a factor of $\sqrt{N}$ in precision.[3,4]

1.1.   *Classical information and computation*

For a discussion of the history of ideas in classical and quantum computation, the reader is directed to the central reference work in the field 'Quantum Computation and Quantum Information,' by Nielsen and Chuang.[5] The brief historical discussion in this and the following section relies heavily on this work.

The theory of computation developed beginning in the 1930's by Turing, Church and others, taught us that one can, in principle, construct universal computational engines which can compute anything that can be computed. (We also learned that not everything is computable.) Any given 'Turing machine' can be programmed to perfectly emulate the operation of any other computer and compile any other programming language. Thus one of the founding principles of computer science is that the actual hardware used in a computation is (formally at least) irrelevant.

A key measure of the difficulty of a computational problem is how the computational cost scales with the size N of the input data set. Skipping over many essential details, the main classification distinction is between polynomial and exponential scaling. For example, solving the Schrödinger equation for N electrons (generically) becomes exponentially harder as N increases. So does the classical problem of the traveling salesman who has to find an optimal route to visit N cities. Turing's ideas taught us that if a problem is exponential on one piece of hardware, it will be exponential on any piece of hardware. All universal computers are 'polynomially equivalent.' That is, the scaling is the same up to possible factors no stronger than polynomial in N. It came as a great shock to the world of physicists and computer scientists that if one could build a computer based on the principles of quantum mechanics, certain classes of problems would change from exponential to polynomial. That is, not all hardware is equivalent. Peter Shor[6] developed a remarkable stochastic quantum algorithm for finding the prime factors of large integers which scales only as $(\ln N)^3$, an enormous exponential speed up over the fastest known classical algorithm. [We should be careful to note here that, despite vast effort, it has not been rigorously proven that this particular example (factoring) is an exponentially hard problem classically.]

There are several threads leading to the conclusion that quantum hardware could be very powerful. One thread goes back to Feynman[7] who argued in the 1980's that since the Schrödinger equation is hard to solve, perhaps small quantum systems could be used to simulate other quantum systems of interest. It is felt in some quarters that Feynman tends to be given a bit too much credit since this idea did not directly lead to the greater understanding of information and computation in quantum mechanics. However he definitely did pose the question of whether quantum hardware is different than classical. There are active attempts underway today to build 'quantum simulators' using optical lattices, trapped ions and other quantum hardware. This effort flows both from Feynman's ideas and from the new ideas of quantum information processing which will be described below.

1.2.   *Quantum information*

A classical bit is some physical system that has two stable states, 0 and 1. Quantum bits are quantum systems that have two energy eigenstates $|0\rangle$ and $|1\rangle$. They derive their extraordinary power from being able to be in coherent quantum superpositions of both states. This 'quantum parallelism' allows them (in some sense) to be both 0 and 1 at the same time.

One additional point worth discussing before we proceed is the following. Classical computers dissipate enormous amounts of energy. (The fraction of electrical power consumed in the U.S. by computers is of order 3% of the total electrical consumption from all sectors of the economy.) Quantum computers work using coherent quantum superpositions and hence fail if *any* energy is dissipated during the computation. That is, the unitary time evolution of a quantum computer is fully reversible.

(I ignore here quantum error correction, described below, in which a small amount of energy must be dissipated to correct imperfections in the hardware operation.) In thinking about the thermodynamics of information, Charles H. Bennett[8] made a major contribution in 1973 by discovering the concept of reversible classical computation. He proved that energy dissipation is not inevitable and thus fully reversible computation is possible in principle. It can be argued that this unexpected idea helped lay the ground work for quantum information processing which, as noted, must of necessity be fully reversible.

The first hint that quantum uncertainty might be a 'feature' and not a 'bug,' came from a remarkably prescient idea of Stephen Wiesner, then a graduate student at Columbia, sometime around 1968 or 1970. The idea was so revolutionary that Wiesner was unable to get it published until 1983.[9] The idea was that one could create 'quantum money'[10] with quantum bits for the serial number and this would make it possible to verify the validity of the serial number, but never duplicate it to make counterfeit money. This was really the first understanding of the 'no cloning theorem' later formally proved by Wooters and Zurek[11] and Dieks.[12] The essential idea is that if we measure the polarization of a two-level system (a quantum bit, say a spin-1/2 particle), there are always only two possible answers, ± 1, no matter what the state of the qubit is. Thus, the result of our measurement yields only one classical bit of information. However unless you know what quantization axis to choose for the measurement, you will most likely destroy the original state during the measurement process. This is because the result of any measurement of the polarization of the spin along some axis always shows that the qubit is maximally polarized with projection ± 1 along the axis you chose to measure (as opposed to the axis along which the state was prepared). The act of measurement itself collapses the state and so no further information can be acquired about the original state. Since it takes an infinite number of classical bits to specify the original orientation of the quantization axis, we cannot learn the quantum state from the one bit of classical information produced by the measurement. On the other hand, if you are told what axis to use in the measurement, then you can determine whether the polarization is $+1$ or -1 along that axis with complete certainty and without destroying the state, since the measurement is QND (quantum non-demolition).

Wiesner realized that all this means that one could make a quantum serial number by randomly choosing the orientation of a spin for each of, say $N = 20$ qubits. For each qubit the quantization axis could be randomly chosen to be X or Z and the sign of the polarization along that axis could be chosen randomly to be ± 1. The agency that originally created the quantum serial number could verify the validity of the money by making QND measurements with the correct quantization axes (which having chosen the quantization axes, it alone knows) thus obtaining the correct polarizations (if the serial number is not counterfeit), but a counterfeiter (who does not have access to the information on the randomly chosen quantization directions) has a low probability (which turns out to be $(3/4)^N$) of being able to make a copy of the serial number that would pass inspection by the agency. This powerful result is a simple consequence of the Heisenberg uncertainty principle and the fact that the X and Z components of the spin are incompatible observables. As such it could have been 'obvious' to practitioners of quantum mechanics anytime after about 1930, but it was not understood until 40 years later, and then by only one person.

Though Wiesner could not initially get his idea published, one person who paid attention was Charles Bennett,[13,14] now at IBM Corporation. Bennett, working with Gilles Brassard, realized in 1984 that the quantum money idea could be used create an unbreakable quantum encryption protocol. The only provably secure encryption technique uses a secret key called a 'one-time pad.' The key is a random string of bits as long as the message to be sent and is used to code and decode the message. The problem is that the sender and receiver must both have a copy of the key and it can only be used once. The difficulty of how to secretly transmit the key to the receiver is identical to the original problem of secretly transmitting the message to the receiver and is called the 'key distribution problem.' Bennett and Brassard's so-called BB84 protocol[15] for quantum key distribution solves the problem by transmitting the key as a string of quantum bits (essentially as the serial number of a piece of quantum money). Because of the no-cloning theorem, it is not possible for an eavesdropper to read the key and pass it on undisturbed in order to avoid detection. From this fact, BB84 provides a protocol for insuring that no eavesdropper has read the key and thus that it is safe to use. The BB84 protocol was a major breakthrough in our understanding of quantum information and its transmission. It too was sufficiently revolutionary that the authors had difficulty getting it published.[14] The BB84 protocol and its descendants have led to practical applications in which quantum key distribution can now be routinely done over long distances at megahertz bit rates using the polarization modes of individual photons as the quantum bits. It has been carried out over optical fiber and through the atmosphere over distances of many tens of kilometers and there are now available commercial systems based on these ideas. Quantum cryptography is reviewed by Gisin *et al.*[16]

Following up on the idea of quantum key distribution, Bennett and Wiesner invented the idea of 'quantum dense coding'.[17] Given that measuring a qubit yields only one classical bit of information, it is not possible to transmit more information with a single quantum bit than with a single classical bit. However if the sender and receiver each share one quantum bit from an entangled pair, something remarkable can occur. There are four standard 'Bell states' for an entangled pair ($|01\rangle \pm |10\rangle, |00\rangle \pm |11\rangle$). Suppose that Alice prepares the Bell state $|00\rangle + |11\rangle$, and then sends one of the two qubits to Bob. So far, no information has been transmitted, because if Bob makes a measurement on his qubit, the result is simply random, no matter what axis he chooses for his measurement. However (as we show in detail in a later section), Bob can now perform one of four local unitary operations on his qubit to either leave the state alone or map it into one of the three other Bell states. He then sends the qubit back to Alice who can make a joint measurement[18] on the two qubits and determine which of the four Bell states results. Since there are four possibilities, Bob has transmitted two classical bits by simply returning one quantum bit to Alice.

Another way to illustrate the peculiar power of entanglement is for Alice to prepare the same Bell state, $|00\rangle + |11\rangle$. She then transmits the first qubit to Bob. After that she performs one of four possible local unitary operations on her remaining qubit to either leave the state alone or map it into one of the three other Bell states (details will be supplied further below). Alice then sends this second qubit to Bob who then can measure which of the four states the qubits are in and hence receive two classical bits of information. This is not surprising since he has received two qubits. Recall however that Alice did not decide which of the four 'messages' to send until *after* she had sent the first qubit to Bob. This result clearly demonstrates the 'spooky action at a distance' that bothered Einstein and has further enhanced our understanding of the novel features of information and its transmission through quantum channels. Following up on the work of Bennett and Brassard, Ekert[19] proposed use of entangled pairs as a resource for quantum encryption, though this is not yet a practical technology.

One of the very important applications of these new ideas in quantum information is to the numerical solution of the Schrödinger equation for strongly correlated many-body systems. I. Cirac, G. Vidal, F. Verstrate, and others have realized that quantum information ideas are very helpful in understanding how to efficiently represent in a classical computer the highly entangled quantum states that occur in strongly correlated many-body systems. This has led to real breakthroughs in our understanding and modeling of complex many-body systems.[20]

1.3.   *Quantum algorithms*

As Feynman noted, simulating quantum systems on a classical computer is exponentially hard. It is clear that a quantum computer can efficiently simulate certain quantum systems (e.g. the computer can at least simulate itself, and a piece of iron can be used to simulate iron!) The next development in going from quantum information to quantum computation was taken by David Deutsch who posed the question of whether there is a quantum extension of the Church-Turing idea that any computation running on a classical computer can be *efficiently* simulated on a universal Turing machine. Deutsch asked if there exists a universal quantum simulator which can efficiently simulate any other quantum system. The answer to this profound question remains unknown, but in 1985 Deutsch[21] opened the door to the world of quantum algorithms. He found a 'toy' computational problem that could be solved on a quantum computer in a manner that is impossible classically. In 1992 Deutsch and Jozsa[22] simplified and extended the earlier result.

The modern formulation of the problem is the following. Consider a function $f(x)$ whose domain is $x = \pm 1$ and whose range is also ± 1. There are exactly four such functions: $f_1(x) = x$, $f_2(x) = -x$, $f_3(x) = +1$, $f_4(x) = -1$. If we measure

$f(-1)$ and $f(+1)$ we acquire two bits of classical information and know which of the four functions we have. Obviously this requires two evaluations of the function. Now, notice that the functions can be divided into two classes, balanced (i.e. $f(-1)+f(+1) = 0$) and constant (i.e. unbalanced). Since there are only two classes, finding out which class the function is in means acquiring one classical bit of information. However, it still requires two evaluations of the function. Hence classically, we are required to learn the full function (by making two evaluations) before we can determine which class it is in. Quantum mechanically we can determine the class with only a single measurement! The trick is to use the power of quantum superpositions. By putting x into a superposition of both $+1$ and -1, a single evaluation of the function can be used to determine the class (but *not* which function in the class since the measurement yields only one bit of classical information).

This idea launched searches for other quantum algorithms. Consider the problem of finding a particular entry in a large unordered database. (For example imagine looking for a friend's phone number in a phone book that was not alphabetically ordered). Classically there is no faster procedure than simply starting at the beginning and examining the database entries one at a time until the desired entry is found. On average this requires $N/2$ 'looks' at the database. Lov Grover[23, 24] found a quantum search algorithm that allows one to find an entry in an unordered database in only $\sim \sqrt{N}$ 'looks,' which is a speed up by a factor $\sim \sqrt{N}$ over the classical search. Again, the superposition and uncertainty principles come into play. A given 'look' can actually be a superposition of single looks at every possible entry at the same time. Since searching is a very generic problem in computer science, this quantum result is quite significant, even though the speedup is not exponential.

The real excitement came in 1994 when Peter Shor[6, 25] found efficient quantum algorithms for the discrete logarithm problem and the Fourier transform problem. The latter can be used to find the prime factors of large integers and hence break RSA public key encryption. The classical Fourier transform is defined by the linear transformation

$$\tilde{f}(k) = \frac{1}{\sqrt{N}} \sum_{j=0}^{N-1} e^{2\pi i k j / N} f(j) \tag{3}$$

and requires of order N^2 operations to evaluate the function $\tilde{f}$. For the special case that $N = 2^n$, the Fast Fourier Transform (FFT) algorithm requires only $\sim N \ln N = n2^n$ steps, but is still exponential in n. Shor's quantum Fourier transform is a unitary transformation in Hilbert space which acts on a state defined by the function f and produces a new state defined by the function $\tilde{f}$. Suppose we have n qubits so that the Hilbert space has dimension 2^n. We can use f to define the following (unnormalized) state

$$|\Psi\rangle = \sum_{j=0}^{2^n-1} f(j)|j\rangle. \tag{4}$$

The quantum Fourier transform applied to this state is defined by

$$|\tilde{\Psi}\rangle = U_{\mathrm{QFT}}|\Psi\rangle = \sum_{j=0}^{2^n-1} \tilde{f}(j)|j\rangle. \tag{5}$$

Remarkably, this operation can be carried out in order n^2 steps (more precisely order $n^2 \ln n \ln \ln \ln n$). This is an *exponential speed up* over the classical FFT procedure.

The bad news is that being in possession of the quantum state $|\tilde{\Psi}\rangle$ does not actually tell you the values of $\tilde{f}(j)$. Determining these would in general require exponentially many measurements. This is because a measurement of which state the qubits are in randomly yields $|j\rangle$ with probability $S(j) \equiv |\tilde{f}(j)|^2$. Suppose however that f is periodic with an integer period and is simple enough that its spectral density $S(j)$ is strongly peaked at some value of j. Then with only a few measurements one would find j with high probability and hence the period could be found. Peter Shor mapped the prime factor problem onto the problem of finding the period of a certain function and hence made it soluble. This result caused a sensation and thereafter the field of quantum information and computation exploded with feverish activity which continues unabated.

With this overview, we are now ready to begin a more detailed discussion. Two interesting topics which will not be covered in this discussion are quantum walks[26] and adiabatic quantum computation.[27,28]

2.   Introduction to quantum information

In recent decades there has been a 'second quantum revolution' in which we have come to much (but still not completely) understand the role of information in quantum mechanics. These notes provides only the briefest introduction to a few of the most basic concepts. For fuller discussions the reader is directed to Nielsen and Chuang[5] and Mermin.[29]

The bit is the smallest unit of classical information. It represents the information we gain when we receive the definitive answer to a yes/no or true/false question about which we had no prior knowledge. A bit of information, represented mathematically as binary digit 0 or 1, can be represented physically inside a computer in the form of two physical states of a switch (on/off) or the voltage state of some transistor circuit (high/low). One of the powerful features of this digital (as opposed to analog) encoding of information is that it is robust against the presence of noise as long as the noise is small compared to the difference in signal strength between the high and low states.

It is useful to note that there are two possible encodings of a mathematical bit in a physical bit. The low voltage state can represent 0 and the high voltage state can represent 1, or we can use the reverse. It does not matter which encoding we use as long as everyone using the information agrees on it. If two parties are using opposite encodings, they can still readily translate from one to the other by performing a NOT operation (which maps 0 to 1 and vice versa) to each bit.

The situation is very different for quantum bits (qubits). A quantum bit is represented physically by an atom or other quantum system which has a two-state Hilbert space. For example the state (in Dirac notation) $|0\rangle$ could be represented physically by the atom being in the ground state and the state $|1\rangle$ could be represented by the atom being in its first excited state. (We assume that the other excited states can be ignored.) All two-state quantum systems are mathematically equivalent to a spin-1/2 particle in a magnetic field with Hamiltonian

$$H = \frac{\hbar}{2}\vec{\omega} \cdot \sigma, \tag{6}$$

where $\vec{\omega}$ is a vector (with units of frequency) representing the strength and direction of the pseudo magnetic field, and $\vec{\sigma} = (\sigma^x, \sigma^y, \sigma^z)$ are the Pauli spin matrices. The transition frequency from ground to excited state is $\omega \equiv |\vec{\omega}|$. Up to an additive constant, this form of Hamiltonian is the most general Hermitian operator on the two-dimensional Hilbert space. That is, the Pauli operators $(\sigma^x, \sigma^y, \sigma^z)$ for the three components of the 'spin' form a basis which spans the set of all possible operators (ignoring the identity operator). Because they are non-commuting, they are mutually incompatible and it is not possible to have a simultaneous eigenstate of more than one of the operators. The most general measurement that can be made is a 'Stern-Gerlach' measurement of the projection of the spin along a single axis

$$\mathcal{O} = \hat{n} \cdot \vec{\sigma}. \tag{7}$$

Since $\mathcal{O}^2 = 1$, the eigenvalues of $\mathcal{O}$ are ± 1 and such a Stern-Gerlach measurement yields precisely one classical bit of information telling us whether the spin is aligned or anti-aligned to the axis $\hat{n}$. It is a strange feature of quantum spins that no matter what axis we choose to measure the spin along, we always find that it is precisely parallel or antiparallel to that axis.

The ideal Stern-Gerlach measurement is quantum non-demolition (QND), that is it is repeatable. If we measure (say) σ^z for an unknown quantum state, the result will be (possibly randomly) ± 1. As long as there are no stray magnetic fields which cause the spin to precess in between measurements,[a] subsequent measurements of σ^z will not be random but will instead be identical to the first. If this is followed by a measurement of (say) σ^x the result will be completely random and unpredictable with $+1$ and -1 each occuring half the time. However subsequent measurements of σ^x will all agree with the first if the measurement is QND.

---

[a]That is, a measurement is QND if the operator being measured and its coupling to the measurement apparatus both commute with the system Hamiltonian.

The spinor eigenfunctions of H are (up to an unimportant global phase factor)

$$|\psi_+\rangle = \begin{pmatrix} \cos\frac{\theta}{2}e^{i\varphi/2} \\ \sin\frac{\theta}{2}e^{-i\varphi/2} \end{pmatrix}, \tag{8}$$

$$|\psi_-\rangle = \begin{pmatrix} -\sin\frac{\theta}{2}e^{i\varphi/2} \\ \cos\frac{\theta}{2}e^{-i\varphi/2} \end{pmatrix}, \tag{9}$$

where θ and φ are respectively the polar and azimuthal angles defining the direction of $\vec{\omega}$. For the case $\theta = \varphi = 0$, the Hamiltonian reduces to

$$H_0 = \frac{\omega}{2}\sigma^z, \tag{10}$$

with eigenfunctions that are simply the 'spin up' and 'spin down' basis states

$$|0\rangle = |\downarrow\rangle = \begin{pmatrix} 0 \\ 1 \end{pmatrix} \tag{11}$$

$$|1\rangle = |\uparrow\rangle = \begin{pmatrix} 1 \\ 0 \end{pmatrix}. \tag{12}$$

Throughout these notes we will use the $0, 1$ and the $\downarrow, \uparrow$ notation interchangeably.

We see from the above that, ignoring any overall phase factor, it takes two real numbers (or equivalently an infinite number of classical bits) to specify the quantum state of a spin. The huge asymmetry between the infinite number of classical bits needed to specify a state and the single classical bit we can obtain by doing a measurement is a key concept in our understanding of quantum information, quantum encryption and quantum information processing. The act of measurement collapses the spin state onto the measurement axis and we know exactly what the state of the spin is after the measurement. However (unless we were told in advance which axis to use) we have no way of knowing what the actual state of the spin was before the measurement. This leads us directly to the no-cloning theorem[11] which states that it is impossible make a copy of an unknown quantum state. If we are given an unknown state the only way we can copy it is to make a measurement to see what the state is and then orient additional spins to match that state. However since the state is unknown and we are only allowed to make one measurement, we are forced to guess which axis to measure along and we cannot guarantee that the state has not been changed by the act of measurement.

The mathematical version of the argument for the no-cloning theorem is the following. Let us start with a product state of a qubit in an unknown superposition state and an ancilla qubit in the $|0\rangle$ state

$$|\Psi\rangle = [\alpha|0\rangle + \beta|1\rangle] \otimes |0\rangle. \tag{13}$$

After the cloning operation we desire to have both qubits in the same state

$$|\Psi_{\text{clone}}\rangle = [\alpha|0\rangle + \beta|1\rangle] \otimes [\alpha|0\rangle + \beta|1\rangle] \tag{14}$$

$$|\Psi_{\text{clone}}\rangle = \alpha^2|00\rangle + \alpha\beta|10\rangle + \alpha\beta|01\rangle + \beta^2|11\rangle]. \tag{15}$$

The only operations which are physically possible in quantum mechanics are represented by unitary transformations. Thus we seek a transformation U obeying

$$|\Psi_{\text{clone}}\rangle = U|\Psi\rangle. \tag{16}$$

U is a linear transformation and so it is impossible for it to produce the non-linear coefficients $\alpha^2, \alpha\beta, \beta^2$ that we see above. Hence cloning an unknown state is impossible. Of course if the state is known, we could use a unitary transformation $U(\alpha, \beta)$ that explicitly depends on the known parameters α and β and successfully achieve cloning. We cannot do so however with a unitary transformation which does not depend on knowledge of α and β. Thus known states are readily cloned, but unknown states cannot be. This result will have profound implications for the encryption of information using quantum states and for quantum error correction.

We can understand the no cloning theorem from another point of view. Quantum bits also encode only a single classical bit (the measurement result), but unlike the classical case where there are only two possible encoding schemes, there are an infinity of different quantum encodings defined by the axis $\hat{n} = (\sin\theta\cos\varphi, \sin\theta\sin\varphi, \cos\theta)$ that was used for the preparation and should be used for the measurement if the information is to be correctly decoded. For the classical case, the transformation from one encoding scheme to the other is simply the NOT operation $0 \rightarrow 1, 1 \rightarrow 0$. For the quantum case the transformation is a unitary operation which rotates the spin states on the Bloch sphere through the appropriate angle connecting the old axis $\hat{n}$ and the new axis $\hat{n}'$

$$U_{\hat{n}'\hat{n}} = e^{i\frac{\Theta}{2}\hat{v}\cdot\vec{\sigma}}, \tag{17}$$

where $\cos\Theta = \hat{n}' \cdot \hat{n}$ defines the angle between the two axes and $\hat{v} = \frac{\hat{n}'\times\hat{n}}{|\hat{n}'\times\hat{n}|}$ is the unit vector perpendicular to the plane formed by the two axes. If we are not given any information about which of the infinitely many encoding schemes have been used we cannot reliably decode the information. If we do know the encoding axis $\hat{n}$, the above transformation law will tell us what the results will be for measurements done in a different 'decoding' basis $\hat{n}'$. In particular it tells us that if we encode on the z axis and decode on the x axis the results of the decoding will be totally random and uncorrelated with the encoding. For example, we can express $|\uparrow\rangle, |\downarrow\rangle$, the ± 1 eigenstates of σ^z as a superposition of the eigenstates of σ^x

$$|\uparrow\rangle = \frac{1}{2}[|\rightarrow\rangle + |\leftarrow\rangle] \tag{18}$$

$$|\downarrow\rangle = \frac{1}{2}[|\rightarrow\rangle - |\leftarrow\rangle], \tag{19}$$

which confirms the statement made above that if we make a σ^x measurement after a σ^z measurement, the results will be ± 1 with equal probability. This is distinctly different than the classical case where, if we accidentally use the wrong decoding scheme, the results are still deterministic and perfectly correlated with the encoded data (just backwards with 0 and 1 being interchanged).

Further below, we will discuss quantum teleportation. This is a process in which an unknown quantum state is perfectly reproduced at a distant location. This is not restricted by the no-cloning theorem because the original state is destroyed during the process of teleportation. For example teleportation could consist of simply swapping two states

$$|\Psi\rangle = [\alpha|0\rangle + \beta|1\rangle] \otimes |0\rangle, \tag{20}$$

$$|\Psi_{\text{teleport}}\rangle = |0\rangle \otimes [\alpha|0\rangle + \beta|1\rangle]. \tag{21}$$

Because the coefficients α and β enter only linearly, there exists a unitary transformation which can accomplish this task. We will explore a particular protocol for how this can be achieved using entangled states further below.

3.   Quantum money and quantum encryption

In the 1970's Steven Wiesner, then a graduate student at Columbia, came up with the idea of 'quantum money' which cannot be counterfeited. This was the very first idea in a chain of ideas that has led to the revolution in quantum information. Legend has it that the idea was so far ahead of its time that Wiesner was unable to get it published until much later.[9] He did however discuss it with Charles Bennett and a decade later it bore fruit in the form of quantum encryption, about which more below.

The idea of quantum money is essentially based on the no-cloning theorem, although that theorem was not formally stated and proved until sometime later.[11] The idea is to create a serial number using quantum bits which encode a random string of 0's and 1's. The quantum wrinkle is that for each bit, one randomly selects the Z encoding ($\hat{n} = \hat{z}$) or the X encoding ($\hat{n} = \hat{x}$). Thus the serial number might look like this: $|\uparrow, \rightarrow, \rightarrow, \downarrow, \leftarrow, \leftarrow, \ldots\rangle$. Because of the no-cloning theorem, it is impossible for a counterfeiter to make copies of the money and its quantum serial number. The counterfeiter would have to guess which measurement to make (X or Z) on each qubit and would have no way of knowing if she had guessed correctly. On the other hand, the treasury department could easily verify whether any given bill is real or counterfeit. If the bill also carries an ordinary classical serial number that uniquely identifies it, then the treasury department can keep a (secret) record of the bit value and the quantization (encoding) axis used for each qubit on that particular bill (labeled by the classical serial number) and so knows which measurement to make for each one. If for example the first bit is in state $|\uparrow\rangle$, a Z measurement will always yield the correct value of $+1$ with no randomness. If the treasury department uses all the correct measurement orientations and does not recover the correct measurements then the bill is counterfeit (or the qubits have decohered!).

Problem 1. *Prove that the probability of a counterfeit 'copy' of a bill with an N qubit quantum serial number succesfully passing the treasury department's scrutiny is $(\frac{3}{4})^N$. Thus for large N, counterfeiting is very unlikely to succeed. Hint: in order for a particular bit to fail the test, the counterfeiter must have chosen the wrong measurement and the treasury has to be unlucky in its measurement result.*

Show that this result remains valid even if the counterfeiter chooses an arbitrary pair of orthogonal axes, X', Z' to make her measurements, as long as they lie in the same plane as the X, Z axes chosen by Bob and Alice.

3.1. *Classical and quantum encryption*

The only provably secure method of classical encryption is the 'one-time pad.' Suppose that Alice has a message which consists of a string of N classical bits. Alice wishes to send this message to Bob in such a way that Eve cannot eavesdrop on the communication by intercepting and deciphering the message. A one-time pad is a string of random bits (also called the encryption 'key') whose length is at least as large as the message. If Bob and Alice each are in possession of identical copies of the one-time pad then the message can be sent via the following simple protocol. Let the jth bit in Alice's message be M_j and the jth bit in the one-time pad be P_j. Then Alice computes the jth bit in the encrypted message via

$$E_j = M_j \oplus P_j, \tag{22}$$

where $\oplus$ means addition modulo 2. Essentially this encryption rule means that if $P_j = 0$, do nothing to the message bit and if $P_j = 1$, then flip the message bit. Because E_j is completely random and uncorrelated with any of the other encrypted bits, Eve is unable to decipher the message even if she intercepts it during the transmission to Bob. Bob however is able to decrypt it by the same operation using the same key (one-time pad)

$$D_j = E_j \oplus P_j. \tag{23}$$

Because $P_j \oplus P_j = 0$, we have $D_j = M_j$ and Bob recovers the original message.

In order to work, the key must be as long as the original message, must be completely random, must be kept secret, and must never be re-used. While perfectly secure, this method suffers from the so-called 'key distribution problem.' Securely transmitting a one-time pad of length N from Alice to Bob has the same difficulty as the problem of securely transmitting the original message! Bennett and Brassard[15]

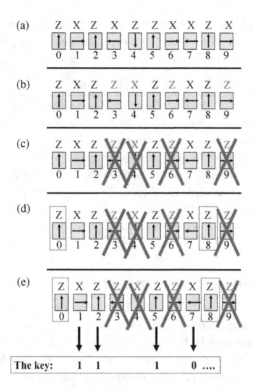

Fig. 1.   BB84 protocol for quantum key distribution. (a) Alice sends Bob a series of quantum bits chosen randomly from among four (non-orthogonal) states: $+Z, -Z, +X, -X$. (b) For each bit Bob randomly guesses whether to measure Z or X. Approximately half of his guesses are correct. (c) Bob publicly announces which measurements he chose but not the results. Alice tells him which ones to throw away because he guessed incorrectly. (d) Bob publicly announces the measurement results from a randomly selected subset of M of the remaining bits (indicated by the extra surrounding boxes). If Alice confirms that all the results are correct, then there is a very low probability, $P_M = (3/4)^M$ that an eavesdropper has intercepted the communication. (e) Alice and Bob know that Bob's remaining bit measurements agree perfectly with what Alice sent and these are used to create the key using the mapping: $+Z \to 1, -Z \to 0, +X \to 1, -X \to 0$.

realized that quantum mechanics provides a solution to this problem. One can simply distribute the one-time pad as the serial numbers of quantum money!

The so-called BB84 protocol developed by Brassard and Bennett illustrated in Fig. 1 works as follows. Alice sends a long string of quantum bits to Bob which have been encoded using randomly selected Z and X encoding (just as for quantum money). Bob does not know what the correct measurements are to make so he guesses randomly. He then announces publicly what measurement axis he chose for each qubit but keeps secret the measurement results. Typically he will have guessed correctly about half the time. Alice then publicly states which measurements Bob performed correctly and Bob discards the others. He now has about half the number of qubits as before but he knows that he made the correct measurements on them. Thus he should have the correct bit values for the key. But, how can he be sure that Eve has not intercepted the message? If she has, she will necessarily corrupt some of the results before passing them along to Bob (which she must do in order to conceal her presence). In essence what she is passing along is a counterfeit bill. How can Bob check that this has or has not occurred? The next step in the protocol is that Bob selects a subset of size M of the (good) qubit measurement results at random and publicly announces them. Alice then publicly announces whether Bob got them right. If he got them right then the probably that Eve was intercepting the message before passing it on is very low, $P_M = (\frac{3}{4})^M$ and it is safe to use the remaining results (which are known to both Bob and Alice) as the encryption key. If the public results do not match, then Eve's presence has been detected and Alice and Bob do not proceed with using the corrupted key. (If they did proceed, neither Eve nor Bob would be able to read the message because the cloned key is necessarily corrupted with high probability.)

4. Entanglement, bell states and superdense coding and grover search and teleportation

So far we have discussed only the states of a single qubit. As we shall see, multi-qubits states will prove to be remarkably rich. A classical register holding N bits, can be in any one of 2^N states. These states can represent, for example, all N-bit binary numbers from 0 to $2^N - 1$. For the quantum case we have studied single qubits which have a Hilbert space spanned by 2 orthogonal states. For N qubits, we have a Hilbert space which is exponentially large, spanned by 2^N orthogonal basis states. We can label these basis states with binary numbers just as for the classical case. However in the quantum case, the Hilbert space includes states which are arbitrary superpositions of all the basis states. While the coefficients for the amplitudes entering such superpositions are continuous complex numbers, let us, for simplicity of counting, restrict our attention to coefficients that are ± 1 so that

$$|\Psi\rangle = \frac{1}{\sqrt{2^N}}\{\pm|000\ldots000\rangle \pm |000\ldots001\rangle \pm |000\ldots010\rangle \pm |000\ldots011\rangle \ldots$$
$$\pm |111\ldots100\rangle \pm |111\ldots101\rangle \pm |111\ldots110\rangle \pm |111\ldots111\rangle\} \quad (24)$$

Because we have 2^N sign choices (ignoring the fact that the overall sign is irrelevant) the total number of states is enormously large: $2^{(2^N)}$. Thus even using only this highly restricted set of states, the storage capacity of quantum memories is truly enormous. While this capacity might be useful at intermediate steps in a quantum computation, let us never forget that when we readout the memory we always obtain only N classical bits of information representing which of the 2^N measurement basis states the memory collapses onto.

Let us focus our attention for the moment on the case of $N = 2$ for which there are four orthogonal basis states. A *separable state* is one which can be written as the product of a state for the first qubit and a state for the second qubit

$$|\Psi\rangle = [\alpha_1|0\rangle + \beta_1|1\rangle] \otimes [\alpha_2|0\rangle + \beta_2|1\rangle]$$

$$= \alpha_1\alpha_2|00\rangle + \alpha_1\beta_2|01\rangle + \beta_1\alpha_2|10\rangle + \beta_1\beta_2|11\rangle. \tag{25}$$

An *entangled state* of two qubits[b] is any state which cannot be written in this form. A convenient basis in which to represent entangled states is the so-called Bell basis

$$|B_0\rangle = \frac{1}{\sqrt{2}}[|\uparrow\downarrow\rangle - |\downarrow\uparrow\rangle] \tag{26}$$

$$|B_1\rangle = \frac{1}{\sqrt{2}}[|\uparrow\downarrow\rangle + |\downarrow\uparrow\rangle] \tag{27}$$

$$|B_2\rangle = \frac{1}{\sqrt{2}}[-|\uparrow\uparrow\rangle + |\downarrow\downarrow\rangle] \tag{28}$$

$$|B_3\rangle = \frac{1}{\sqrt{2}}[-|\uparrow\uparrow\rangle - |\downarrow\downarrow\rangle]. \tag{29}$$

Each of these is a maximally entangled state, but they are mutually orthogonal and span the full Hilbert space. Therefore linear superpositions of them can represent product states. For example,

$$|\uparrow\rangle|\uparrow\rangle = -\frac{1}{\sqrt{2}}[|B_2\rangle + |B_3\rangle]. \tag{30}$$

Entanglement is very mysterious and entangled states have many peculiar and counter-intuitive properties. In an entangled state the individual spin components have zero expectation value and yet the spins are strongly correlated. For example,

$$\langle B_0|\vec{\sigma}_1|B_0\rangle = \vec{0} \tag{31}$$

$$\langle B_0|\vec{\sigma}_2|B_0\rangle = \vec{0} \tag{32}$$

$$\langle B_0|\sigma_1^x\sigma_2^x|B_0\rangle = -1 \tag{33}$$

$$\langle B_0|\sigma_1^y\sigma_2^y|B_0\rangle = -1 \tag{34}$$

$$\langle B_0|\sigma_1^z\sigma_2^z|B_0\rangle = -1. \tag{35}$$

This means that the spins have quantum correlations which are stronger than is possible classically. In particular,

$$\langle B_0|\vec{\sigma}_1 \cdot \vec{\sigma}_2|B_0\rangle = -3 \tag{36}$$

despite the fact that in any single-spin (or product state) $|\langle \vec{\sigma}\rangle| \leq 1$.

To explore this further, let us consider the concept of *entanglement entropy*.[20, 30–32] In classical statistical mechanics, entropy is a measure of

---

[b]The concept of entanglement is more difficult to uniquely define and quantify for $N > 2$ qubits.

randomness or disorder. It is proportional to the logarithm of the number of microstates a thermodynamic system can be in given the observed macrostate. Formally, the entropy of a microstate probability distribution $\{P_k; k = 1, N\}$ is given by

$$S = -k_{\mathrm{B}} \sum_{j=1}^{N} P_j \ln P_j, \tag{37}$$

(with the convention that $0 \ln 0$ is replaced by $\lim_{x \to 0} x \ln x = 0$) where k_{B} is the Boltzmann constant. In the canonical ensemble of statistical mechanics one assumes that all microstates of a given energy are equally likely. If there are N such microstates then $P_j = \frac{1}{N}$ and we obtain $S = k_{\mathrm{B}} \ln N$. The logarithmic form is important because it means that in thermodynamics, the entropy is an extensive quantity. Thus for example if we have a system consisting of two subparts with N_1 and N_2 microstates, the total number of microstates is the product $N_{\mathrm{TOT}} = N_1 N_2$ and the probability of a particular joint microstate is $P_j = \frac{1}{N_1 N_2}$, but because of its logarithmic form, the entropy is the sum of the two entropies: $S = S_1 + S_2$.

In information theory, the Shannon entropy of a message drawn from an ensemble of N possible code words is

$$S = -\sum_{j=1}^{N} P_j \ln_2 P_j, \tag{38}$$

where P_j is the probability that the message will consist of the jth codeword and the logarithm is conventionally measured in base 2 so that the entropy (information content) is measured in bits. For example, if Alice sends Bob a message containing a single physical bit (a 0 or 1), and the a priori probability of the bit being 0 is $P_0 = \frac{1}{2}$ and of being 1 is $P_1 = \frac{1}{2}$, then $S = 1$ bit. On the other hand if Bob knows ahead of time that Alice almost always sends him a 0 (because he knows $P_0 = 0.999$, say), then there will rarely be a surprise in the message and the Shannon information entropy (which is a measure of how much Bob learns on average from the message) is much lower, $S = -0.999 \ln_2 0.999 - 0.001 \ln_2 0.001 \approx 0.01$ bits. If Alice sends Bob two messages (drawn from the same or different ensembles of code words) the logarithmic form guarantees that the total entropy (and hence information content) is additive, $S = S_1 + S_2$. We can connect this to statistical mechanics by noting that if Bob and Alice both look at a *macrostate* of a thermodynamics system and Alice sends Bob the actual *microstate*, the information content of the message is precisely the thermodynamic entropy (modulo the factor of the Boltzmann constant and the change from base e to base 2 in the logarithm).

In quantum mechanics, the concept of the classical probability distribution is replaced by the density matrix

$$\rho = \sum_{j} P_j |\psi_j\rangle\langle\psi_j|. \tag{39}$$

Here we imagine a situation in which state $|\psi_j\rangle$ in the Hilbert space is drawn at random from the ensemble with classical probability P_j. Note that the different $|\psi_j\rangle$ are normalized but need not be orthogonal. The ensemble average of any observable $\mathcal{O}$ is given by

$$\langle \mathcal{O} \rangle = \mathrm{Tr}\{\mathcal{O}\rho\}. \tag{40}$$

Problem 2. *Prove the identity in Eq. (40) by computing the trace in some basis and rearranging the sum on all states so that it becomes an insertion of the identity. Do not assume that the different $|\psi_j\rangle$ are orthogonal.*

The density matrix for a system with a Hilbert space of dimension M is always an $M \times M$ positive semi-definite (i.e. having only zero or positive real eigenvalues) Hermitian matrix with unit trace. In thermal equilibrium, the density matrix is given by

$$\rho = \frac{1}{Z}e^{-\beta H} \tag{41}$$

with $Z \equiv \mathrm{Tr}\, e^{-\beta H}$.

For a single spin-1/2, the density matrix has the following convenient general representation

$$\rho = \frac{1}{2}\sum_{j=0}^{3}\langle Q_j\rangle Q_j, \tag{42}$$

where $Q_0 = \hat{I}, Q_1 = \sigma^x, Q_2 = \sigma^y, Q_3 = \sigma^z$ and $\hat{I}$ is the identity operator. This can be rewritten

$$\rho = \frac{1}{2}[\hat{I} + \langle \vec{\sigma}\rangle \cdot \vec{\sigma}]. \tag{43}$$

Thus the density matrix for a single spin is determined solely by the spin polarization of the ensemble. The ensemble could represent a large collection of spins (as in NMR) or the average results of measuring a single qubit many times (allowing the qubit to come to equilibrium with its environment before each measurement is repeated).

Problem 3. *Prove the identity in Eq. (43) by first showing that $\mathrm{Tr}\, Q_j = 2\delta_{j,0}$ and $\mathrm{Tr}\, Q_j Q_k = 2\delta_{j,k}$.*

The analog of the Shannon entropy for a classical probability distribution is the von Neuman entropy of the density matrix

$$S = -\mathrm{Tr}\{\rho \ln_2 \rho\}. \tag{44}$$

Evaluating the trace in the basis of eigenstates of $\vec{m} \cdot \vec{\sigma}$, where $\vec{m} = \langle \vec{\sigma} \rangle$ is the Bloch vector (spin polarization), it is straigtforward to show that

$$S = -\left\{ \frac{1+m}{2} \ln_2 \left[\frac{1+m}{2} \right] + \frac{1-m}{2} \ln_2 \left[\frac{1-m}{2} \right] \right\} \tag{45}$$

where $m = |\vec{m}|$. From this it follows that $S(m)$ decreases monotonically from $S(0) = 1$ to $S(1) = 0$. Zero von Neuman entropy means that a system is in a single pure state

$$\rho = |\phi\rangle\langle\phi|, \tag{46}$$

from which it follows that

$$\rho^2 = \rho \tag{47}$$

for a pure state. Equivalently, the density matrix of a pure state has one eigenvalue of unity and $M - 1$ zero eigenvalues.

Before turning finally to the concept of entanglement entropy, let us generalize the above results to the case of two qubits. The density matrix is 4×4 and the analog of Eq. (42) is

$$\rho = \frac{1}{4} \sum_{j,k=0}^{3} \langle Q_j^{(1)} Q_k^{(2)} \rangle Q_j^{(1)} Q_k^{(2)}, \tag{48}$$

Problem 4. *Prove the identity in Eq. (48) by first showing that* $\operatorname{Tr} Q_j^{(a)} Q_k^{(b)} = 4\delta_{j,k}\delta_{a,b}.$

We can get a clear picture of the Bell states by examining the two-spin correlators $\langle Q_j^{(1)} Q_k^{(2)} \rangle$ in the so-called 'Pauli bar plot' for the state $|B_0\rangle$ shown in Fig. 2.

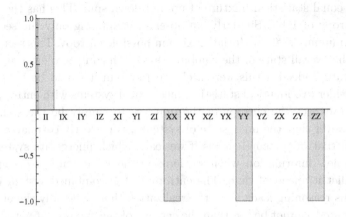

Fig. 2.   'Pauli bar plot' of one and two spin correlators in the Bell state $|B_0\rangle$.

We see that all the single-spin expectation values vanish. Because of the entanglement, each spin is on average totally unpolarized. Yet three of the two-spin correlators, XX, YY, ZZ, are all -1 indicating that the two spins are pointing in exactly opposite directions. This is becasue $|B_0\rangle$ is the rotationally invariant spin-singlet state.

With this background, we are now ready to tackle the concept of entanglement entropy. Consider product state $|\uparrow\uparrow\rangle$ with density matrix

$$\rho = |\uparrow\uparrow\rangle\langle\uparrow\uparrow|. \tag{49}$$

Suppose now that we only had access to the first qubit and not the second. We can compute the expectation value of any observable associated with the first qubit using Eq. (40), but it will be convenient to carry out the trace in two steps, first over the states of the second qubit state and then over those of the first. We can do this by defining the so-called 'reduced density matrix' for the first spin

$$\bar{\rho}_1 \equiv \mathrm{Tr}_2\rho. \tag{50}$$

This is an ordinary 2×2 density matrix for the first spin. In this simple case

$$\bar{\rho} = \mathrm{Tr}_2|\uparrow\uparrow\rangle\langle\uparrow\uparrow| = |\uparrow\rangle_{11}\langle\uparrow| \tag{51}$$

corresponding to a pure state with the first qubit in the up state. In a product state, individual qubits can be viewed as being separately in their own pure states.

Consider now what happens when we compute the reduced density matrix for the first qubit in the entangled Bell state $|B_0\rangle$. This is a pure state with zero entropy. However computation of the reduced density matrix yields

$$\bar{\rho} = \frac{1}{2}\{|\uparrow\rangle\langle\uparrow| + |\downarrow\rangle\langle\downarrow|\} = \frac{1}{2}\begin{pmatrix} 1 & 0 \\ 0 & 1 \end{pmatrix}, \tag{52}$$

which is an equal statistical mixture of up and down spin. This has the maximum possible entropy of 1 bit. Similarly, an observer examining only the second qubit would see an impure state with the maximum possible entropy. This seems to be at odds with the overall state of the combined system having zero entropy and being in a pure state. Indeed if this were a classical system it would be impossible. To see why consider two identical statistical mechanical systems with entropy $S_1 = S_2$. Normally we expect the entropy of the universe to simply be the sum $S_1 + S_2$. Suppose however that the microstate of system 2 is perfectly correlated with (e.g. identical to) that of system 1. Thus if we learn which microstate system 1 is in, we gain no new information when we look at the microstate of system 2 since we can predict it ahead of time. The entropy of the combined system is just S_1, not $2S_1$. This reasoning leads us to the realization that classically the entropy of a composite system cannot be less than the entropy of the component with the largest entropy.

Problem 5. *Derive Eq. (52).*

The quantum result derived above, that each component has finite entropy but the overall system has zero entropy is in clear contradiction to the classical result.[c] Evidently there is a kind of 'negative entropy' associated with the non-classical correlations which cancels out the entropy of the separate components in an entangled system. The 'entanglement entropy' of a bipartite system which is in a pure state is defined by von Neumann entropy of the reduced density matrix of each of the subsystems

$$S_E = S(\bar{\rho}_1) = S(\bar{\rho}_2). \tag{53}$$

For the Bell states, $S_E = 1$ bit, while for any product state, S_E vanishes. Notice that the fact that we found the polarization of the individual qubits in the Bell state vanishes is consistent with Eq. (45) which says that the entropy is a maximum when the polarization vanishes.

Entanglement entropy is one quantitave measure of the entanglement between two halves of a bipartite system and in recent years has (along with the 'entanglement spectrum'[33]) become a valuable theoretical tool in the analysis of correlations in many-body systems[20, 30–32] and has even forged connections between condensed matter physics and quantum gravity.[34] It is also clear from the discussion above that if a qubit in a quantum computer becomes entangled with an external reservoir due to spurious coupling, the coherence needed to carry out computations is lost.

Entanglement also enters the modern picture of quantum measurement.[35–38] In the Stern-Gerlach experiment, one does not measure the spin directly. Rather the magnetic field gradient entangles spin with position, and then one measures the position. From the position one infers the spin projection. Suppose that we start with an initial wave function in which the spin is pointing in some arbitrary direction

$$|\Psi\rangle = [\alpha|\uparrow\rangle + \beta|\downarrow\rangle]\Phi(\vec{r}), \tag{54}$$

where Φ is the initial spatial wave packet of the spin as it approaches the measurement apparatus. After passing through the Stern-Gerlach magnet (which has its field gradient in the z direction), the up spin component's trajectory is deflected upward and the down spin component is deflected downward resulting in

$$|\Psi\rangle = [\alpha|\uparrow\rangle\Phi_\uparrow(\vec{r}) + \beta|\downarrow\rangle\Phi_\downarrow(\vec{r})]. \tag{55}$$

---

[c] A situation that Charles Bennett has summarized by stating that 'A classical house is at least as dirty as its dirtiest room, but a quantum house can be dirty in every room, yet still perfectly clean overall.'

If the deflection is sufficiently strong that the two spatial states cease to have common support ($\Phi_\uparrow(\vec{r})\Phi_\downarrow(\vec{r}) = 0$ for all $\vec{r}$), then a measurement of position unambiguously determines the spin projection.[d]

Notice that prior to the measurement, the reduced density matrix for the spin degree of freedom is simply

$$\bar{\rho} = [\alpha|\uparrow\rangle + \beta|\downarrow\rangle][\alpha^*\langle\uparrow| + \beta^*\langle\downarrow|] = \begin{pmatrix} \alpha \\ \beta \end{pmatrix} \begin{pmatrix} \alpha^* & \beta^* \end{pmatrix} = \begin{pmatrix} |\alpha|^2 & \alpha\beta^* \\ \alpha^*\beta & |\beta|^2 \end{pmatrix}. \tag{56}$$

The non-zero off-diagonal terms ('cohences') imply that there will be transverse components of the spin polarization in addition to the z component

$$\langle\sigma^x\rangle = \text{Tr}\,\sigma^x\bar{\rho} = 2\,\text{Re}[\alpha^*\beta] \tag{57}$$

$$\langle\sigma^y\rangle = \text{Tr}\,\sigma^y\bar{\rho} = 2\,\text{Im}[\alpha^*\beta] \tag{58}$$

$$\langle\sigma^z\rangle = \text{Tr}\,\sigma^z\bar{\rho} = |\alpha|^2 - |\beta|^2. \tag{59}$$

After the spin passes through the magnet, the full density matrix for spin and position is

$$\rho(\vec{r},\vec{r}') = \langle\vec{r}|\rho|\vec{r}'\rangle = \begin{pmatrix} |\alpha|^2\Phi_\uparrow(\vec{r})\Phi_\uparrow^*(\vec{r}') & \alpha\beta^*\Phi_\uparrow(\vec{r})\Phi_\downarrow^*(\vec{r}') \\ \alpha^*\beta\Phi_\uparrow^*(\vec{r})\Phi_\downarrow(\vec{r}') & |\beta|^2\Phi_\downarrow(\vec{r})\Phi_\downarrow^*(\vec{r}') \end{pmatrix}. \tag{60}$$

and the unconditional reduced density matrix for the spin (tracing out the position by integrating over $\vec{r} = \vec{r}'$ and therefore ignoring the measurement result!) yields

$$\bar{\rho}_{\text{uc}} = \begin{pmatrix} |\alpha|^2 & \alpha\beta^*\langle\Phi_\uparrow|\Phi_\downarrow\rangle \\ \alpha^*\beta\langle\Phi_\downarrow|\Phi_\uparrow\rangle & |\beta|^2 \end{pmatrix}. \tag{61}$$

We see that as the measurement pointer states become orthogonal, the off-diagonal elements decay away and the density matrix becomes impure. This phenomenon is referred to as measurement-induced dephasing. One can show that the rate of dephasing is directly equivalent to the rate of information gain in the measurement.[37] (We emphasize again however that we are here referring to the ensemble average over the measurement results which ignores the particular measurement result.)

Instead of averaging over all measurement results, we can ask the following question: What is the spin density matrix conditioned on the measurement result being $\vec{R}$? This is given by

$$\bar{\rho}_c(\vec{R}) = \frac{1}{P(\vec{R})}\langle\vec{R}|\rho|\vec{R}\rangle = \frac{1}{P(\vec{R})}\begin{pmatrix} |\alpha|^2\Phi_\uparrow(\vec{R})\Phi_\uparrow^*(\vec{R}) & \alpha\beta^*\Phi_\uparrow(\vec{R})\Phi_\downarrow^*(\vec{R}) \\ \alpha^*\beta\Phi_\uparrow^*(\vec{R})\Phi_\downarrow(\vec{R}) & |\beta|^2\Phi_\downarrow(\vec{R})\Phi_\downarrow^*(\vec{R}) \end{pmatrix}, \tag{62}$$

---

[d]Technically to achieve perfect distinguishability, one only needs a weaker condition, namely that the two spatial states are orthogonal. However in this more general case you might need to measure some more complex operator to fully distinguish the two measurement 'pointer states.'

where we have introduced the factor

$$P(\vec{R}) \equiv |\alpha|^2 |\Phi_\uparrow(\vec{R})|^2 + |\beta|^2 |\Phi_\downarrow(\vec{R})|^2, \tag{63}$$

to satisfy the probability normalization condition $\mathrm{Tr}\,\bar{\rho}_c(\vec{R}) = 1$. Notice that $P(\vec{R})$ is precisely the probability density that the position measurement (unconditioned on the spin state) will yield the value $\vec{R}$. This normalization factor also makes sense because if we ensemble average over all measurement results (weighted by the probability of occurrence of each measurement result) we correctly recover the unconditional reduced density matrix of Eq. (61)

$$\bar{\rho}_{uc} = \int d^3R\, P(\vec{R}) \bar{\rho}_c(\vec{R}), \tag{64}$$

because the normalization factor cancels out.

Remarkably, the reduced density matrix conditioned on the measurement result corresponds to a pure state. This can be easily seen by noting that we can write the conditional density matrix in terms of a conditional state vector (wave function)

$$\bar{\rho}_c = |\Psi_c\rangle\langle\Psi_c|, \tag{65}$$

where

$$|\Psi_c\rangle \equiv \frac{1}{\sqrt{P(\vec{R})}} [\alpha_c|\uparrow\rangle + \beta_c|\downarrow\rangle], \tag{66}$$

and the new coefficients depend on the particular measurement result, $\vec{R}$

$$\alpha_c = \alpha \Phi_\uparrow(\vec{R}) \tag{67}$$

$$\beta_c = \beta \Phi_\downarrow(\vec{R}). \tag{68}$$

The change in the state induced by the act of measurement is known as the 'back action.' This change in state modifies the spin polarization

$$\langle \sigma^x \rangle = \frac{2}{P(\vec{R})} \mathrm{Re}[\alpha_c^* \beta_c] \tag{69}$$

$$\langle \sigma^y \rangle = \frac{2}{P(\vec{R})} \mathrm{Im}[\alpha_c^* \beta_c] \tag{70}$$

$$\langle \sigma^z \rangle = \frac{1}{P(\vec{R})} [|\alpha_c|^2 - |\beta_c|^2]. \tag{71}$$

For a weak measurement $\Phi_\uparrow$ and $\Phi_\downarrow$ do not differ by much and the backaction is small. Experimental progress in developing nearly ideal quantum limited amplifiers is such that it is now possible to directly observe this backaction and how it varies with the strength of the measurement.[39] For a strong measurement, $\Phi_\uparrow$ and $\Phi_\downarrow$ are fully separated and have no common support. Thus $\Phi_\uparrow(\vec{R})\Phi_\downarrow(\vec{R}) = 0$ and we have complete collapse of the spin polarization to

$$\langle \sigma^x \rangle = 0 \tag{72}$$

$$\langle \sigma^y \rangle = 0 \tag{73}$$

$$\langle \sigma^z \rangle = \pm 1. \tag{74}$$

where the sign of $\langle \sigma^z \rangle$ is determined by $\vec{R}$ and is positive with probability $|\alpha|^2$ and negative with probability $|\beta|^2$.

Our discussion so far has been of the entanglement between a spin and the measurement apparatus pointer. We have seen that a strong measurement of the pointer variable (which in the example above collapses the pointer variable to a particular value of $\vec{R}$) leads to a weak or strong back action on the spin which partially or completely collapses it to a definite state depending on the degree of entanglement (the measurement strength). We turn now to further consideration of the 'spooky' correlations in entangled states.

We have already seen for the Bell state $|B_0\rangle$ that the components of the two spins are perfectly anti-correlated. Suppose now that Alice prepares two qubits in this Bell state and then sends one of the two qubits to Bob who is far away (say one light-year). Alice now chooses to make a measurement of her qubit projection along some arbitrary axis $\hat{n}$. For simplicity let us say that she chooses the $\hat{z}$ axis. Let us further say that her measurement result is -1. Then she immediately knows that if Bob chooses to measure his qubit along the same axis, his measurement result will be the opposite, $+1$. It seems that Alice's measurement has collapsed the state of the spins from $|B_0\rangle$ to $|\downarrow\uparrow\rangle$. This 'spooky action at a distance' in which Alice's measurement seems to instantaneously change Bob's distant qubit was deeply troubling to Einstein.[1]

Upon reflection one can see that this effect cannot be used (in violation of special relativity) for superluminal communication. Even if Alice and Bob and agreed in advance on what axis to use for measurements, Alice has no control over her measurement result and so cannot use it to signal Bob. It is true that Bob can immediately find out what Alice's measurement result was, but this does not give Alice the ability to send a message. In fact, suppose that Bob's clock was off and he accidentally made his measurement slightly before Alice. Would either he or Alice be able to tell? The answer is no, because each would see a random result just as expected. This must be so because in special relativity, the concept of simultaneity is frame-dependent and not universal.

Things get more interesting when Alice and Bob choose different measurement axes. Einstein felt that quantum mechanics must somehow not be a complete description of reality and that there might be 'hidden variables' which if they could be measured would remove the probabilistic nature of the quantum description. However in 1964 John S. Bell proved a remarkable inequality[40] showing

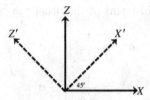

Fig. 3. Measurement axes used by Alice (solid lines) and Bob (dashed lines) in establishing the Clauser-Horn-Shimoni-Holt (CHSH) inequality.

that when Alice and Bob use certain different measurement axes, the correlations between their measurement results are stronger than any possible classical correlation that could be induced by (local) hidden variables. Precision experimental measurements which violate the Bell inequalities are now the strongest proof that quantum mechanics is correct and that local hidden variable theories are excluded. Perhaps the simplest way to understand this result is to consider the CHSH inequality developed by Clauser, Horn, Shimoni and Holt[41] following Bell's ideas. Consider the measurement axes shown in Fig. 3. The experiment consists of many trials of the following protocol. Alice and Bob share a pair of qubits in an entangled state. Alice randomly chooses to measure the first qubit using X or Z while Bob randomly chooses to measure the second qubit using X' or Z'. After many trials (each starting with a fresh copy of the entangled state), Alice and Bob compare notes on their measurement results and compute the following correlation function:

$$S = \langle XX' \rangle + \langle ZZ' \rangle - \langle XZ' \rangle + \langle ZX' \rangle, \tag{75}$$

which can be rewritten

$$S = \langle (X + Z)X' \rangle - \langle (X - Z)Z' \rangle. \tag{76}$$

Alice and Bob note that their measurement results are random variables which are always equal to either $+1$ or -1. In a particular trial Alice chooses randomly to measure either X or Z. Surely however, the variable not measured still has a value of either $+1$ or -1. If this is true, then either $X = Z$ or $X = -Z$. Thus either $X + Z$ vanishes or $X - Z$ vanishes in any given realization of the random variables. The combination that does not vanish is either $+2$ or -2. Hence it follows immediately that S is bounded by the CHSH inequality

$$-2 \leq S \leq +2. \tag{77}$$

It turns out however that the quantum correlations between the two spins in the entangled pair violate this classical bound. They are stronger than can ever be possible in any classical local hidden variable theory. Because $\vec{\sigma}$ is a vector we can resolve its form in one basis in terms of the other via

$$\sigma^{x'} = \frac{1}{\sqrt{2}}[\sigma^z + \sigma^x] \tag{78}$$

$$\sigma^{z'} = \frac{1}{\sqrt{2}}[\sigma^z - \sigma^x]. \tag{79}$$

Thus we can express S in terms of the 'Pauli bar' correlations

$$S = \frac{1}{\sqrt{2}}[\langle XX + XZ + ZX + ZZ \rangle - \langle XZ - XX - ZZ + ZX \rangle]. \tag{80}$$

For Bell state $|B_0\rangle$, these correlations are shown in Fig. 2 and yield

$$S = -2\sqrt{2}, \tag{81}$$

in clear violation of the CHSH inequality. Strong violations of the CHSH inequality are routine in modern experiments.[e] This teaches us that in our quantum world, observables do not have values if you do not measure them. There are no hidden variables which determine the random values. The unobserved spin components simply do not have values. Recall that σ^x and σ^z are incompatible observables and when we choose to measure one, the other remains not merely unknown, but unknowable.

It is ironic that Einstein's arguments that quantum mechanics must be incomplete because of the spooky properties of entanglement have led to the strongest experimental tests verifying the quantum theory and falsifying all local hidden variable theories. We are forced to give up the idea that physical observables have values before they are observed.

4.1.   *Quantum dense coding*

We saw above that quantum correlations are strong enough to violate certain classical bounds, but the spooky action at a distance seemed unable to help us send signals. Actually it turns out that by using a special 'quantum dense coding' protocol, we can use entanglement to help us transmit information in a way that is impossible classically.[17] As background let us recall that for a single qubit there are only two orthogonal states which are connected by a rotation of the spin through an angle of π. For rotation by π around the y axis the unitary rotation operator is

$$R_\pi^y = e^{-i\frac{\pi}{2}\sigma^y} = -i\sigma^y, \tag{82}$$

---

[e] Although strictly speaking, there are loopholes associated with imperfections in the detectors and the fact that Alice and Bob typically do not have a space-like separation.

and we map between the two orthogonal states

$$R_\pi^y|\uparrow\rangle = +|\downarrow\rangle \tag{83}$$

$$R_\pi^y|\downarrow\rangle = -|\uparrow\rangle \tag{84}$$

while any other rotation angle simply produces linear general combinations of the two basis states. For example, for rotation by $\pi/2$ we have

$$R_{\frac{\pi}{2}}^y = e^{-i\frac{\pi}{4}\sigma^y} = \frac{1}{\sqrt{2}}[1 - i\sigma^y], \tag{85}$$

and we have

$$R_{\frac{\pi}{2}}^y|\uparrow\rangle = \frac{1}{\sqrt{2}}[|\uparrow\rangle + |\downarrow\rangle] = |\rightarrow\rangle \tag{86}$$

$$R_{\frac{\pi}{2}}^y|\downarrow\rangle = -\frac{1}{\sqrt{2}}[|\uparrow\rangle - |\downarrow\rangle] = -|\leftarrow\rangle. \tag{87}$$

These are linearly independent of the starting state but never orthogonal to it except for the special case of rotation by π.

The situation is very different for two-qubit entangled states. We take as our basis the four orthogonal Bell states. Suppose that Alice prepares the Bell state $|B_0\rangle$ and sends one of the qubits to Bob who is in a distant location. Using a remarkable protocol called quantum dense coding,[17] Alice can now send Bob two classical bits of information by sending him the remaining qubit. The protocol relies on the amazing fact that Alice can transform the initial Bell state into any of the others by purely local operations on her remaining qubit without communicating with Bob. The four possible unitary operations Alice should perform are I, X, iY, Z which yield

$$I|B_0\rangle = |B_0\rangle \tag{88}$$

$$Z|B_0\rangle = |B_1\rangle \tag{89}$$

$$X|B_0\rangle = |B_2\rangle \tag{90}$$

$$iY|B_0\rangle = |B_3\rangle. \tag{91}$$

It seems somehow miraculous that without touching Bob's qubit, Alice can reach all four orthogonal states by merely rotating her own qubit. This means that there are four possible two-bit messages Alice can send by associating each with one of the four operations I, X, iY, Z according to the following

message	operation
00	X
01	I
10	iY
11	Z

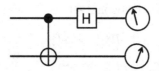

Fig. 4.   Bell measurement circuit with a CNOT and Hadamard gate. This circuit permits measurement of which Bell state a pair of qubits is in by mapping the states to the basis of eigenstates of σ_1^z and σ_2^z.

After encoding her message by carrying out the appropriate operation on her qubit, Alice physically transmits her qubit to Bob. Bob then makes a joint measurement (details provided further below) on the two qubits which tells him which of the four Bell states he has and thus recovers two classical bits of information even though Alice sent him only one quantum bit after deciding what the message was. The pre-positioning of the entangled pair has given them a resource which doubles the number of classical bits that can be transmitted with one (subsequent) quantum bit. Of course in total Alice transmitted two qubits to Bob. The key point is that the first one was sent in advance of deciding what the message was. How weird is that!?

This remarkable protocol sheds considerable light on the concerns that Einstein raised in the EPR paradox.[1] It shows that the special correlations in Bell states can be used to communicate information in a novel and efficient way by 'prepositioning' entangled pairs shared by Alice and Bob. However causality and the laws of special relativity are not violated because Alice still has to physically transmit her qubit(s) to Bob in order to send the information.

The above protocol requires Bob to determine which of the four Bell states he has. The quantum circuit shown in Fig. 4 uniquely maps each of the Bell states onto one of the four computational basis states (eigenstates of Z_1 and Z_2). The first symbol indicates the CNOT gate which flips the target qubit (in this case qubit 2) if and only if the control qubit (qubit 1) is in the excited state. This can be explicitly written in matrix form as

$$\text{CNOT}_{12} = \frac{\sigma_1^0 + \sigma_1^z}{2}\sigma_2^x + \frac{\sigma_1^0 - \sigma_1^z}{2}\sigma_2^0, \tag{92}$$

where σ_j^0 is the identity operator for qubit j. The second gate in the circuit (denoted H) is the Hadamard gate acting on the first qubit

$$\text{H} = \frac{1}{\sqrt{2}}[\sigma^z + \sigma^x]. \tag{93}$$

The Hadamard gate obeys $\text{H}^2 = 1$ and interchanges the X and Z components of a spin:

$$H|\uparrow\rangle = |\rightarrow\rangle \tag{94}$$

$$H|\downarrow\rangle = |\leftarrow\rangle \tag{95}$$

$$H|\rightarrow\rangle = |\uparrow\rangle \tag{96}$$

$$H|\leftarrow\rangle = |\downarrow\rangle \tag{97}$$

Problem 6. *Prove the following identities for the circuit shown in Fig. 4*

$$H_1 \, \text{CNOT}_{12}|B_0\rangle = +|ge\rangle \tag{98}$$

$$H_1 \, \text{CNOT}_{12}|B_1\rangle = +|ee\rangle \tag{99}$$

$$H_1 \, \text{CNOT}_{12}|B_2\rangle = -|gg\rangle \tag{100}$$

$$H_1 \, \text{CNOT}_{12}|B_3\rangle = -|eg\rangle \tag{101}$$

Once Bob has mapped the Bell states onto unique computational basis states, he makes a separate measure of the state of each qubit thereby gaining two bits of classical information and effectively reading the message Alice has sent. Note that the overall sign in front of the basis states produced by the circuit is irrelevant and not observable in the measurement process. Also note that to create Bell states in the first place, Alice can simply run the circuit in Fig. 4 backwards.

4.2. *No-cloning theorem revisited*

Now that we have understood the EPR paradox and communication via quantum dense coding, we can gain some new insight into the no-cloning theorem. It turns out that if cloning of an unknown quantum state were possible, then we could use an entangled pair to communicate information at superluminal speeds in violation of special relativity. Consider the following protocol. Alice and Bob share an entangled Bell pair in state $|B_0\rangle$. Alice chooses to measure her qubit in either the Z basis or the X basis. The choice she makes defines one classical bit of information. The result of the measurement collapses the entangled pair into a simple product state. If Alice chooses the Z basis for her measurement, then Bob's qubit will be either $|\uparrow\rangle$ or $|\downarrow\rangle$. If Alice chooses to measure in the X basis, then Bob's qubit will be either $|\rightarrow\rangle$ or $|\leftarrow\rangle$. Bob can distinguish these cases by cloning his qubit to make many copies. If he measures a large number of copies in the Z basis and always gets the same answer, he knows that his qubit is almost certainly in a Z eigenstate. If even a single measurement result is different from the first, he knows his qubit cannot be in a Z eigenstate and so must be in an X eigenstate. (Of course he could also measure a bunch of copies in the X basis and gain the same information.) This superluminal communication would violate special relativity and hence cloning must be impossible.

In fact, cloning would make it possible for Alice to transmit an unlimited number of classical bits using only a single Bell pair. Alice could choose an arbitrary measurement axis $\hat{n}$. The specification of $\hat{n}$ requires two real numbers (the polar and azimuthal angles). It would take a very large number of bits to represent these real numbers to some high accuracy. Now if Bob can make an enormous number of copies of his qubit, he can divide the copies in three groups and measure the vector spin polarization $\langle \vec{\sigma} \rangle$ to arbitrary accuracy. From this he knows the polarizaton axis $\hat{n} = \pm \langle \vec{\sigma} \rangle$ Alice chose (up to an unknown sign since he does not know the sign of Alice's measurement result for $\hat{n} \cdot \sigma$). Hence Bob has learned a large number of classical bits of information. The accuracy (and hence the number of bits) is limited only by the statistical uncertainties resulting from the fact that his individual measurement results can be random, but these can be reduced to an arbitrarily low level with a sufficiently large number of copies of the state.

4.3.   *Quantum teleportation*

We noted earlier than even though it is impossible to clone an unknown quantum state, it is possible for Alice to teleport it to Bob as long as her copy of the original is destroyed in the process.[42] Just as for quantum dense coding, teleportation protocols also take advantage of the power of 'pre-positioned' entangled pairs. However unlike quantum dense coding where Alice ultimately sends her qubit to Bob, teleportation only requires Alice to send two classical bits to Bob. A simple protocol is as follows: Alice creates a $|B_0\rangle$ Bell state and sends one of the qubits to Bob. Alice has in her possession an additional qubit in an unknown state

$$|\psi\rangle = \alpha|g\rangle + \beta|e\rangle \tag{102}$$

which she wishes to transmit to Bob. Alice applies the Bell state determination protocol illustrated in Fig. 4 to determine the joint state of the unknown qubit and her half of the Bell pair she shares with Bob. She then transmits two classical bits giving her measurement result to Bob. To see how Bob is able to reconstruct the initial state, note that we can rewrite the initial state of the three qubits in the basis of Bell states for the two qubits that Alice will be measuring as follows

$$\begin{aligned}
|\Psi\rangle &= [\alpha|e\rangle + \beta|g\rangle]|B_0\rangle \\
&= \frac{1}{2}|B_0\rangle[-\alpha|e\rangle - \beta|g\rangle] + \frac{1}{2}|B_1\rangle[-\alpha|e\rangle + \beta|g\rangle] \\
&\quad + \frac{1}{2}|B_2\rangle[-\beta|e\rangle - \alpha|g\rangle] + \frac{1}{2}|B_3\rangle[+\beta|e\rangle - \alpha|g\rangle]
\end{aligned} \tag{103}$$

From this representation we see that when Alice tells Bob which Bell state she found, Bob can find a local unitary operation to perform on his qubit to recover the original unknown state (up to an irrelevant overall sign). The appropriate operations are

Alice's Bell state	Bob's operation
$|B_0\rangle$	X
$|B_1\rangle$	$-iY$
$|B_2\rangle$	I
$|B_3\rangle$	Z

5. Quantum error correction

Now that we understand entanglement, we are in a position to tackle quantum error correction.

To overcome the deleterious effects of electrical noise, cosmic rays and other hazards, modern digital computers rely heavily on error correcting codes to store and correctly retrieve vast quantities of data. Classical error correction works by introducing extra bits which provide redundant encoding of the information. Error correction proceeds by measuring the bits and comparing them to the redundant information in the auxiliary bits. Another benefit of the representation of information as discrete bits (with 0 and 1 corresponding to a voltage for example) is that one can ignore small noise voltages. That is, $V = 0.99$ volts can be safely assumed to represent 1 and not 0.

All classical (and quantum) error correction codes are based on the assumption that the hardware is good enough that errors are rare. The goal is to make them even rarer. For classical bits there is only one kind of error, namely the bit flip which maps 0 to 1 or vice versa. We will assume for simplicity that there is probability $p \ll 1$ that a bit flip error occurs, and that error occurrences are uncorrelated among different bits. One of the simplest classical error correction codes to understand involves repetition and majority rule. Suppose we have a classical bit carrying the information we wish to protect from error and we have available two ancilla bits (also subject to errors). The procedure consists copying the state of the first bit into the two ancilla bits. Thus a 'logical' 1 is represented by three 'physical' bits in state 111, and a 'logical' 0 is represented by three 'physical' bits in state 000. Any other physical state is an error state outside of the logical state space. Suppose now that one of the three physical bits suffers an error. By examining the state of each bit it is a simple matter to identify the bit which has flipped and is not in agreement with the 'majority.' We then simply flip the minority bit so that it again agrees with the majority. This procedure succeeds if the number of errors is zero or one, but it fails if there is more than one error. Of course since we have replaced one imperfect bit with three imperfect bits, this means that the probability of an error occurring has increased considerably. For three bits the probability P_n of n errors is given by

$$P_0 = (1 - p)^3 \tag{104}$$

$$P_1 = 3p(1 - p)^2 \tag{105}$$

$$P_2 = 3p^2(1 - p) \tag{106}$$

$$P_3 = p^3. \tag{107}$$

Because our error correction code only fails for two or more physical bit errors the error probability for our logical qubit is

$$p_{\text{logical}} = P_2 + P_3 = 3p^2 - 2p^3. \tag{108}$$

If $p < 1/2$, then the error correction scheme reduces the error rate (instead of making it worse). If for example $p = 10^{-6}$, then $p_{logical} \sim 3 \times 10^{-12}$. Thus the lower the raw error rate, the greater the improvement. Note however that even at this low error rate, a petabyte (8×10^{15} bit) storage system would have on average 24,000 errors. Futhermore, one would have to buy three petabytes of storage since 2/3 of the disk would be taken up with ancilla bits!

We are now ready to enter the remarkable and magic world of quantum error correction. Without quantum error correction, quantum computation would be impossible and there is a sense in which the fact that error correction is possible is even more amazing and counter intuitive than the fact of quantum computation itself. Naively, it would seem that quantum error correction is completely impossible. The no-cloning theorem does not allow us to copy an unknown state of a qubit onto ancilla qubits. Furthermore, in order to determine if an error has occurred, we would have to make a measurement, and the back action (state collapse) from that measurement would itself produce random unrecoverable errors.

Part of the power of a quantum computer derives from its analog character—quantum states are described by continuous real (or complex) variables. This raises the specter that noise and small errors will destroy the extra power of the computer just as it does for classical analog computers. Remarkably, this is *not* the case! This is because the quantum computer also has characteristics that are digital. Recall that any measurement of the state of qubit always yields a binary result. Amazingly this makes it possible to perform quantum error correction and keep the calculation running even on an imperfect and noisy computer. In many ways, this discovery (by Shor[43, 44] and by Steane[45, 46]) is even more profound and unexpected than the discovery of efficient quantum algorithms that work on ideal computers. For an introduction to the key concepts behind quantum error correction and fault-tolerant quantum computation, see the reviews by Raussendorf[47] and by Gottesman.[48, 49]

It would seem obvious that quantum error correction is impossible because the act of measurement to check if there is an error would collapse the state, destroying any possible quantum superposition information. Remarkably however, one can encode the information in such a way that the presence of an error can be detected by measurement, and if the code is sufficiently sophisticated, the error can be corrected, just as in classical computation. Classically, the only error that exists is the bit flip. Quantum mechanically there are other types of errors (e.g. phase flip, energy decay, erasure channels, etc.). However codes have been developed[5, 43–46, 48, 49] (using a minimum of 5 qubits) which will correct all possible quantum errors. By concatenating these codes to higher levels of redundancy, even small imperfections in the error correction process itself can be corrected. Thus quantum superpositions can in principle be made to last essentially forever even in an imperfect noisy system. It is this remarkable insight that makes quantum computation possible. Many other ideas have been developed to reduce error rates. Alexei Kitaev at Caltech in particular has developed novel theoretical ideas for topologically protected qubits which are impervious to local perturbations.[50, 51] The goal of realizing such topologically protected logical qubits is being actively pursued for trapped ions[52, 53] and super-conducting qubits.[54, 55] Certain strongly correlated condensed matter systems may offer the possibility of realizing non-abelian quasiparticle defects which could also be used for topologically protected qubits. For a review of this vast field see Das Sarma *et al.*[56]

As an entré to this rich field, we will consider a simplified example of one qubit in some state $\alpha|0\rangle + \beta|1\rangle$ plus two ancillary qubits in state $|0\rangle$ which we would like to use to protect the quantum information in the first qubit. As already noted, the simplest classical error correction code simply replicates the first bit twice and then uses majority voting to correct for (single) bit flip errors. This procedure fails in the quantum case because the no-cloning theorem prevents replication of an unknown qubit state. Thus there does not exist a unitary transformation which takes

$$[\alpha|0\rangle + \beta|1\rangle] \otimes |00\rangle \longrightarrow [\alpha|0\rangle + \beta|1\rangle]^{\otimes 3}. \tag{109}$$

As was mentioned earlier, this is clear from the fact that the above transformation is not linear in the amplitudes α and β and quantum mechanics is linear. One can however perform the repetition code transformation:

$$[\alpha|0\rangle + \beta|1\rangle] \otimes |00\rangle \longrightarrow [\alpha|000\rangle + \beta|111\rangle], \tag{110}$$

since this is in fact a unitary transformation. Just as in the classical case, these three physical qubits form a single logical qubit. The two logical basis states are

$$|0\rangle_{\log} = |000\rangle$$
$$|1\rangle_{\log} = |111\rangle. \tag{111}$$

The analog of the single-qubit Pauli operators for this logical qubit are readily seen to be

$$X_{\log} = X_1 X_2 X_3$$
$$Y_{\log} = i X_{\log} Z_{\log} \tag{112}$$
$$Z_{\log} = Z_1 Z_2 Z_3.$$

We see that this encoding complicates things considerably because now to do even a simple single logical qubit rotation we have to perform some rather non-trivial three-qubit joint operations. It is not always easy to achieve an effective Hamiltonian that can produce such joint operations, but this is an essential price we must pay in order to carry out quantum error correction.

It turns out that this simple code cannot correct all possible quantum errors, but only a single type. For specificity, let us take the error operating on our system to be a single bit flip, either X_1, X_2, or X_3. These three together with the identity operator, I, constitute the set of operators that produce the four possible error states of the system we will be able to correctly deal with. Following the formalism developed by Daniel Gottesman,[47–49] let us define two *stabilizer* operators

$$S_1 = Z_1 Z_2 \tag{113}$$

$$S_2 = Z_2 Z_3. \tag{114}$$

These have the nice property that they commute both with each other and with all three of the logical qubit operators listed in Eq. (112). This means that they can both be measured simultaneously and that the act of measurement does *not* destroy the quantum information stored in any superposition of the two logical qubit states. Furthermore they each commute or anticommute with the four error operators in such a way that we can uniquely identify what error (if any) has occurred. Each of the four possible error states (including no error) is an eigenstate of both stabilizers with the eigenvalues listed in the table below

error	S_1	S_2
I	+1	+1
X_1	−1	+1
X_2	−1	−1
X_3	+1	−1

Thus measurement of the two stabilizers yields two bits of classical information (called the 'error syndrome') which uniquely identify which of the four possible error states the system is in and allows the experimenter to correct the situation by applying the appropriate error operator, I, X_1, X_2, X_3 to the system to cancel the original error.

We now have our first taste of fantastic power of quantum error correction. We have however glossed over some important details by assuming that either an error has occurred or it hasn't (that is, we have been assuming we are in a definite error state). At the next level of sophistication we have to recognize that we need to be able to handle the possibility of a quantum superposition of an error and no error. After all, in a system described by smoothly evolving superposition amplitudes, errors can develop continuously. Suppose for example that the correct state of the three physical qubits is

$$|\Psi_0\rangle = \alpha|000\rangle + \beta|111\rangle, \tag{115}$$

and that there is some perturbation to the Hamiltonian such that after some time there is a small amplitude ϵ that error X_2 has occurred. Then the state of the system is

$$|\Psi\rangle = [\sqrt{1 - |\epsilon|^2} I + \epsilon X_2]|\Psi_0\rangle. \tag{116}$$

(The reader may find it instructive to verify that the normalization is correct.)

What happens if we apply our error correction scheme to this state? The measurement of each stabilizer will always yield a binary result, thus illustrating the dual digital/analog nature of quantum information processing. With probability

$P_0 = 1 - |\epsilon|^2$, the measurement result will be $S_1 = S_2 = +1$. In this case the state collapses back to the original ideal one and the error is removed! Indeed, the experimenter has no idea whether ϵ had ever even developed a non-zero value. All she knows is that if there was an error, it is now gone. This is the essence of the Zeno effect in quantum mechanics that repeated observation can stop dynamical evolution. Rarely however (with probability $P_1 = |\epsilon|^2$) the measurement result will be $S_1 = S_2 = -1$ heralding the presence of an X_2 error. The correction protocol then proceeds as originally described above. Thus error correction still works for superpositions of no error and one error. A simple extension of this argument shows that it works for an arbitrary superposition of all four error states.

There remains however one more level of subtlety we have been ignoring. The above discussion assumed a classical noise source modulating the Hamiltonian parameters. However in reality, a typical source of error is that one of the physical qubits becomes entangled with its environment. We generally have no access to the bath degrees of freedom and so for all intents and purposes, we can trace out the bath and work with the reduced density matrix of the logical qubit. Clearly this is generically not a pure state. How can we possibly go from an impure state (containing the entropy of entanglement with the bath) to the desired pure (zero entropy) state? Ordinary unitary operations on the logical qubit preserve the entropy so clearly will not work. Fortunately our error correction protocol involves applying one of four possible unitary operations *conditioned on the outcome of the measurement of the stabilizers.* The wave function collapse associated with the measurement gives us just the non-unitarity we need and the error correction protocol works even in this case. Effectively we have a Maxwell demon which uses Shannon information entropy (from the measurement results) to remove an equivalent amount of von Neumann entropy from the logical qubit!

To see that the protocol still works, we generalize Eq. (116) to include the bath

$$|\Psi\rangle = [\sqrt{1 - |\epsilon|^2}|\Psi_0, \text{Bath}_0\rangle + \epsilon X_2]|\Psi_0, \text{Bath}_2\rangle. \tag{117}$$

For example, the error could be caused by the second qubit having a coupling to a bath operator $\mathcal{O}_2$ of the form

$$V_2 = g\, X_2 \mathcal{O}_2, \tag{118}$$

acting for a short time $\epsilon\hbar/g$ so that

$$|\text{Bath}_2\rangle \approx \mathcal{O}_2|\text{Bath}_0\rangle. \tag{119}$$

Notice that once the stabilizers have been measured, then either the experimenter obtained the result $S_1 = S_2 = +1$ and the state of the system plus bath collapses to

$$|\Psi\rangle = |\Psi_0, \text{Bath}_0\rangle, \tag{120}$$

or the experimenter obtained the result $S_1 = S_2 = -1$ and the state collapses to

$$|\Psi\rangle = X_2|\Psi_0, \text{Bath}_2\rangle. \qquad (121)$$

Both results yield a product state in which the logical qubit is unentangled with the bath. Hence the algorithm can simply proceed as before and will work.

Problem 7. *Compute the Shannon entropy gained by the measurement and show that it is precisely the entanglement entropy which has been removed from the system by the act of measurement.*

Finally, there is one more twist in this plot. We have so far described a measurement-based protocol for removing the entropy associated with errors. There exists another route to the same goal in which purely unitary multi-qubit operations are used to move the entropy from the logical qubit to some ancillae, and then the ancillae are reset to the ground state to remove the entropy. The reset operation could consist, for example, of putting the ancillae in contact with a cold bath and allowing the qubits to spontaneously and irreversibly decay into the bath. Because the ancillae are in a mixed state with some probability to be in the excited state and some to be in the ground state, the bath ends up in a mixed state containing (or not containing) photons resulting from the decay. Thus the entropy ends up in the bath. It is important for this process to work that the bath be cold so that the qubits always relax to the ground state and are never driven to the excited state. We could if we wished, measure the state of the bath and determine which error (if any) occurred, but in this protocol, no actions conditioned on the outcome of such a measurement are required. The first error correction circuit for superconducting qubits[57] used a minimalist version of such a non-measurement based protocol.

Quantum error correction was first realized experimentally some time ago in NMR,[58] then in trapped ions,[59] quantum optics[60,61] and more recently in superconducting qubits.[57] Despite considerable progress,[60–64] we have not yet achieved the goal of a logical qubit which is effectively immortal, but we are moving closer.

6. Introduction to circuit QED

In the last decade, there has been truly amazing experimental progress in realizing superconducting electrical circuits which operate at microwave frequencies and which behave quantum mechanically. These circuits exhibit quantized energy levels and can be placed in quantum superpositions of these levels. An important component of this progress has been the realization that it is extremely useful to apply the ideas of non-linear quantum optics to such circuits. Quantum electrodynamics is the study of the effect of quantum fluctuations of the electromagnetic field on electrons and atoms. Quantum fluctuation effects include spontaneous decay (by photon emission) of excited atomic states, Lamb shifts of energy levels, etc. The

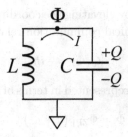

Fig. 5. Simple LC oscillator whose coordinate is the flux Φ and whose momentum is the charge $Q = C\dot{\Phi} = CV$.

spectrum and strength of these quantum fluctuations can be modified by engineering the available electromagnetic modes by placing the system under study in a cavity. By analogy with this 'cavity QED', we will here study 'circuit QED'. In addition to being a novel test bed for non-linear quantum optics in a completely new regime, circuit QED represents an important architecture for implementing the theoretical ideas about quantum information processing discussed above.

6.1.   *Electromagnetic oscillators*

The simplest electrical circuit to understand is the LC oscillator illustrated in Fig. 5. Let us define the node flux variable by the time integral of the voltage[65]

$$\Phi(t) \equiv \int^t d\tau\, V(\tau). \tag{122}$$

The relation $V(t) = \dot{\Phi}(t)$ shows us (via the Faraday induction relation) that the node flux is indeed the magnetic flux threading the inductor. The Lagrangian is therefore

$$L = \frac{1}{2}C\dot{\Phi}^2 - \frac{1}{2L}\Phi^2, \tag{123}$$

In this representation, the kinetic energy term in the Lagrangian represents the electrostatic energy stored in the capacitor and the potential energy term represents the kinetic (magnetic) energy stored in the electron motion through the inductor.

The momentum canonically conjugate to the flux coordinate is given by

$$Q = \frac{\delta L}{\delta\dot{\Phi}} = C\dot{\Phi} = CV, \tag{124}$$

and is the charge on the capacitor. The Hamiltonian is given by

$$H(Q,\Phi) = Q\dot{\Phi} - L(\Phi,\dot{\Phi}) = \frac{Q^2}{2C} + \frac{\Phi^2}{2L}. \tag{125}$$

This is the Hamiltonian of a simple harmonic oscillator with 'mass' C and 'spring constant' $1/L$. The resonance frequency is therefore $\Omega = \frac{1}{\sqrt{LC}}$ as expected.

We quantize this oscillator by elevating the coordinate and conjugate momentum to operators obeying the canonical commutation relation

$$[\hat{Q}, \hat{\Phi}] = -i\hbar. \tag{126}$$

These operators can in turn be represented in terms of raising and lowering operators

$$\hat{\Phi} = \Phi_{\text{ZPF}}(a + a^\dagger) \tag{127}$$

$$\hat{Q} = -iQ_{\text{ZPF}}(a - a^\dagger) \tag{128}$$

where the zero-point fluctuations of the flux and charge are given by

$$\Phi_{\text{ZPF}} \equiv \sqrt{\frac{\hbar Z}{2}} \tag{129}$$

$$Q_{\text{ZPF}} \equiv \sqrt{\frac{\hbar}{2Z}} \tag{130}$$

and $Z \equiv \sqrt{\frac{L}{C}}$ is called the characteristic impedance of the resonator. Notice that if the characteristic impedance on the order of the quantum of resistance h/e^2, then Q_{ZPF} is on the order of the electron charge and Φ_{ZPF} is on the order of the flux quantum.

To get a better feeling for the meaning of the characteristic impedance consider the input admittance of the parallel LC resonator

$$Y(\omega) = j\omega C + \frac{1}{j\omega L} \tag{131}$$

where we are using the electrical engineer's convention for the square root of minus one: j $= -$i. This can be rewritten

$$Y(\omega) = \frac{1}{jZ} \left(\frac{\Omega}{\omega} \right) \left[1 - \left(\frac{\omega}{\Omega} \right)^2 \right]. \tag{132}$$

We see that the collective mode frequency is determined by the zero of the admittance as a function of frequency and the slope with which the admittance crosses zero determines the characteristic impedance of the resonance.

With these two quantities in hand, we have everything we need to quantize this normal mode. The Hamiltonian (ignoring the zero-point energy) is given by the usual expression

$$H = \hbar\Omega\, a^\dagger a, \tag{133}$$

and the physical observables $\hat{\Phi}$ and $\hat{Q}$ are given by Eq. (128). In the case of a general 'black box' containing an arbitrary linear (purely reactive) circuit with a single input port, the admittance at that port will have a series of zeros corresponding to each internal collective mode as shown in Fig. (6).

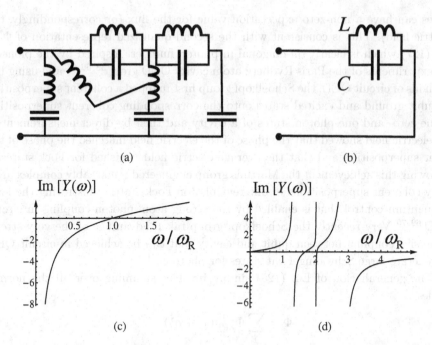

Fig. 6. a-b) Arbitrary one-port devices containing purely reactive elements. c) Frequency dependence of the admittance for device (b) with a single zero of the admittance and hence a single collective mode. d) Admittance of a device with multiple zeros of the admittance each separated by a pole. The zeros correspond to a separate normal mode of the system under open circuit conditions at the input port.

We should understand the raising and lowering operators as photon creation and destruction operators. We are used to thinking of photons in free space, but here we are in a lumped element circuit. The electric field of the photon mode lives between the capacitor plates and the magnetic field lives in a physically separate space inside the inductor. We are also used to seeing photons introduced only in second-quantized notation. However it can be instructive to use an ordinary first quantization approach to this simple harmonic oscillator. Since we have chosen the flux Φ as the coordinate, we can easily write the ground state wave function for the oscillator as a function of this coordinate

$$\Psi_0(\Phi) = \frac{1}{\sqrt{F}} e^{-\frac{\Phi^2}{4\Phi_{ZPF}^2}} \tag{134}$$

$$\Psi_1(\Phi) = \frac{1}{\sqrt{F}} \frac{\Phi}{\Phi_{ZPF}} e^{-\frac{\Phi^2}{4\Phi_{ZPF}^2}} \tag{135}$$

where $F \equiv \sqrt{2\pi\Phi_{ZPF}^2}$ is the normalization factor.

From this first-quantized representation we see that photon Fock states wave functions have definite reflection parity in Φ and hence Fock states (eigenstates of photon number) obey $\langle n|\hat{\Phi}|n\rangle = 0$. Only a coherent superposition of different Fock states can have a non-zero expectation value for the flux (or correspondingly, the electric field). This is consistent with the second-quantized representation of $\hat{\Phi}$ in Eq. (127) which is clearly off-diagonal in photon number. Inspired by the pioneering experiments of the Paris Rydberg atom cavity QED group[38, 66, 67] and using the methods of circuit QED, the Schoelkopf group first mapped a coherent superposition of qubit ground and excited states onto the corresponding coherent superposition of the zero- and one-photon states of a cavity and then by direct measurement of the electric field showed that the phase of the electric field matched the phase of the qubit superposition and that the average electric field vanished for Fock states.[68] Following this achievement the Martinis group engineered remarkably complex arbitrary coherent superpositions of different photon Fock states, illustrating the level of quantum control that is enabled by the strong atom-photon coupling in circuit QED.[69, 70] Very recently the Schoelkopf group taken advantage of the very strong dispersive coupling between qubit and cavity that can be achieved in circuit QED to generate large Schrödinger cat states for photons.[71]

The generaliation of Eq. (128) simply involves summing over all the normal modes

$$\hat{\Phi} = \sum_j \Phi_{\text{ZPF}}^{(j)}(a_j + a_j^\dagger) \tag{136}$$

$$\hat{Q} = -i \sum_j Q_{\text{ZPF}}^{(j)}(a_j - a_j^\dagger) \tag{137}$$

where

$$\Phi_{\text{ZPF}}^{(j)} \equiv \sqrt{\frac{\hbar Z_j}{2}} \tag{138}$$

$$Q_{\text{ZPF}}^{(j)} \equiv \sqrt{\frac{\hbar}{2Z_j}} \tag{139}$$

involves the impedance Z_j of the jth mode as viewed from the input port.

6.2.   *Superconducting qubits*

In order to go beyond the simple LC harmonic oscillator to create a qubit, we need a non-linear element to produce anharmonicity in the spectrum. The non-linear circuit element of choice in superconducting systems is the Josephson tunnel junction. The first evidence that Josephson tunneling causes the Cooper pair box to exhibit coherent superpositions of different charge states was obtained by Bouchiat et al.[72] This was followed in 1999 by the pioneering experiment of the NEC group[73] demonstrating time-domain control of the quantum state of the CPB using very rapid control pulses to modulate the offset charge.

The remarkable recent progress in creating superconducting quantum bits and manipulating their states has been summarized in several reviews.[74–82] Nearly 30 years ago Leggett discussed the fundamental issues concerning the collective degrees of freedom in superconducting electrical circuits and the fact that they themselves can behave quantum mechanically.[83] As noted earlier, the essential collective variable in a Josephson junction[84] is the phase difference of the superconducting order parameter across the junction. The first experimental observation of the quantization of the energy levels of the phase 'particle' was made by Martinis, Devoret and Clarke in 1985.[85,86]

A number of different qubit designs, illustrated in Figs. 6 and 7 have been developed around the Josephson junction including the Cooper pair box (CPB)[72,73,87–92] based on charge, the flux qubit,[93–95] and the phase qubit.[96,97] Devoret and co-workers have recently introduced the fluxonium qubit[98,99] in which the small Josephson junction is shunted by a very high inductance created from a string of larger Josephson junctions. Figure 8 shows an 'evolutionary phylogeny' for these different types of qubits and Fig. 9 classifies them into a 'periodic table' of the

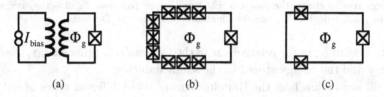

(a)   (b)   (c)

Fig. 7. Inductively shunted qubits. a) Phase qubit with a transformer flux bias circuit driven by current I_{bias}. Josephson junction is indicated by box with cross. b) Fluxoniumqubit. The shunt inductor has been replaced by an array of a large number of Josephson junctions. The array junctions are chosen to have a sufficiently large ratio of Josephson energy E_J to charging energy E_C that phase slips can be neglected and the array is a good approximation to a very large inductor. Flux bias circuit not shown. c) Flux qubit consisting of a superconducting loop with three Josephson junctions. Flux bias circuit not shown.

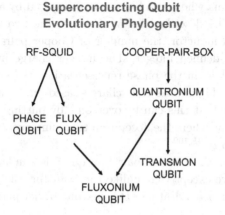

Fig. 8. Evolutionary phylogeny of superconducting qubits. (Courtesy M. Devoret.)

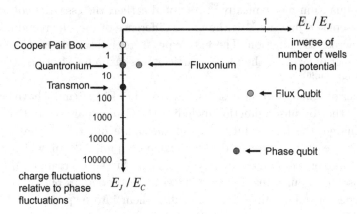

Fig. 9. 'Periodic Table' of superconducting qubits. E_J is the tunneling Josephson energy, $4E_C$ is the energy cost to charge the junction with one Cooper pair, and $E_L/2$ is the energy cost to 'charge' the shunt inductor with one flux quantum. (Courtesy M. Devoret.)

'elements' according to the relative size of the Coulomb charging energy, the Josephson energy and the energy stored in the shunt inductor.

We will not review here the Hamiltonians of these different types of qubits and their relative merits in terms of their sensitivity to noise perturbations but instead focus on the Cooper pair and the so-called 'transmon qubit' which holds the current records for phase coherence time.

6.3.   *The cooper pair box and the transmon qubit*

The Cooper pair box (CPB)[65] is topologically distinct from the other designs shown in Fig. 7 in that it has no wire closing the loop around the junction. It consists very simply of a small antenna whose two halves are connected by a Josephson junction. The Hamiltonian will be described below. Because the two sides of the junction are not shunted by an inductor, the number of Cooper pairs transferred through the junction is a well-defined integer. The integer charge implies the conjugate phase is compact; that is, in the phase representation, the system obeys periodic boundary conditions. This implies that charge-based qubits are sensitive to stray electric field noise, but that this can be overcome by putting the Cooper pair box in the 'transmon' regime where the Josephson tunneling energy dominates over the Coulomb charging energy.[91, 100]

The Lagrangian for the Cooper pair box qubit is similar to that of the LC oscillator studied above except that unlike the inductor, the energy stored in the Josephson junction is not quadratic in the flux but rather periodic

$$L = \frac{1}{2}C_\Sigma \dot{\Phi}^2 + E_J \cos\varphi, \tag{140}$$

where $C_\Sigma = C_{\text{geom}} + C_J$ is the sum of the geometric capacitance of the antenna plus the Josephson tunnel junction capacitance, and φ is the phase difference of the superconducting order parameter across the Josephson junction and is related to the usual flux variable by

$$\varphi = 2\pi \frac{\Phi}{\Phi_0} = \frac{2e}{\hbar}\Phi, \tag{141}$$

and Φ_0 is the superconducting flux quantum. Because there is no shunting inductor, the physical state of the Josephson junction is invariant under shifts of the phase of the order parameter by 2π. Hence φ is an angular variable and the wave function obeys periodic boundary conditions. It is convenient to work with φ rather than Φ and so we write

$$L = \frac{\hbar^2}{8e^2}C_\Sigma\dot{\varphi}^2 + E_J\cos\varphi. \tag{142}$$

This is the Lagrangian of a quantum rotor in a gravitational field, where E_J plays the role of the strength of the gravitational field and C_Σ plays the role of the moment of inertia of the rotor. The angular momentum conjugate to the angle φ is

$$n = \frac{\delta L}{\delta\dot{\varphi}} = \frac{\hbar^2}{4e^2}C_\Sigma\dot{\varphi}, \tag{143}$$

and the quantum Hamiltonian for the rotor may be written

$$H = 4E_C[\hat{n} - n_g]^2 - E_J\cos\hat{\varphi}, \tag{144}$$

and $E_C \equiv \frac{e^2}{2C_\Sigma}$ is the charging energy of a single electron (half a Cooper pair) stored on the total capacitance. $\hat{n} \equiv -i\frac{\partial}{\partial\varphi}$ is the integer-valued angular momentum conjugate to the angle φ and is the operator representing the integer number of Cooper pairs that have tunneled through the Josephson junction relative to the equilibrium classical charge state. Note that we have included in the Hamiltonian the so-called offset charge n_g which represents external bias voltage as shown in Fig. 6. In addition to any intentionally applied external bias, n_g can also contain random fluctuations due to stray charges jumping around in the underlying substrate or the Josephson junction barrier. It turns out that in the limit $E_J \gg E_C$ the low frequency noise in the offset charge can be neglected[91, 100] resulting in very long coherence times for the transmon qubit and so we will henceforth neglect this term.

In the limit $E_J \gg E_C$, the 'gravitational' force is very strong and the 'moment of inertia' is very large so the phase angle undergoes only very small quantum fluctuations around zero. In this limit one can safely expand the cosine in a power series when studying the low-lying excitations[91]

$$H = H_0 + V, \tag{145}$$

$$H_0 = 4E_C\hat{n}^2 + \frac{E_J}{2}\hat{\varphi}^2, \tag{146}$$

$$V = E_J\left[-\frac{1}{4!}\hat{\varphi}^4 + \frac{1}{6!}\hat{\varphi}^6 + \cdots\right] \tag{147}$$

We see that H_0 is the Hamiltonian of a simple LC harmonic oscillator with frequency $\hbar\Omega = \sqrt{8E_JE_C}$ and with the leading order effect of the Josephson junction being to play the role of a (relatively large) linear inductor whose energy is quadratic in the phase (flux) across it. In the limit $E_J \gg E_C$ we are effectively ignoring the fact that φ is an angular variable and taking the harmonic oscillator coordinate φ to be non-compact and imposing vanishing boundary conditions at infinity on the wave functions. It is this assumption which makes conjugate momentum $\hat{n}$ continuous rather than integer-valued and also allows us to neglect the offset charge term.[91] Essentially the large charge fluctuations associated with the small phase fluctuations wash out the discreteness of the charge.

In terms of second quantization, we have

$$\hat{\varphi} = \varphi_{ZPF}[b + b^\dagger] \tag{148}$$

with $\varphi_{ZPF} = \frac{\hbar\Omega}{2E_J}$. The Hamiltonian becomes

$$H_0 = \hbar\Omega b^\dagger b \tag{149}$$

$$V = -\frac{\alpha}{2}(b + b^\dagger)^4 + \cdots \approx -\alpha b^\dagger b - \frac{\alpha}{2}b^\dagger b^\dagger bb \tag{150}$$

where we have made the rotating wave approximation in the last term of V. The anharmonicity is given by[91]

$$\alpha \approx E_C \tag{151}$$

and the first term in V leads to a small renormalization of the bare transition frequency

$$\hbar\tilde{\Omega} \approx \hbar\Omega - E_C. \tag{152}$$

The perturbation term V represents the fact that the Josephson energy is not simply quadratic in flux as in a linear inductor. We are effectively dealing with a non-linear inductor which causes the system to have the anharmonic spectrum we need to use it as a qubit. The transition frequency from ground to first excited state is $\hbar\Omega_{01} = \hbar\tilde{\Omega}$ while the next transition in the ladder is $\hbar\Omega_{12} = \hbar\Omega_{01} - E_C$. In a typical transmon qubit[91,100] with $E_J/E_C \sim 50 - 100$, this corresponds to a $\sim$3–5% negative anharmonicity of $\sim$200 MHz. This anharmonicity is large enough that smooth microwave pulses with durations on the scale of a few nanoseconds will be sufficiently frequency selective that leakage into higher levels of the transmon can be neglected and we can treat it as a two-level qubit.

6.4. *Circuit QED: qubits coupled to resonators*

We have seen that the presence of the Josephson junction causes the transmon to become a weakly anharmonic oscillator. If we couple this qubit to a lumped element LC resonator, a coplanar waveguide resonator or a 3D cavity, we will have a system analogous to that studied in cavity QED. In this 'circuit QED' setup the coupling to the microwave mode(s) of the resonator can be very strong. Focusing on the case where only a single mode of the resonator is important and assuming that the qubit does not strongly perturb the resonator mode, our Hamiltonian becomes

$$H = \hbar\tilde{\Omega}b^\dagger b - \frac{\alpha}{2}b^\dagger b^\dagger bb + \hbar\omega_r a^\dagger a + \hbar g[a^\dagger b + ab^\dagger], \tag{153}$$

where g represents the transition dipole coupling strength between the qubit and the resonator. We have noted that the transmon anharmonicity is weak. If despite this, the coupling to the cavity mode is sufficiently weak relative to the anharmonicity, $g \ll \alpha$, then we can limit our Hilbert space to the ground and first excited states of the qubit and treat it as a two-level system by making the substitutions

$$b^\dagger b \longrightarrow \frac{1+\sigma^z}{2} \tag{154}$$

$$b \longrightarrow \sigma^- \tag{155}$$

$$b^\dagger \longrightarrow \sigma^+ \tag{156}$$

which leads us (up to irrelevant constants and making a rotating wave approximation) to the standard Jaynes-Cummings Hamiltonian of cavity QED[38]

$$H = \hbar\omega_r a^\dagger a + \frac{\hbar\tilde{\Omega}}{2}\sigma^z + \hbar g[a\sigma^+ + a^\dagger\sigma^-]. \tag{157}$$

In the dispersive limit where the qubit is detuned by a distance $\Delta \gg g$ from the cavity, applying a unitary transformation which diagonalizes the Hamiltonian to lowest order in g yields the dispersive Hamiltonian[38, 101]

$$H = \hbar\omega_r a^\dagger a + \frac{\hbar\tilde{\Omega}}{2}\sigma^z + \hbar\chi a^\dagger a\sigma^z, \tag{158}$$

where the dispersive coupling is given by $\chi = -g^2/\Delta$. We see that the dispersive coupling causes the frequency of the qubit to depend on the photon occupation of the cavity[102, 103] and the frequency of the cavity to depend on the state of the qubit.[101] Thus this term permits QND readout of the state of the qubit by measuring the cavity frequency (e.g. by measuring the phase shift of photons reflected from the cavity) or QND readout of the state of cavity by measurement of the transition frequency of the qubit.[103]

In practice the coupling $g \sim 100$–200 MHz is typically sufficiently strong and the qubit anharmonicity is sufficiently small that the above derivation (in which the two-level qubit approximation is made first) is not quantitatively accurate. In addition the approximation that the cavity mode is not distorted by the presence of the qubit is often inaccuracte. Nigg *et al.*[104] have developed a much more careful treatment which does not treat the qubit cavity coupling perturbatively and explicitly takes advantage of the weak anharmonicity of the qubit(s). This so-called 'Black Box Quantization' (BBQ) scheme uses a commercial large-scale numerical finite-element Maxwell equation solver to ('exactly') solve for the normal modes of the linear system consisting of the cavity strongly coupled to the qubit (treated as a simple harmonic oscillator by replacing the Josephson cosine term by its leading quadratic approximation). The normal modes computed this way can be arbitrarily distorted by strong coupling between the (multiple) cavity modes and the qubit and thus form a highly efficient basis in which to express the weak anharmonic term in the Josephson Hamiltonian. The harmonic Hamiltonian for the system treats the qubit and the cavity modes all on an equal footing and is given by

$$H_0 = \sum_j \hbar\omega_j a_j^\dagger a_j, \tag{159}$$

where the frequencies of the normal modes are determined by the locations of the numerically computed of the zeros of the admittance $Y(\omega)$ at a port across the Josephson inductance.

The Josephson phase variable φ (or equivalently the flux at a network port defined across the Josephson inductance) can (by analogy with Eq. (136)) be precisely expressed in terms of the jth normal mode operators as

$$\hat{\varphi} = \sum_j \varphi_{\mathrm{ZPF}}^{(j)}[a_j^\dagger + a_j]. \tag{160}$$

Because the quadratic term in the expansion of the cosine has already been included in the harmonic part of the Hamiltonian, the next leading term is

$$V = -\frac{E_J}{4!}\hat{\varphi}^4 = -\frac{E_J}{4!}\left\{\sum_j \varphi_{\mathrm{ZPF}}^{(j)}[a_j^\dagger + a_j]\right\}^4. \tag{161}$$

Normal ordering and making rotating wave approximations (RWA) leads to corrections to normal mode frequencies analogous to Eq. (152). In addition every mode develops some anharmonicity (self-Kerr) inherited from the Josephson junction. The most anharmonic mode can be identified as 'the qubit' and the remainder as 'cavity modes.' In addition to self-Kerr terms, there will be cross-Kerr terms in which the frequency of a given mode depends on the occupation numbers of other modes. The general form of the non-linear Hamiltonian within the RWA is

$$V = \sum_j \delta\omega_j \hat{n}_j + \frac{1}{2}\sum_{j,k} \chi_{jk}\hat{n}_j\hat{n}_k \tag{162}$$

where $\hat{n}_j = a_j^\dagger a_j$ is the occupation number operator for the jth mode. At this point it is now generally safe and accurate to map the most anharmonic component (aka the qubit) onto a spin-1/2 as done above. The procedure outlined here is much more accurate than making the two-level approximation first and treating the coupling to the cavity perturbatively, especially in the case where the anharmonicity is weak and the coupling g is strong. It should not be forgotten that there may be contributions from the sixth and higher-order terms in the expansion of the cosine potential which have been neglected here.

One of the remarkable aspects of circuit QED is that it is easy to reach the strong-dispersive limit where some elements of the χ matrix are $\sim 10^1$–10^3 times larger than the line widths of both the qubit and the cavity. This makes it possible for example to create photon cat states using the self-Kerr non-linearity which the cavity inherits from the qubit.[71] This strong dispersive coupling also opens up a new toolbox for mapping qubit states onto cavity cat states[105] and possibly even for autonomous error correction using cavity states as memories.[106] It is also worth noting from the form of the cross-Kerr Hamiltonian that both qubits and cavities can be readily dephased by stray photons hopping into and out of higher cavity modes that would otherwise be considered irrelevant. Hence very careful filtering to remove these stray photons is important.

7.   Summary

These notes have attempted to convey some of the basic concepts in quantum information as well as provide a basic introduction to circuit QED. A much more detailed review of circuit QED is presented in the author's lecture notes in the Proceedings of the 2011 Les Houches Summer School on Quantum Machines.[2]

Acknowledgments

The author is grateful for helpful conversations with numerous colleagues including among others Charles Bennett, Liang Jiang, Sreraman Muralidharan, Uri Vool, Claudia De Grandi, Simon Nigg, Matt Reed, Michael Hatridge, Shyam Shankar, Gerhard Kirchmair, Robert Schoelkopf and Michel Devoret. The author's research is support by Yale University, the NSF, the ARO/LPS and IARPA.

References

1. A. Einstein, B. Podolsky, and N. Rosen, Can quantum-mechanical description of physical reality be considered complete?, *Phys. Rev.* **47**, 777–780, (1935).
2. S. M. Girvin, *Proceedings of the 2011 Les Houches Summer School on Quantum Machines*, chapter Circuit QED: Superconducting Qubits Coupled to Microwave Photons. Oxford University Press, (2012).
3. I. D. Leroux, M. H. Schleier-Smith, and V. Vuletić, Implementation of cavity squeezing of a collective atomic spin, *Phys. Rev. Lett.* **104**, 073602, (2010).
4. I. D. Leroux, M. H. Schleier-Smith, and V. Vuletić, Orientation-dependent entanglement lifetime in a squeezed atomic clock, *Phys. Rev. Lett.* **104**, 250801, (2010).

5. I. L. Chuang and M. A. Nielsen, *Quantum Information and Quantum Computation.* (Cambridge University Press, 2000).

6. P. W. Shor, Algorithms for quantum computation: Discrete logarithms and factoring. In *Proc. 35th Annual Symp. on Foundations of Computer Science.* IEEE Press, (1994).

7. R. P. Feynman, Simulating physics with computers, *Int. J. of Theor. Phys.* **21**, 467–488, (1982).

8. C. H. Bennett, Logical reversibility of computation, *IBM J. Res. Develop.* **17**, 525, (1973).

9. S. Wiesner, Conjugate coding, *Sigact News* **15**, 78, (1983).

10. S. Aaronson and P. Christiano, Quantum money from hidden subspaces. In *Proceedings of the 44th symposium on Theory of Computing, STOC '12*, pp. 41–60, (2012).

11. W. Wootters and W. Zurek, A single quantum cannot be cloned, *Nature* **299**, 802-803, (1982).

12. D. Dieks, Communication by epr devices, *Physics Letters A* **92**, 271272, (1982).

13. C. Bennett, (private communication).

14. D. Divincenzo, (private communicaton).

15. C. H. Bennett and G. Brassard, Quantum cryptography: Public key distribution and coin tossing. In *Proc. IEEE Int. Conf. on Computers, Systems, and Signal Processing*, p. 175, (1984).

16. N. Gisin, G. Ribordy, W. Tittel, and H. Zbinden, Quantum cryptography, *Rev. Mod. Phys.* **74**, 145–195, (2002).

17. C. H. Bennett and S. J. Wiesner, Communication via one- and two-particle operators on einstein-podolsky-rosen states, *Phys. Rev. Lett.* **69**, 2881–2884, (1992).

18. Alice can perform a unitary transformation which un-entangles the four Bell states, mapping them into four distinct states in the measurement basis ($|00\rangle, |01\rangle, |10\rangle, |11\rangle$). Then simply measuring both qubits tells her which of the four Bell states she had received.

19. A. K. Ekert, Quantum cryptography based on bell's theorem, *Phys. Rev. Lett.* **67**, 661–663, (1991).

20. M. Rizzi, S. Montangero, and G. Vidal, Simulation of time evolution with multiscale entanglement renormalization ansatz, *Phys. Rev. A* **77**, 052328, (2008).

21. D. Deutsch, Quantum theory, the church-turing principle and universal quantum computer, *Proc. R. Soc. Lond. A* **400**, 97, (1985).

22. D. Deutsch and R. Jozsa, Rapid solution of problems by quantum computation, *Proc. R. Soc. Lond. A* **439**, 553, (1992).

23. L. Grover, A fast quantum mechanical algorithm for database search. In *Proc. 28th Annual ACM Symp. on the Theory of Computation*, p. 212. ACM Press, (1996).

24. L. K. Grover, Quantum mechanics helps in searching for a needle in a haystack, *Phys. Rev. Lett.* **79**, 325–328, (1997).

25. P. W. Shor, Polynomial-time algorithms for prime factorization and discrete logarithms on a quantum computer, *SIAM J. Comp.* **26**, 1484, (1997).

26. J. Kempe, Quantum random walks: An introductory overview, *Contemporary Physics* **44**, 307–327, (2003).

27. E. Farhi, J. Goldstone, S. Gutmann, J. Lapan, A. Lundgren, and D. Preda, A quantum adiabatic evolution algorithm applied to random instances of an np-complete problem, *Science* **292**, 472–475, (2001).

28. E. Farhi, D. Gosset, I. Hen, A. W. Sandvik, P. Shor, A. P. Young, and F. Zamponi, Performance of the quantum adiabatic algorithm on random instances of two optimization problems on regular hypergraphs, *Phys. Rev. A* **86**, 052334, (2012).

29. N. D. Mermin, *Quantum Computer Science: An Introduction.* (Cambridge University Press, 2007).

30. J. Eisert, M. Cramer, and M. B. Plenio, *Colloquium*: Area laws for the entanglement entropy, *Rev. Mod. Phys.* **82**, 277–306, (2010).

31. R. Horodecki, P. Horodecki, M. Horodecki, and K. Horodecki, Quantum entanglement, *Rev. Mod. Phys.* **81**, 865–942, (2009).

32. J. C. F. Verstraete and V. Murg, Matrix product states, projected entangled pair states, and variational renormalization group methods for quantum spin systems, *Advances in Physics* **57**, 143–224, (2008).

33. H. Li and F. D. M. Haldane, Entanglement spectrum as a generalization of entanglement entropy: Identification of topological order in non-abelian fractional quantum hall effect states, *Phys. Rev. Lett.* **101**, 010504, (2008).

34. S. Sachdev, What can gauge-gravity duality teach us about condensed matter physics?, *Annual Review of Condensed Matter Physics* **3**, 9, (2012). [arXiv: 1108.1197].

35. T. A. Brun, A simple model of quantum trajectories, *Am. J.Phys.* **70**, 719, (2002).

36. K. Jacobs and D. A. Steck, A straightforward introduction to continuous quantum measurement, *Contemporary Physics* **47**, 279, (2007).

37. A. A. Clerk, M. H. Devoret, S. M. Girvin, F. Marquardt, and R. J. Schoelkopf, Introduction to quantum noise, measurement and amplification, *Rev. Mod. Phys.* **82**, 1155–1208, (2010). (Longer version with pedagogical appendices available at: arXiv.org:0810.4729).

38. S. Haroche and J.-M. Raimond, *Exploring the Quantum: Atoms, Cavities and Photons.* (Oxford University Press, 2006).

39. M. Hatridge, S. Shankar, M. Mirrahimi, F. Schackert, K. Geerlings, T. Brecht, K. M. Sliwa, B. Abdo, L. Frunzio, S. M. Girvin, R. J. Schoelkopf, and M. H. Devoret, Quantum back-action of an individual variable-strength measurement, *Science* **339**, 178–181, (2013).

40. J. S. Bell, On the einstein podolsky rosen paradox, *Physics* **1**, 195–200, (1964).

41. J. F. Clauser, M. A. Horne, A. Shimony, and R. A. Holt, Proposed experiment to test local hidden-variable theories, *Phys. Rev. Lett.* **23**, 880–884, (1969).

42. C. H. Bennett, G. Brassard, C. Crépeau, R. Jozsa, A. Peres, and W. K. Wootters, Teleporting an unknown quantum state via dual classical and einstein-podolsky-rosen channels, *Phys. Rev. Lett.* **70**, 1895–1899, (1993).

43. P. W. Shor, Scheme for reducing decoherence in quantum computer memory, *Phys. Rev. A* **52**, R2493–R2496, (1995).

44. P. W. Shor, Fault-tolerant quantum computation. In *Proc. 37th Annual Symp. on Fundamentals of Computer Science*, pp. 56–65. IEEE Press, (1996).

45. A. M. Steane, Error correcting codes in quantum theory, *Phys. Rev. Lett.* **77**, 793–797, (1996).

46. A. M. Steane, Simple quantum error-correcting codes, *Phys. Rev. A* **54**, 4741–4751, (1996).

47. R. Raussendorf, Key ideas in quantum error correction, *Phil. Trans. R. Soc. A* **370**, 4541–4565, (2012).

48. D. Gottesman, *Stabilizer Codes and Quantum Error Correction.* Phd thesis, Caltech, (1997).

49. D. Gottesman, An introduction to quantum error correction. In ed. J. S. J. Lomonaco, *Quantum Computation: A Grand Mathematical Challenge for the Twenty-First Century and the Millennium*, pp. 221–235, (2002). [arXiv:quant-ph/0004072].

50. A. Kitaev, Fault-tolerant quantum computation by anyons, *Annals of Physics* **303**, 2, (2003).

51. D. P. DiVincenzo, Fault-tolerant architectures for superconducting qubits, *Phys. Scr.* **T137**, 014020, (2009).

52. Y. L. Z. C. F. R. M. Müller, K. Hammerer, and P. Zoller, Simulating open quantum systems: from many-body interactions to stabilizer pumping, *New. J. Phys.* **13**, 085007, (2011).

53. J. T. Barreiro, M. Müller, P. Schindler, D. Nigg, T. Monz, M. Chwalla, M. Hennrich, C. F. Roos, P. Zoller, and R. Blatt, An open-system quantum simulator with trapped ions, *Nature* **470**, 486, (2011).

54. S. Gladchenko, D. Olaya, E. Dupont-Ferrier, B. Doucot, L. B. Ioffe, and M. E. Gershenson, Superconducting nanocircuits for topologically protected qubits, *Nature Physics* **5**, 48–53, (2009).

55. S. E. Nigg and S. M. Girvin, Stabilizer quantum error correction toolbox for superconducting qubits. [arXiv:1212.4000].

56. C. Nayak, S. H. Simon, A. Stern, M. Freedman, and S. Das Sarma, Non-abelian anyons and topological quantum computation, *Rev. Mod. Phys.* **80**, 1083–1159, (2008).

57. M. D. Reed, L. DiCarlo, S. E. Nigg, L. Sun, L. Frunzio, S. M. Girvin, and R. J. Schoelkopf, Realization of three-qubit quantum error correction with superconducting circuits, *Nature* **482**, 382–385, (2012).

58. D. G. Cory, M. D. Price, W. Maas, E. Knill, R. Laflamme, W. H. Zurek, T. F. Havel, and S. S. Somaroo, Experimental quantum error correction, *Phys. Rev. Lett.* **81**, 2152–2155, (1998).

59. J. Chiaverini, D. Leibfried, T. Schaetz, M. D. Barrett, R. B. Blakestad, J. Britton, W. M. Itano, J. D. Jost, E. Knill, C. Langer, R. Ozeri, and D. J. Wineland, Realization of quantum error correction, *Nature* **432**, 602–605, (2004).

60. T. B. Pittman, B. C. Jacobs, and J. D. Franson, Demonstration of quantum error correction using linear optics, *Phys. Rev. A* **71**, 052332, (2005).

61. T. Aoki, G. Takahashi, T. Kajiya, J. ichi Yoshikawa, S. L. Braunstein, P. van Loock, and A. Furusawa, Quantum error correction beyond qubits, *Nature Physics* **5**, 541–546, (2009).

62. E. Knill, R. Laflamme, R. Martinez, and C. Negrevergne, Benchmarking quantum computers: The five-qubit error correcting code, *Phys. Rev. Lett.* **86**, 5811–5814, (2001).

63. N. Boulant, L. Viola, E. M. Fortunato, and D. G. Cory, Experimental implementation of a concatenated quantum error-correcting code, *Phys. Rev. Lett.* **94**, 130501, (2005).

64. P. Schindler, J. T. Barreiro, T. Monz, V. Nebendahl, D. Nigg, M. Chwalla, M. Hennrich, and R. Blatt, Experimental repetitive quantum error correction, *Science* **332**, 1059–1061, (2011).

65. M. Devoret, *Les Houches Session LXIII, 1995*, chapter Quantum Fluctuations in Electrical Circuits, pp. 351–385. Elsevier, Amsterdam, (1997).

66. J. Raimond, M. Brune, and S. Haroche, Manipulating quantum entanglement with atoms and photons in a cavity, *Rev. Mod. Phys.* **73**, 565, (2001).

67. S. Deleglise, I. Dotsenko, C. Sayrin, J. Bernu, M. Brune, J.-M. Raimond, and S. Haroche, Reconstruction of non-classical cavity field states with snapshots of their decoherence, *Nature* **455**, 510–514, (2008). ISSN 0028-0836.

68. A. A. Houck, D. I. Schuster, J. M. Gambetta, J. A. Schreier, B. R. Johnson, J. M. Chow, L. Frunzio, J. Majer, M. H. Devoret, S. M. Girvin, and R. J. Schoelkopf, Generating single microwave photons in a circuit, *Nature* **449**, 328–331, (2007).

69. M. Hofheinz, E. M. Weig, M. Ansmann, R. C. Bialczak, E. Lucero, M. Neeley, A. D. O'Connell, H. Wang, J. M. Martinis, and A. N. Cleland, Generation of Fock states in a superconducting quantum circuit, *Nature* **454**, 310–314, (2008).

70. M. Hofheinz, H. Wang, M. Ansmann, R. C. Bialczak, E. Lucero, M. Neeley, A. D. O'Connell, D. Sank, J. Wenner, J. M. Martinis, and A. N. Cleland, Synthesizing arbitrary quantum states in a superconducting resonator, *Nature* **459**, 546–549, (2009).

71. G. Kirchmair, B. Vlastakis, Z. Leghtas, S. E. Nigg, H. Paik, E. Ginossar, M. Mirrahimi, L. Frunzio, S. M. Girvin, and R. J. Schoelkopf, Observation of quantum state collapse and revival due to the single-photon kerr effect. arXiv:1211.2228.

72. V. Bouchiat, D. Vion, P. Joyez, D. Esteve, and M. H. Devoret, Quantum coherence with a single cooper pair, *Phys. Scr.* **T76**, 165, (1998).

73. Y. Nakamura, Y. A. Pashkin, and J. S. Tsai, Coherent control of macroscopic quantum states in a single-cooper-pair box, *Nature* **398**, 786–788, (1999). ISSN 0028-0836.

74. M. Devoret and J. Martinis, Superconducting qubits. In ed. D. Esteve, J. M. Raimond, and J. Dalibard, *Quantum Entanglement and Information Processing*, Vol. 79, *Les Houches Summer School Session*, pp. 443–485, (2004). Les Houches Session 79th on Quantum Entanglement and Information Processing, 2003.

75. D. Esteve and D. Vion, Solid state quantum bit circuits. In ed. H. Bouchiat, Y. Gefen, S. Gueron, G. Montambaux, and J. Dalibard, *Nanophysics: Coherence and Transport*, Vol. 81, *Les Houches Summer School Session*, pp. 537+, (2005). Les Houches Session 81st on Nanophysics — Coherence and Transport, 2004.

76. G. Wendin and V. Shumeiko, *Handbook of Theoretical and Computational Nanotechnology*, Vol. 3, chapter 'Superconducting circuits, qubits and computing', pp. 223–309. American Scientific Publishers, Los Angeles, (2006).

77. G. Wendin and V. S. Shumeiko, Quantum bits with Josephson junctions, *J. Low Temperature Physics* **33**, 724, (2007).

78. J. Clarke and F. K. Wilhelm, Superconducting quantum bits, *Nature* **453**, 1031–1042, (2008). ISSN 0028-0836.

79. R. Schoelkopf and S. Girvin, Wiring up quantum systems, *Nature* **451**, 664, (2008).

80. J. You and F. Nori, Superconducting circuits and quantum information, *Physics Today* **58**, 42–47, (2005).

81. F. Nori, Superconducting qubits: Atomic physics with a circuit, *Nature Physics* **4**, 589–590, (2008).

82. A. N. Korotkov, Special issue on quantum computing with superconducting qubits, *Quantum Information Processing* **8**, 51–54, (2009).

83. A. Leggett, Macroscopic quantum systems and the quantum theory of measurement, *Progress of Theoretical Physics Suppl.* **69**, 80–100, (1980).

84. M. H. Devoret and J. M. Martinis, *Experimental Aspects of Quantum Computing*, Vol. 3, chapter Implementing Qubits with Superconducting Integrated Circuits, pp. 163–203. Springer, (2005).

85. J. Martinis, M. Devoret, and J. Clarke, Energy level quantization in the zero-voltage state of a current-biased josephson junction, *Phys. Rev. Lett.* **55**, 1543–1546, (1985).

86. J. Clarke, A. N. Cleland, M. H. Devoret, D. Esteve, and J. M. Martinis, Quantum mechanics of a macroscopic variable: The phase difference of a Josephson junction, *Science* **239**, 992–997, (1988).

87. D. Averin, A. Zorin, and K. Likharev, Bloch oscillations in small josephson junctions, *Sov. Phys. JETP* **61**, 407, (1985).

88. M. Büttiker, Zero-current persistent potential drop across small-capacitance josephson junctions, *Phys. Rev. B* **36**, 3548–3555, (1987). doi: 10.1103/PhysRevB.36.3548.

89. P. Lafarge, P. Joyez, D. Esteve, C. Urbina, and M. H. Devoret, Two-electron quantization of the charge on a superconductor, *Nature* **365**, 422–424, (1993).

90. D. Vion, A. Aassime, A. Cottet, P. Joyez, H. Pothier, C. Urbina, D. Esteve, and M. H. Devoret, Manipulating the quantum state of an electrical circuit, *Science* **296**, 886–889, (2002).

91. J. Koch, T. M. Yu, J. Gambetta, A. A. Houck, D. I. Schuster, J. Majer, A. Blais, M. H. Devoret, S. M. Girvin, and R. J. Schoelkopf, Charge-insensitive qubit design derived from the Cooper pair box, *Phys. Rev. A* **76**, 042319, (2007). ISSN 1050-2947.

92. J. A. Schreier, A. A. Houck, J. Koch, D. I. Schuster, B. R. Johnson, J. M. Chow, J. M. Gambetta, J. Majer, L. Frunzio, M. H. Devoret, S. M. Girvin, and R. J. Schoelkopf, Suppressing charge noise decoherence in superconducting charge qubits, *Phys. Rev. B* **77**, 180502, (2008).

93. J. E. Mooij, T. P. Orlando, L. Levitov, L. Tian, C. H. van der Wal, and S. Lloyd, Josephson persistent-current qubit, *Science* **285**, 1036–1039, (1999).

94. C. H. van der Wal, A. C. J. ter Haar, F. K. Wilhelm, R. N. Schouten, C. J. P. M. Harmans, T. P. Orlando, S. Lloyd, and J. E. Mooij, Quantum superposition of macroscopic persistent-current states, *Science* **290**, 773–777, (2000).

95. I. Chiorescu, Y. Nakamura, C. J. P. M. Harmans, and J. E. Mooij, Coherent quantum dynamics of a superconducting flux qubit, *Science* **299**, 1869–1871, (2003).

96. J. Martinis, S. Nam, J. Aumentado, and C. Urbina, Rabi oscillations in a large Josephson-junction qubit, *Phys. Rev. Lett.* **89**, 117901, (2002).

97. A. J. Berkley, H. Xu, R. C. Ramos, M. A. Gubrud, F. W. Strauch, P. R. Johnson, J. R. Anderson, A. J. Dragt, C. J. Lobb, and F. C. Wellstood, Entangled macroscopic quantum states in two superconducting qubits, *Science* **300**, 1548–1550, (2003).

98. V. E. Manucharyan, J. Koch, L. Glazman, and M. Devoret, Fluxonium: Single cooper-pair circuit free of charge offsets, *Science* **326**, 113–116, (2009).

99. V. E. Manucharyan, J. Koch, M. Brink, L. I. Glazman, and M. H. Devoret, Coherent oscillations between classically separable quantum states of a superconducting loop, *arXiv:0910.3039*, (2009).

100. A. Houck, J. Koch, M. Devoret, S. Girvin, and R. Schoelkopf, Life after charge noise: recent results with transmon qubits, *Quantum Information Processing* **8**, 105, (2009).

101. A. Blais, R.-S. Huang, A. Wallraff, S. M. Girvin, and R. J. Schoelkopf, Cavity quantum electrodynamics for superconducting electrical circuits: an architecture for quantum computation, *Phys. Rev. A* **69**, 062320, (2004).

102. D. Schuster, A. Houck, J. Schreier, A. Wallraff, J. Gambetta, A. Blais, L. Frunzio, B. Johnson, M. Devoret, S. Girvin, and R. Schoelkopf, Resolving photon number states in a superconducting circuit, *Nature* **445**, 515–518, (2007).

103. B. R. Johnson, M. D. Reed, A. A. Houck, D. I. Schuster, L. S. Bishop, E. Ginossar, J. M. Gambetta, L. DiCarlo, L. Frunzio, S. Girvin, and R. Schoelkopf, Quantum non-demolition detection of single microwave photons in a circuit, *Nature Physics* **6**, 663–667, (2010).

104. S. E. Nigg, H. Paik, B. Vlastakis, G. Kirchmair, S. Shankar, L. Frunzio, M. H. Devoret, R. J. Schoelkopf, and S. M. Girvin, Black-box superconducting circuit quantization, *Phys. Rev. Lett.* **108**, 240502, (2012).

105. Z. Leghtas, G. Kirchmair, B. Vlastakis, M. Devoret, R. Schoelkopf, and M. Mirrahimi, Deterministic protocol for mapping a qubit to coherent state superpositions in a cavity. arXiv:1205.2401.

106. Z. Leghtas, G. Kirchmair, B. Vlastakis, R. Schoelkopf, M. Devoret, and M. Mirrahimi, Hardware-efficient autonomous quantum error correction. arXiv:1207.0679.

Chapter 6

Cavity Polaritons: Crossroad Between Non-Linear Optics and Atomic Condensates

Alberto Amo* and Jacqueline Bloch†

Laboratoire de Photonique et Nanostructures, LPN/CNRS
Route de Nozay, 91460 Marcoussis, France
*alberto.amo@lpn.cnrs.fr †jacqueline.bloch@lpn.cnrs.fr

Polaritons are mixed light-matter particles arising from the strong coupling between excitons and photons confined in a two-dimensional semiconductor microcavity. Their bosonic nature along with the possibility to directly manipulate them in a semiconductor chip makes them an excellent workbench to study non-linear properties of interacting degenerate bosons in a dissipative system. In this chapter we review the basic properties of semiconductor microcavities in the strong coupling regime including the methods used to achieve polariton condensation and to observe quantum fluid effects. We then describe how the engineering of the planar cavity using different technological approaches allows confining polaritons in low dimensional structures with a large variety of geometries. We illustrate the fascinating physics that arise from these low dimensional microstructures choosing a few examples: periodically modulated wires, coupled micropillars and polariton lattices. Cavity polaritons thus appear as a versatile platform to implement non-linear Hamiltonians and to study non-linear topological excitations.

1. Introduction

The spontaneous emission of light is not an intrinsic property of the emitter but can be dramatically modified when changing its electromagnetic environment. This idea has been first proposed by Purcell in 1946 in a short paper,[1] where he predicted that spontaneous emission can be enhanced or inhibited depending of the density of electromagnetic modes at the emitter location. In particular if an emitter is placed in an optical cavity and spectrally resonant with a confined optical mode, spontaneous emission into the cavity mode can be strongly enhanced as compared to emission into other modes.[2] Photons are funnelled into the cavity mode and this effect has been implemented to realize efficient single photon emitters.[3-5] If the radiative coupling between the emitter and the cavity mode is stronger than both the cavity loss (escape of the photon out of the cavity) and the emitter damping rate, a very interesting regime can be achieved named strong coupling regime. The spontaneous emission of light becomes reversible: photons emitted within the cavity can oscillate long enough to be reabsorbed, reemitted again and so on. As a result, the energy undergoes coherent Rabi oscillations between light and matter. The strong coupling regime gives rise to the formation of mixed light matter eigenstates.

Excitonic polaritons are bosonic quasi-particles arising from the strong coupling regime between excitons, electron-hole pairs bound by Coulombic interactions, and photons propagating in a medium. As described in details in Chapter 2, excitonic polaritons are the fundamental excitations of bulk direct semiconductors at low temperatures.[6,7] They have been extensively investigated in the 70's.[8,9]

Excitonic polaritons have known a spectacular resurgence of interest when revisited in 1992 by C. Weisbuch in two-dimensional semiconductor microcavities.[10] Indeed, cavity excitonic polaritons rapidly appeared as a fascinating system to investigate the physics of quantum fluids in the presence of interactions and in controlled geometries. As we will illustrate in this chapter, cavity polaritons provide a physical system at the frontier between non-linear optics and atomic physics. Their bosonic nature along with their lightness makes polaritons an extraordinary workbench to study bosonic condensation, the macroscopic accumulation of particles in a single quantum state, at temperatures much higher than those required in ultracold atomic gases (5–300 K vs <1 μK). Polariton condensation was first achieved in a CdTe planar microcavity in 2005–2006,[11,12] and it is now routinely obtained in GaAs-based structures,[13,14] GaN[15], ZnO[102,103] and organic microcavities.[16] The excitonic component of polaritons adds interactions to the condensates, resulting in remarkable non-linear fluid effects, like superfluid-like behavior,[17,18] or the nucleation of quantized vortices[19] and oblique dark solitons[20] when the condensate moves against an obstacle, phenomena characteristic of atomic condensates. Additionally, polariton condensates present out-of-equilibrium features due to the continuous escape of photons out of the cavity.[21] The steady state of the system is not necessarily determined by thermodynamics but by the interplay of pumping, relaxation and decay.

From a complementary perspective, polaritons can be simply viewed as photonic particles in which the excitonic component provides a very large $\chi^{(3)}$ non-linearity. This gives rise to phenomena like optical parametric oscillation[22,23] or squeezing,[24] with the advantage of having extremely low thresholds. Polaritons thus share non-linear properties both with standard Kerr non-linear optical systems[25,26] and with ultracold atomic Bose-Einstein condensates, with novel properties arising from their short lifetime.

In this chapter we review the concept of strong light matter coupling in semiconductor microcavities and the phenomenon of polariton condensation in 2D microcavities. We then show different techniques to reduce the dimensionality of the system and to engineer the polariton landscape. In this way, bosonic condensation can be studied in a wide variety of geometries, ranging from periodic potentials to double and multiple well structures, or two-dimensional lattices. In the presence of interactions, the geometry of the potential radically changes the properties of the condensate. We show several examples, ranging from the formation of gap solitons to non-linear Josephson oscillations and spontaneous spin currents. The manipulation of polariton condensates in engineered potentials opens several avenues of study. The strong non-linearities allow the implementation of polariton functionalities in integrated photonic circuits with low optical thresholds. The design of at-will geometries can be used to create two-dimensional polariton lattices, where engineered bands and synthetic gauge fields for light can be foreseen.

2. Cavity polaritons: exciton-photon strong coupling regime

In 1992, C. Weisbuch and collaborators investigated by low temperature reflectivity a GaAs based microcavity containing seven quantum wells[10] (see Fig. 1(a)). When tuning the cavity mode across the quantum well excitonic resonance, they observed an anti-crossing, signature of the exciton-photon strong coupling regime. One year later, R. Houdré *et al.* reported angle resolved photoluminescence measurements evidencing the characteristic polariton dispersions.[27] An example of these dispersions is shown on Fig. 1(b).

These dispersion relations can be easily derived considering the coupling between a 2D continuum of exciton states, and a 2D continuum of photon modes. Indeed in the optical cavity, photon confinement along the cavity axis (z direction, perpendicular to the (xy) plane of cavity) results in a quantification of the photon wave-vector component k_z along the (z) direction: $k_z = p\frac{\pi}{L_c}$. Here p denotes the order of the considered Fabry Perot optical mode and L_c the thickness of the cavity layer. Thus we can define a two dimensional continuum of photon modes, labelled by their in-plane wavevector $k_{//}$ and with energy $E_c(k_{//}) = \frac{\hbar c}{n_{cav}}\sqrt{k_{//}^2 + (\frac{\pi}{L_c})^2}$, where n_{cav} is the cavity layer refractive index. Close to $k_{//} = 0$, the cavity mode dispersion can be approximated with a parabola. Thus the confinement along the (z) direction allows defining an effective mass for the photon given by the following expression:

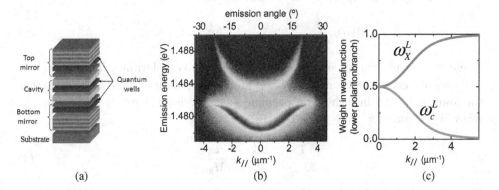

(a)   (b)   (c)

Fig. 1.   (a) Schematic of a semiconductor cavity sample (b) Polariton dispersion measured from the angle and energy resolved emission of a GaAs/AlGaAs planar microcavity under low power non-resonant excitation. (c) Exciton (ω_X^L) and photon (ω_C^L) weight in the eigenfunctions corresponding to the lower polariton branch at zero exciton-cavity detuning, $\delta(k_{//} = 0) = 0$.

$M_{phot} = \frac{\hbar p \pi n_{cav}}{c L_c}$. Typically M_{phot} is 5 orders of magnitude smaller than the free electron mass m_e.

We now consider the electronic states of the quantum well, which can be excited when promoting electrons from the valence to the conduction band.[28] The lowest energy states are excitonic states, describing the quantum states of an electron hole pair bound by Coulombic interaction and free of motion in the plane of the quantum well. The 1s excitonic states can be labelled by their in-plane wavevector $k_{//}$, and have an energy of the form: $E_X(k_{//}) = E_X^0 + \frac{\hbar^2 k^2}{2 M_X}$, where $M_X = m_e + m_h$ is the exciton effective mass (of the order of $0.6\,m_e$ in GaAs based quantum wells). This energy dispersion describes the kinetic energy related to the exciton centre of mass motion in the plane of the quantum well.

Excitons and photons are coupled via dipolar interaction.[28] Because of in-plane translation invariance, each exciton of a given in-plane wave-vector $k_{//}$ is coupled to the photon mode of same $k_{//}$. As a result, for each $k_{//}$, the hamiltonian describing the coupled system reads:

$$H(k_{//}) = \begin{bmatrix} E_X(k_{//}) & \Omega/2 \\ \Omega/2 & E_C(k_{//}) \end{bmatrix} \tag{1}$$

where $\Omega/2$ is the exciton-photon radiative coupling. For each value of $k_{//}$, the eigenstates $/UP>$ and $/LP>$ of the system are linear superposition of the exciton and the photon states, with relative exciton $w_X^{U/L}$ and photon weights $w_C^{U/L}$ which depend on the exciton-photon detuning $\delta(k_{//}) = E_C(k_{//}) - E_X(k_{//})$, as follows:

$$w_C^U = w_X^L = \sqrt{\frac{\Delta + \delta}{2\Delta}} \quad \text{and} \quad w_C^L = w_X^U = \sqrt{\frac{\Delta - \delta}{2\Delta}} \tag{2}$$

where $\Delta = \sqrt{\delta^2 + g^2}$.

The energy of the upper [lower] eigenstates is:

$$E^{U[L]}(k_{//}) = \frac{(E_X(k_{//}) + E_C(k_{//}))}{2} + [-]\frac{\Delta}{2} \tag{3}$$

These exciton-photon mixed states are named cavity polaritons. They form two branches of eigenstates, namely the lower and upper polariton branches. Notice that the lower polariton branch has a peculiar s-shaped dispersion,[27] with an effective mass M_{pol} close to $k_{//} = 0$ given by:

$$\frac{1}{M_{pol}} = \frac{w_X^L}{M_X} + \frac{w_C^L}{M_{phot}} \tag{4}$$

Since M_{phot} is typically three orders of magnitude smaller than M_{exc}, a small photon weight in the polariton state is sufficient to give an effective mass to the polariton state much smaller than that of the bare exciton. This was illustrated in ZnO nanowires where polaritons with a photon weight of only 0.03, were shown to have already a significantly reduced effective mass as compared to the bulk ZnO exciton and to present very interesting properties related to this small photon fraction.[29]

The nature of the polariton states strongly changes along the polariton branches. This is illustrated in Fig. 1(c), where the exciton and photon components of the lower branch polariton are shown as a function of $k_{//}$. Close to $k_{//} = 0$, the lower polariton state is a half exciton-half-photon quasi-particle. As $k_{//}$ increases, the polariton state evolves toward a more and more exciton like state. At very large $k_{//}$, the eigenstates are excitonic states, with a large density of states reflecting the large exciton effective mass. This region in k-space is named the excitonic reservoir. It is highly populated in experiments where the polariton branch is excited via non-resonant pumping, and will play a key role in such experiments.

All the physical properties of cavity polaritons reflect their dual light-matter nature. On the one hand, from their photonic component, polaritons present many properties similar to those of light:

— they propagate like photons with a group velocity which can be very close to the speed of light;[30]
— polaritons are coupled to the free space optical modes, via the escape of their photon part through the cavity mirrors. Thus polariton states can be probed and excited via their photonic component. Moreover photons emitted by polaritons fully reflect their quantum properties (spin, coherence, density, etc....);
— finally, as explained above, polaritons inherit a very small effective mass because of their photonic nature.

On the other hand, from their excitonic component, polaritons interact with their environment. They interact with phonons,[31] with free electrons,[32] or with excitons.[33] Moreover, because of their excitonic component, polaritons present strong polariton-polariton interactions.[34,35] This is a very important property which will be further addressed in the chapter and which is responsible for enhanced non-linearities.

To summarize this introduction, we would like to stress that because of their dual light-matter nature, cavity polaritons behave as photons, but with strong non-linearities inherited from their excitonic component.

3. Laterally confining cavity polaritons

A semiconductor microcavity operating in the strong coupling regime can be grown in a single step by epitaxy deposition. Thus, the as-grown samples are two dimensional. Nevertheless it can be interesting to further reduce the dimensionality of the system or to engineer the potential in which polaritons are created. Many fascinating physical effects are expected when considering such non-linear systems in lower dimensionality and in the presence of an engineered potential landscape.

One approach, which is not controllable, relies on the natural disorder potential present in as grown cavity layers. Several groups have reported polariton localization in the minima of this potential landscape.[38-40] Still real engineering of the potential in which polaritons are generated, requires development of techniques which allow full control of the cavity geometry. For this purpose, a variety of ingenious methods have been developed to produce lower dimensionality resonators. The lateral confinement can be obtained by acting on the exciton part of the polariton and/or on its photon part.

Let us address some of these techniques:

- A local change in the dielectric constant of a cavity sample can be obtained by depositing **a patterned thin metallic layer** on top of the samples. This is sufficient to slightly modify the cavity mode energy and thus to induce a confinement potential on the polariton states. The group of Y. Yamamoto in Stanford used this technique to create periodic potential in one or two dimensions.[41,42] Polariton lattices of various geometries using this method will be discussed further at the end of this chapter. More recently other promising hybrid techniques have been implemented to shape the polariton potential landscape: a slab photonic crystal (PC) based on a sub-wavelength grating can be used for the top mirror[43]; a nanowire cavity can be positioned on a patterned substrate.[44]

- Another way to laterally confine polariton states is to **locally change the thickness of the cavity layer**. This was done prior to the growth of the top mirror in the group of B. Deveaud[36,45] (see Fig. 2(a)). Figure 2(b) shows polariton states measured on a cavity which has been processed into 0D mesas using this technique.

- **Surface acoustic waves** have also been used to modulate the polariton energy by the group of P. Santos.[46,47] With this technique, both the photon mode is affected via a local change in the medium refractive index, and the excitonic

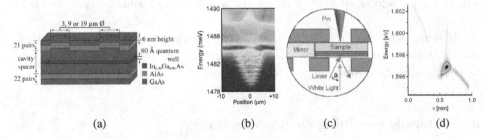

(a)   (b)   (c)   (d)

Fig. 2.   (a) Scheme of the mesa structures grown in the group of B. Deveaud by locally changing the cavity width. (b) Spatial and energy resolved emission from a mesa of $9\,\mu m$ width, showing a ladder of confined polariton states. From Ref. 36. (c) Experimental method developed in the group of D. Snoke to induce local strain in a planar microcavity by using a pin. (d) Spatial and energy resolved photoluminescence of the lower polariton branch at $k_{//} = 0$ (grey line) for a large area excitation showing a local minimum. The colored area shows the emission when a localized excitation spot is placed slightly off the position of the potential minimum, showing localisation in the trap. From Ref. 37.

energy via the local strain induced by the acoustic wave. This last effect is shown to be the dominant one. Periodic modulation of the polariton energy using this technique has been used with one or two trains of surface acoustic waves.[46,47]

- **A local strain** can be also applied locally on a cavity sample to change the exciton energy. Using a tip which pushes on the back side of the sample, the group of D. Snoke has successfully demonstrated the generation of a local trap in real space for polaritons[13,37] (see Fig. 2(c)–(d)).
- Finally the technique which will be used in all experiments on microstructured resonators described below, relies on the **deep etching** of the cavity layer.[48–50] Using electron beam lithography, any design can be transferred into a mask deposited on top of the cavity. Then dry etching allows imprinting the desired geometry into the cavity sample. This technique allows inducing strong lateral confinement because of the large refractive index difference between air and semiconductor. It does not impose any restriction on the shape of the structures except that their lateral dimension should not be too small (all lateral dimensions should be larger than typically 1.5 μm) to avoid non-radiative recombinations on the side walls of the etched structures, which would strongly degrade the optical properties of the system.

In the next subsections we show several examples of reduced dimensionality geometries, illustrating the potential of deep etching for polariton band engineering.

3.1. *One dimensional microcavities*

Cavity samples can be processed in the shape of elongated photonic wires.[50,51] Typically the lateral size L_x of the wires is of a few microns, while their extension along the (y) direction exceeds a few hundreds of microns. As a result, the k_x component of the polariton wave-vector becomes quantized because of lateral confinement of the photon mode, while free motion along the (y) direction is preserved. Thus $k_x = \frac{p_x \pi}{L_x}$, where p_x is an integer and L_x the lateral width of the wire.

Figure 3(a) shows angle resolved emission measurements performed on a single microwire evidencing the formation of polariton 1D sub-bands corresponding to different values of p_x. Notice that because of strain induced by the lateral patterning, a splitting develops between the photon mode with linear polarization parallel to the wire (TM) and the photon mode with polarization orthogonal to the wire axis (TE mode polarized along the (x) direction). Thus each sub-band of given index p_x is split into two bands, one TE and the other one TM linearly polarized.

The lateral confinement due to the finite dimension along the x direction induces a blueshift proportional to $\frac{1}{L_x^2}$. Thus if the lateral dimension is locally varied, an effective potential barrier or potential dip can be designed. Examples of such lateral shaping of one dimensional cavities are shown on Fig. 3. Figure 3(b) illustrates how a static periodic potential can be engineered by periodically modulated the wire lateral dimension.[52,53] Angularly resolved measurements on such modulated

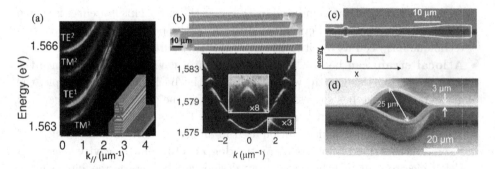

Fig. 3.   (a) Dispersion of polariton subbands arising from the lateral confinement in a 1D microwire (sketched in the inset). Note that the subbands are split in TE and TM linearly polarized modes. (b) Scanning electron microscopy image of polariton microwires with periodic lateral confinement. The corresponding dispersion shows the opening of gaps and the formation of minibands. From Ref. 52. (c) Microwire with an engineered localized obstacle and the corresponding potential profile. (d) Polariton interferometer. Image credits: CNRS-Laboratoire de Photonique et Nanostructures.

wire demonstrate the formation of mini-Brillouin zones together with the opening of mini-gaps. The widths of both the mini-bands and mini-gaps are fully controlled by the geometrical dimension of the wire.

Other polaritonic circuits are shown on Fig. 3: in (c) a single defect is created, defined by a single potential dip of fully controlled width and height. In (d) a more complex circuit defines a polariton interferometer.[54]

3.2.   *Zero dimensional cavities: single and coupled micropillars*

Cavity polaritons can be fully discretized when confined in micropillars with typical lateral dimensions of a few microns.[49,55,56] In this condition, both k_x and k_y the in-plane components of the polariton states, are quantized. For instance, if square shaped micropillars are considered (Fig. 4(a)), $k_x = \frac{p_x \pi}{L_x}$, and $k_y = \frac{p_y \pi}{L_y}$, with p_x and p_y being integers.[57] The confined polariton states are spectrally more separated as the pillar size is reduced as illustrated in Fig. 4(b)–(c).

Such zero dimensional resonators are the building blocks for more complex systems named photonic molecules, which are made of an assembly of coupled micropillars.[58] The simplest molecule is defined by two micropillars with centre to centre distance smaller than their diameter.[59] The discrete modes of each micropillar hybridize into coupled modes delocalized over the entire molecule. As illustrated in Fig. 4(g), the spatial shape of the polariton modes in the case of the two coupled micropillars can be directly imaged by photoluminescence measurements. Hybridization of the modes of the two pillars is evidenced, in very strong analogy with the case of diatomic molecules. Another interesting example of photonic molecules is made of a ring of six coupled micropillars[60] (Fig. 11(a)). In such structure the polariton states are delocalized over the entire structure, with quantized phase jumps between

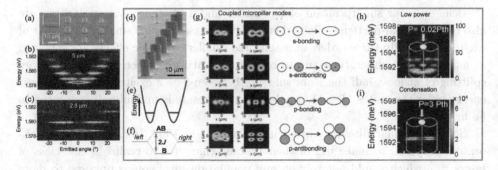

Fig. 4.   (a) Single micropillars of different sizes and shapes. (b)–(c) Emission spectra from squared micropillars with lateral size of 5 μm (b) and 2.8 μm (c). They show quantised energy levels with larger gaps in the second case due to the tighter confinement. (d) Array of coupled micropillars with different couplings J determined by the centre-to-centre distances. (e) The coupled pillars can be described as two coupled quantum wells, giving rise to bonding and antibonding eigenmodes (f) with spatial distributions extended over the whole molecule, as shown in (g) — left column: experiment, central column: calculation. When pumping the centre of the molecule (h), at high excitation density condensation is observed simultaneously in the s-bonding and s-antibonding states (i). From Ref. 61.

adjacent micropillars. This is similar to the hybridization of P_z orbitals in a benzene molecule, which shares a similar geometry. We will describe in more details further in the chapter the profound analogy between the photonic molecule and the real benzene molecule, and in particular we will discuss the existence of the analogous of a spin-orbit coupling in the photonic system.

These examples of microstructured resonators illustrate the huge potentiality of the polariton platform as a quantum simulator. They allow implementing different Hamiltonians with various potential geometries and strong non-linearities.

4.   Polariton condensation

As already mentioned above, excitons are quasi-particles made of an electron-hole pair bound by Coulombic interactions. They are issued from the pairing of two fermions and thus present a quasi-bosonic nature. Their effective mass, typically of the order of the electron mass, is several orders of magnitude smaller than that of, for instance, Rubidium atoms. As a result, exciton condensation is expected to be possible at temperatures of the order of a few tenths of Kelvin,[62] thus much higher than for cold atoms. Exciton condensation has been the subject of intense research but only recently long range order has been demonstrated with indirect excitons.[63] Nevertheless a lot of open questions remain concerning the precise interpretation of these experiments. In particular because of the strong exchange coupling, exciton condensates are expected to be dark.[64] Using the fact that the measurement of the emission energy allows to evaluate the local exciton density,[65] the group of F. Dubin recently brought experimental evidence of a dark exciton condensate.[66]

Being a linear superposition of an exciton and a photon, cavity polaritons are also bosonic quasi-particles. Since their effective mass is much smaller than that of excitons (typically three orders of magnitude smaller), bosonic effects are expected at much higher temperatures, up to room temperature.[67] Since there is an energy splitting of at least half the Rabi splitting between $J = 2$ excitons and the lowest energy polariton states ($J = 1$), a polariton condensate is undoubtedly expected to be bright.

Soon after their discovery of cavity polaritons in 1992, Imamoglu and collaborators proposed to make use of their bosonic character to realize a new class of lasers,[68,69] which would not rely on any population inversion as opposed to more conventional lasers. The basic principle of the polariton laser is the bosonic stimulation of polariton scattering induced by the macroscopic population of a polariton quantum state. First proposed for polariton-phonon scattering,[69] it was then generalized to polariton-polariton scattering.[33] In this framework, as soon as the occupancy of a polariton state exceeds unity, bosonic stimulation of the scattering toward this state results in the build-up of a macroscopic population of the considered state, and the appearance of spontaneous coherence.[70]

Evidence for the bosonic character of cavity polaritons was first shown in pair scattering processes (also named parametric scattering).[22,23] Pumping a polariton state resonantly close the inflexion point of the lower polariton branch, both parametric oscillation[22] and parametric amplification[23,71,72] were demonstrated with the build-up of a large occupancy of the signal (at $k_{//} = 0$) and idler (at twice the $k_{//}$ of the pump beam) polariton states. Thanks to the giant non-linearity of cavity polaritons, inherited from their excitonic content, the polariton based Optical Parametric Oscillator was shown to have a record low threshold[73,74] and has been put forward as a promising source of correlated photon beams.[75-77]

Two other methods have been used to generate a macroscopically occupied polariton state. These are resonant excitation, and non-resonant pumping. In the following we will address the specificity of each of these methods, illustrated by experimental examples.

4.1.　*Resonantly driven polariton condensates*

Tuning both the energy and the angle of incidence of the excitation laser beam, polaritons can be resonantly excited in a well-defined quantum state. In this excitation scheme, the polariton field is coherently driven: the in-plane wavevector and the energy of the polariton condensate are optically controlled, as well as its polarization. In a mean field approximation, the polariton field can be described by a set of Ginzburg Landau equations for the exciton and photon field, taking into account polariton-polariton interactions and the decay γ of the polariton state due to the finite lifetime of the photon inside the cavity.[17] These equations can be reformulated for the polariton field as follows:

$$i\hbar\partial_t\psi(r,t) = \left[D - \frac{i\hbar\gamma}{2} + V(r) + \hbar g|\psi(r,t)|^2\right]\psi(r,t) + F_p e^{i(k_p r - \omega t)}, \quad (5)$$

where D describes the polariton kinetic energy taking into account the non-parabolicity of the polariton dispersion, $V(r)$ is the static potential defining the cavity geometry, g the polariton-polariton interaction constant, and F_p the amplitude of the driving laser field.

This non-linear equation gives rise to a rich variety of phenomena such as optical bistability,[78,79] multistability[80,81] and optical switching.[82,83] An example of such non-linear effects is shown on Fig. 5(a). A cavity is resonantly pumped on the lower branch and the transmitted laser beam is monitored. Its intensity shows a clear hysteresis cycle when swiping up and down the excitation power. The existence of such bistable state is induced by polariton-polariton interactions.[79]

Polariton hydrodynamics can be explored in this resonant configuration. For instance, scattering by disorder has been investigated, sending a controlled coherent polariton flow onto a defect.[17,18] When the interaction energy is larger than the kinetic energy, polariton scattering around the defect disappears, signature of superfluidity. Polariton interactions renormalize the polariton dispersion in such a way that the elastic ring in k-space corresponding to the energy of the incident polariton flow entirely shrinks.[84] As a result no final state is available for elastic scattering and polaritons flow straight without being affected by the local defect. A very different situation is obtained when the kinetic energy is larger than the interaction energy. In this case, nucleation of topological defects such as integer vortices[85,86] or dark solitons[20,87] are observed in the wake of the obstacle. A review on polariton quantum hydrodynamics can be found in Ref. 88. An example of such experiments is shown on Fig. 5(b)–(c), where a dark oblique soliton is revealed on the wake of a defect in real space imaging and interferometry experiments.[20]

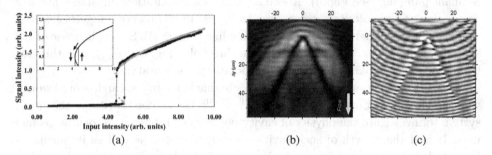

(a) (b) (c)

Fig. 5. (a) Bistable behavior of the transmitted signal when exciting a microcavity with an energy and momentum close to the inflexion point. From Ref. 79. (b) Real space emission from a polariton fluid encountering an obstacle in its flowpath. Two dark solitons nucleate in the wake of the obstacle, evidenced by a phase jump of π in an interferometric measurement (c). From Ref. 20.

Polariton hydrodynamic effects have also been studied in original optical parametric configurations that allow for the propagation of polariton bullets[72] and the controlled nucleation of vortices of different angular momenta.[89,90]

4.2.   *Polariton condensation under non resonant excitation*

Polariton condensation under non resonant excitation appears as the experimental scheme the closest to seminal experiments with cold atoms.[91,92] A laser beam, strongly blue detuned with respect to the polariton resonances, injects hot electron-hole pairs. After ultra-fast energy relaxation assisted by LO phonon emission, part of these electron-hole pairs bind and form excitons. They populate the so-called excitonic reservoirs (j = 1 states at large in-plane wave-vectors and j = 2 states). The lower polariton branch gets also populated, according to the balance between scattering rates and polariton decay.[93,94] Different scattering mechanisms contribute to the population of the polariton branch: exciton-phonon scattering,[31] exciton-exciton scattering,[33] and in some cases electron-exciton scattering.[32,95–97] When the occupancy of a given polariton state reaches unity, then scattering into this state becomes stimulated and a highly non-linear increase of its population is triggered. This population dynamics can be simulated by time dependent rate equations describing the population evolution of each polariton state along the different branches.[33,98] Depending on the relative strength between scattering rates and polariton lifetime, the steady state population distribution along the polariton branches can be thermalized or highly out of equilibrium.[21,99,100] In any case, one of the signatures of stimulated scattering and macroscopic occupation of a polariton state is the spontaneous appearance of spatial and temporal coherence. This coherence can be probed experimentally by interferometry experiments.

Polariton condensation under non-resonant optical excitation was observed in 2006 by the groups of B. Deveaud and D. Le Si Dang on a CdTe based microcavity.[12] In their seminal paper, these researchers reported the measurement of the polariton state occupancy in k-space as a function of excitation power under cw non resonant pumping (see Fig. 6). Above a well-defined excitation threshold, massive occupation of the polariton states close to $k_{//} = 0$ was observed, together with a significant spectral narrowing of the emission line (Fig. 6(a)). Strong increase of the spatial coherence was demonstrated above threshold (Fig. 6(b)–(c)), and this was highlighted as a smoking gun feature for polariton condensation.

Polariton condensation is now routinely obtained in many research groups worldwide, and in different material systems.[13–16,101–103] GaAs based cavities is the model system to investigate the physics of cavity polaritons in different resonator geometries. Indeed the growth of such cavities is well mastered as well as its engineering into lower dimension resonators. Nevertheless GaAs condensation is limited to low temperature range (typically below 40 K). Intense research is also dedicated to the development of high quality cavities with different materials hosting much more robust excitons and thus offering the possibility to observe polariton lasing

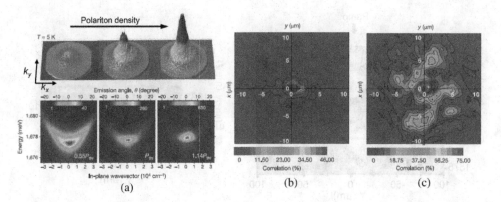

Fig. 6. Polariton condensation in a CdTe microcavity under out of resonance excitation. Condensation is evidenced by the macroscopic accumulation of polaritons at k = 0 above a given threshold density (a), and via the establishment of long range coherence (b), (c). Below threshold, coherence is restricted to the autocorrelation of individual polaritons (b). Above threshold, coherence extends over a macroscopic region (c). From Ref. 12.

up to room temperature. Impressive results have been obtained in GaN based samples,[15,101] in ZnO wires[29,104,105] or ZnO planar cavities[102,103] and also using organic emitters such as anthracene.[16]

4.3.   *Polariton interaction with the excitonic reservoir*

In the case of non-resonant excitation, the description of the polariton condensate wavefunction has to take into account the excitonic reservoir.[106,107] Indeed the excitonic reservoir not only populates the condensate, but also renormalizes its energy and its spatial distribution because of exciton-polariton interactions. A Ginzburg Landau equation can be adapted to the non-resonant excitation case as follows[106,107]:

$$i\hbar\partial_t\psi(r,t) = \left[D + \frac{i\hbar}{2}[R(n_R(r,t)) - \gamma] + V(r) + V_R(r,t) + \hbar g|\psi(r,t)|^2\right]\psi(r,t)$$

(6)

where $n_R(r,t)$ describes the excitonic population within the reservoir, $V_R(r,t)$ the potential induced by this reservoir and $R(n_R(r,t))$ the feeding of the polariton population induced by scattering of excitons from the excitonic reservoir. A coupled time dependent equation describes the dynamics of the excitonic reservoir which is pumped by the excitation laser beam.

   Equation (6), which describes the evolution of the polariton condensate amplitude, is analogous to the equation of a laser. The term $R(n_R(r,t))$ provides the gain of the polariton laser, arising from exciton and polariton relaxation in the lasing mode. The threshold for polariton lasing is achieved when the gain overcomes the losses (escape of the photon). Notice that the threshold does not require

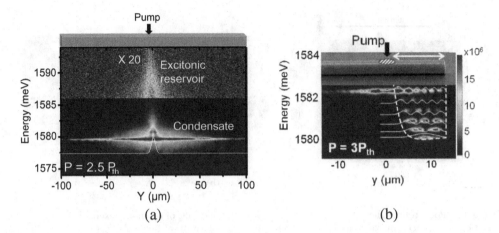

Fig. 7.   (a) Propagation of a polariton condensate in a 1D microwire under non-resonant excitation with a spot of $2\,\mu m$ in diameter. The injected excitonic reservoir creates a potential localised under the excitation spot that accelerates the polariton condensates away from it. From Ref. 108. (b) The potential induced reservoir can be used to confine polaritons in 0D traps when the excitation spot is close to the end of the wire. From Ref. 109.

any population inversion,[69] contrary to a conventional laser where the gain term corresponds to absorption below transparency.

Under non-resonant excitation conditions, peculiar physical properties of polariton condensates arise from their repulsive interactions with the highly populated excitonic reservoir. For instance, the resulting blueshift can be used to put a polariton condensate in motion and induce its expansion over macroscopic distances[108] as illustrated in Fig. 7(a). In this experiment, a one-dimensional microcavity is non-resonantly excited with a laser beam which is tightly focused on a micron sized region of the cavity. Polariton condensation occurs at $k_{//} = 0$ in the excitation region, where the polariton states are blueshift because of the presence of the excitonic reservoir. Because of this optically induced local potential, polaritons tend to be expelled from the excitation region, thus converting this interaction energy into kinetic energy. This is why spatially resolved emission measurements reveal a polariton condensate which spreads over the entire wire. Interferometry experiments show that the spontaneous coherence of the polariton condensate is preserved during this propagation and expansion along the wire. The potential induced by the excitonic reservoir can be also used to engineer a fully reconfigurable trap, in which polaritons are confined and can macroscopically occupy a quantum state. Such optical trapping has been obtained first when positioning the excitation spot close to the end of a wire cavity[108] (see Fig. 7(b)) or using several excitation spots in a planar cavity.[110,111] More complex geometries are now envisaged, sculpting a complex potential landscape with light[112,113] via the creation of several localized excitonic reservoirs.

4.4. *Polariton condensation in a periodic potential: interplay between reservoir interaction and self-interaction*

As we have discussed above, polariton condensates undergo both strong self-interactions and interactions with the excitonic reservoir. This is a specificity of this non-linear photonic system which can manifest itself in particular spatial features. An illustration of the interplay between self-interaction and reservoir interaction is found when considering polariton condensation in a static periodic potential.[52] In this case, the equation describing the polariton wavefunction reads as follows:

$$i\hbar\partial_t\psi(r,t) = \left[D + \frac{i\hbar}{2}[R(n_R(r,t)) - \gamma] + V_{per}(r) + V_R(r,t) + \hbar g|\psi(r,t)|^2\right]\psi(r,t)$$

(7)

V_{per} (r) describes a periodic potential which can be obtained for instance by modulating the width of a 1D microcavity.[52] As previously described, $V_R(r,t)$ describes the repulsive potential due to the excitonic reservoir which, in the case we want to consider here, is a Gaussian potential with a width of the order of the periodic potential period. Depending on the relative strength of the reservoir potential as compared to self-interaction, different localized solutions exist (Fig. 8(a)). For instance, if we

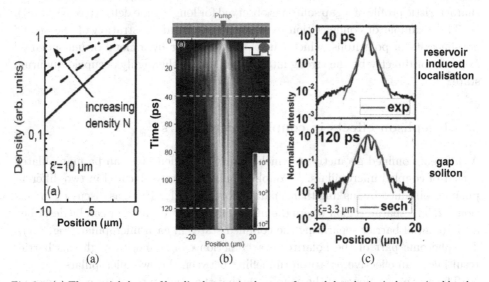

(a)   (b)   (c)

Fig. 8. (a) The spatial shape of localised states in the gap of a modulated wire is determined by the interplay between polariton self-interaction (sech^2 (x), large N, dot-dashed line) and interactions with the reservoir (exponential decay, solid line). (b) Time resolved image of the emission from the gap state under pulsed excitation. At short times, interactions with the reservoir dominate the shape of the emission, (c) upper panel. At later times, (c) lower panel, the reservoir is emptied and a gap soliton shape emerges. From Ref. 52.

neglect the self-interaction term, the localized potential induced by the reservoir can be considered as a defect in the periodic potential. Thus it creates a localized mode within the forbidden gap, centered onto the reservoir and with exponential decay on each side of it. The opposite case is the one where the reservoir potential can be neglected as compared to the self-interaction energy. Then non-linear localized modes exist, which are named gap solitons[114–117] and have a characteristic profile in the shape of a sech^2 function. Thus their profile is much smoother than the exponential profile of a defect state. Gap solitons, localized modes induced by a Kerr type non-linearity, have been observed both for non-linear optical waveguides[115–118] and also with cold atoms in a periodic potential.[161] If one computes the spatial profile of the polariton gap states solutions of Eq. (7), when increasing the ratio between the reservoir potential and the self-interaction energy, a continuous change in the spatial profile from an exponential defect state to a solitonic solution is observed (see Fig. 8(a)). This qualitative change in the spatial profiles can be directly monitored when performing a condensation experiment on a periodically modulated wire cavity under pulsed non resonant excitation. Indeed under such experimental conditions, the reservoir potential dominates self-interaction energy at short time delays. As a result, when a polariton condensate forms within the gap, its spatial profile is exponential, characteristic of a defect-like state. Then because of stimulated exciton and polariton relaxation into the condensate, the reservoir is depleted and its potential is strongly attenuated. As a result, self-interaction becomes predominant and the characteristic profile of a gap soliton is observed for longer time delays (see Fig. 8(c)).

This example of condensation experiment is a clear illustration of the strong non-linearity of polaritons, which can be driven either by a cloud of uncondensed excitons, or directly by the strong interactions within the highly occupied quantum state.

5.   Josephson effects in coupled micropillars

A different confined geometry in which polariton condensates can be manipulated is that of coupled micropillars.[61] Coupling polariton states located in two different pillars is obtained by making them spatially overlap (Fig. 4(d)), as discussed in Section 3.2. The spatial region where the micropillars meet each other can be described as a potential barrier separating the confined states in each micropillar (Fig. 4(e)). The photonic part of the polariton wavefunction can tunnel through this barrier resulting in an effective polariton tunnelling between the two micropillars.

The first consequence of the coupling is that the eigenstates of the system extend over the whole molecule. If we concentrate on the ground states $|\psi_{L/R}\rangle$ of two identical round micropillars, of "s" geometry, the coupling gives rise to a bonding (B) and an antibonding (AB) state which, in the tight binding approach, can be expressed as:

$$|\psi_B\rangle = \frac{1}{\sqrt{2}}(|\psi_L\rangle + |\psi_R\rangle) \quad |\psi_{AB}\rangle = \frac{1}{\sqrt{2}}(|\psi_L\rangle - |\psi_R\rangle), \tag{8}$$

whose energy splitting is given by twice the coupling energy J (Fig. 4(f)). Both states are characterised by amplitudes with maxima in the centre of each micropillar, in-phase for the bonding state and in anti-phase for the antibonding state. Under non-resonant excitation and above a given threshold, simultaneous condensation in both states has been observed, as shown in Fig. 4(i). Note that condensation in several states at the same time is possible in pump dissipative systems. The steady state is not determined by the thermodynamics but by the interplay between pumping, relaxation and losses. The mode competition between different relaxation channels may result in the formation of multiple condensates, as shown in Fig. 4(i) and also as described in Refs. 38, 40.

This very simple system made out of two coupled quantum states shows remarkable dynamical phenomena. If we restrict ourselves to the bonding and antibonding modes, the double well can be viewed as an example of a bosonic junction in which Josephson effects can be studied. In the original proposal of Josephson, two super-conductors, described as a two macroscopic bosonic states, are coupled via a piece of "normal" material which serves as a weak tunnel link. Josephson showed that different current and voltage oscillatory regimes can be reached, the so-called a.c. and d.c. Josephson effects.[119] If we take into account particle interactions, non-linear oscillations and self-trapping are expected[120] and have been studied using two component atomic condensates.[121,122] This dynamics is difficult to access in usual photonic systems due to the required photon lifetimes (larger than the oscillation period) and the need for significant interactions. The polaritonic molecules are in this sense an excellent photonic platform to study these non-linear phases. The most straightforward method is to prepare the system in a given initial state and let it evolve in time, a situation that can be modelled by two coupled non-linear Schrödinger equations:

$$i\hbar\frac{d\psi_L}{dt} = \left(E_L^0 + \hbar g|\psi_L|^2 - i\frac{\hbar}{2\tau} \right) \psi_L - J\psi_R \tag{9}$$

$$i\hbar\frac{d\psi_R}{dt} = \left(E_R^0 + \hbar g|\psi_R|^2 - i\frac{\hbar}{2\tau} \right) \psi_R - J\psi_L \tag{10}$$

where $E_{L(R)}^0$ is the ground state energy of the left (right) micropillar in the absence of coupling and τ is the polariton lifetime, given by the escape of photons out of the cavity. Experimentally, the initialisation of the system can be done with a short laser pulse in resonance with both the bonding and antibonding states, in a Gaussian spot size covering the whole molecule. By selecting the position of the laser with respect to the centre of the molecule, we can prepare a predefined linear combination of B and AB modes resulting in an initial state with more particles on one pillar than the other (for instance, all particles in the left site). In the absence of interactions and decay, this will not be an eigenstate, and we expect the system to show coherent oscillations between the two micropillars. Decay does not qualitatively modify the picture, and allows monitoring the oscillations by looking

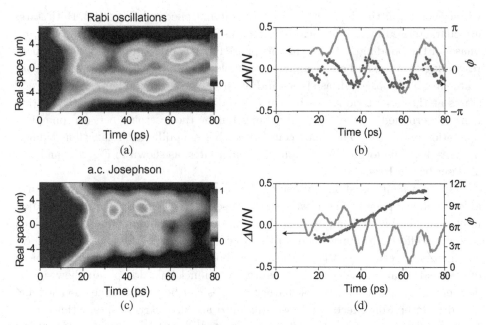

Fig. 9.   (a) Intensity oscillations in the light emitted by two coupled micropillars at low excitation power, in the regime of coherent Rabi oscillations $(E_L^0 - E_R^0 = 0)$. (b) Measured population imbalance $\Delta N/N$ and phase difference ϕ. (c) Oscillations in the a.c. Josephson regime $(E_L^0 - E_R^0 \neq 0)$ and corresponding $\Delta N/N$ and ϕ (d). From Ref. 123.

at the time resolved emission from each of the micropillars of the molecule using a streak camera.

Intensity oscillations in the non-interacting regime are shown in Fig. 9 under low power excitation. The relative population $\Delta N/N = (|\psi_R|^2 - |\psi_L|^2)/(|\psi_R|^2 + |\psi_L|^2)$ oscillates with a period of 21 ps, fully determined by the B-AB splitting $(T = h/2J)$. It is interesting to note that while the populations continuously decrease due to the escape of photons, the magnitude $\Delta N/N$ in the non-interacting regime is not affected by this decay.[123] We can follow in time the phase difference between the emission coming from each pillar by performing an interferometric measurement.[124] The solution of Eqs. (9) and (10) in these conditions predicts an oscillation of the phase difference around a value $\phi = 0$, and identical period as for the intensity oscillations. This behaviour is experimentally observed (Fig. 9(b) red points), showing the power of time resolved photoluminescence to study the dynamics of coupled micropillars.

The regime we have just shown corresponds to the usual situation of Rabi oscillations in a two level system when prepared in a coherent superposition of eigenstates. A different situation takes place when the two micropillars have different ground state energies $(\Delta E = E_L^0 - E_R^0 \neq 0)$. In this case the wavefunctions corresponding to the B and AB modes have different amplitudes in each micropillar (contrary to the case of $\Delta E = 0$). Thus, even if the initial state is perfectly centred in the molecule, the system will start to oscillate as it does not correspond to an eigenstate. This regime is shown in Fig. 9(c). In order to induce a $\Delta E \neq 0$, a non-resonant laser beam is added to the right micropillar. This beam, of very weak intensity, creates a population of non-condensed excitons that blueshifts the ground state energy E_R^0 of that micropillar, leaving unchanged the energy of the other. A different strategy would be to change the energy of the confined ground state of one of the micropillars by changing its size. The oscillations observed in Fig. 9(c) are still given by the B-AB splitting, which is now larger and, thus, the period is shorter. Additionally, if we look at the evolution of the phase difference (Fig. 9(d) red points), we observe a monotonous increase.[124] This situation corresponds to the a.c. effect originally predicted by Josephson between two superconductors. In that case, the role of $\Delta E \neq 0$ is played by a voltage difference externally set across the junction, resulting in the appearance of an a.c. current.

More interesting phenomena appear when the excitation density is increased and the polariton-polariton interactions start to play a role. This is the case shown in Fig. 10, with excitation conditions similar to those of the above described Rabi oscillations but with a strong initial injected density. Polariton self-interactions give rise to a blueshift proportional to the population of each micropillar. When the system oscillates, the energy of each micropillar will dynamically change, resulting in an acceleration of the oscillations. This situation is explored experimentally in Fig. 10(a), where the observed oscillation period at short times after the arrival of the pulse is smaller than the nominal period of 21 ps expected from the tunnel coupling. At later time, particles escape out of the system and interactions dim away recovering the expected single particle oscillation period.[123]

The anharmonic oscillations just discussed take place for moderate population imbalances. If the initial population imbalance is above the threshold $\Delta N/N(t = 0) > (4J/\hbar gN)[(\hbar gN/2J) - 1]^{1/2}$, the system enters the regime of self-trapping: the highly populated micropillar gets so blueshifted with respect to the other one that the tunnel energy is not high enough to compensate for the interaction energy (Fig. 10(c)). Particles in the highly populated micropillar cannot lose energy to

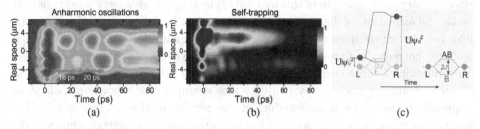

Fig. 10. (a) Anharmonic intensity oscillations observed at high excitation densities and an initial $\Delta N/N = 0.62$. (b) Self trapping obtained at high excitation density and initial $\Delta N/N = 0.98$. In this case, the initial blueshift of the right pillar decouples the system and polaritons remain in that pillar — (c) left panel. At later times, the usual B-AB splitting is recovered and oscillations set in — (c) right panel. From Ref. 123.

tunnel to the other micropillar and stay trapped indefinitely. This is the picture for equilibrium systems, but in the case of polaritons the continuous escape of particles gradually reduces the blueshift and the system recovers the oscillatory regime[123] at long times (Fig. 10(b),(c)).

The observation of non-linear Josephson effects shows that the high confinement achieved in micropillars can enhance the polariton non-linear effects. For instance, it has been shown in a double well for atomic condensates, that non-linear bifurcations[122] and spin-squeezing[125] can be achieved. In the case of polaritons this configuration could be used to generate non-classical states of light.

6.   Ring structures: engineering geometric phases

6.1.   *The polariton benzene molecule*

Confined coupled structures do not only enhance non-linear effects, they can also be used to study the properties of systems with periodic boundary conditions. These geometries are ubiquitous in physics, from the subnanometric scales of the benzene C_6H_6 molecule to the Bohm-Aharonov rings in mesoscopic systems.[126] The optical properties of polaritons provide a unique opportunity to directly visualise the wavefunctions in this kind of geometries and, for instance, gain insights on the fine structure effects that arises when the spin degree of freedom is taken into account.

One of the archetypical ring structures is the benzene-like hexagonal molecule. A benzene molecule is formed by six carbon atoms in a hexagonal structure. The $2s$ and two of the $2p$ orbitals of carbon hybridize to form three sp^2 orbitals arranged in the plane of the molecule. They give rise to the covalent bonding chaining the six carbon atoms and linking each of them to a hydrogen atom. The remaining non-hybridized "π" orbital of each carbon atom contains the valence electrons, which can tunnel from carbon to carbon (Fig. 11(b)). These electrons are responsible for most of the physical properties of the molecule.

It has been speculated that the ring shape of benzene is responsible for the extraordinary diamagnetism of this molecule. The ring-like geometry would not only enable the formation of diamagnetic charge currents in the presence of a magnetic field, but the nature itself of the eigenstates of the molecule would contain permanent spin currents.[127] Questions have arisen on the effects of spin-orbit coupling in the formation of these currents. The engineering of a benzene molecule analogue in a semiconductor microcavity is a very nice example of the power of polaritons to provide hints to the understanding of the physics of the original system. For instance, in this case, we will see that the important role of geometric phases in the electron currents in this molecule can be directly viewed using polaritons.

A benzene molecule analogue can be fabricated in a semiconductor microcavity by etching a hexagonal structure formed by six coupled micropillars (Fig. 11(a)). If we consider only the ground state of the micropillars and first neighbour tunnelling,

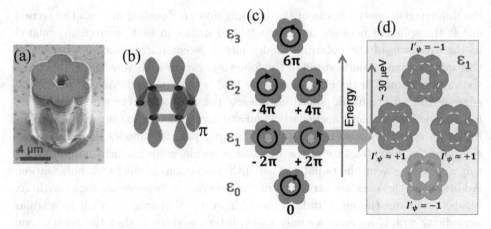

Fig. 11. (a) Scanning electron microscope image of a benzene polariton molecule. (b) Scheme of an actual benzene molecule: black dots represent carbon atoms and blue lines the covalent bonds. The green volumes are the π orbitals from which valence electrons can tunnel. (c) Representation of the vortical eigenstates obtained from a tight binding calculation without spin. (d) Fine structure of the ε_1 level including the polariton pseudospin and spin-orbit coupling effects.

a scalar tight binding model predicts four energy levels $\varepsilon_{|l|}$. The eigenfunctions $|l\rangle = \sum_k 6^{-1/2} \exp(il\pi k/3)|k\rangle$ have maxima in the centre of each of the micropillars $|k\rangle$ ($k = 0, \ldots, 5$), with phase $l\pi k/3$ where $l = 0, \pm1, \pm2, 3$. The eigenstates with $l = \pm1$ and $l = \pm2$ present phase vortices of charge l, that is, their phase increases by $2\pi l$ when circumventing the molecule (Fig. 11(c)). This particular form of the eigenstates is determined by the periodic boundary conditions. The intensity and phase structure of the eigenstates can be directly extracted from interferometric measurements performed on the light emitted by the polariton molecule under low power excitation density, as shown in Ref. 60.

The vortical phase structure of the eigenstates already announces that the diamagnetic properties of benzene are indeed related to the establishing of permanent currents, inbuilt in the form of the eigenstates. In order to get a full description of the properties of the molecule, we need to consider the spin degree of freedom. Polaritons are bosons in the low density regime with only two possible values of the third component of their intrinsic angular momentum ($j_z = \pm1$), mapping into $\sigma+$ and $\sigma-$ circular polarisation of the emitted light. The intrinsic angular momentum can thus be understood as a pseudospin $1/2$ which doubles the degeneracy of each energy level: $|l, \sigma\pm\rangle$.

6.2. *Spin-orbit coupling*

The basis $|l, \sigma\pm\rangle$ is not the basis of eigenstates of our benzene analogue molecule. In order to find it, we need to take into account the effect of the confined geometry on the polariton polarisation. We can consider two different effects. The first one is the dependence on linear polarisation tunnelling between two micropillars,[59] related to the difference in spatial shape of the bonding and antibonding modes. The second one is the splitting between linearly polarised modes in each micropillar, related to the extension of the polariton modes into adjacent micropillars. The origin of this splitting can be understood in the following way. In a 1D waveguide, like those shown in Section 3.1, the electromagnetic field boundary conditions on the lateral edges of the wire result in polaritons linearly polarised along the wire (TM) being less confined than polaritons linearly polarised perpendicular to the wire (TE). This induces a splitting between TE and TM linearly polarised modes. If we now consider the ring-like structure of the benzene molecule, a confinement-induced splitting appears between the radial and azimuthal directions of the linear polarisation. Additionally, the necklace structure of the polaritonic benzene-analogue with six beads, modulates the magnitude of the radial-azimuthal splitting with an angular periodicity $\pi/3$. If we consider only this splitting, and we neglect the polarisation dependent tunnelling (treated elsewhere[60]), a new term appears in the Hamiltonian of the form of a spin-orbit interaction:[60]

$$H_{l_+ l_-}^{SO} = -\frac{\eta}{2}(\delta_{l_+ - l_-, -2} + \delta_{l_+ - l_-, +4}), \tag{11}$$

where η quantifies the splitting and l_+, l_- express the quantum numbers associated to the basis $|l, \sigma\pm\rangle$.

The effect of the spin-orbit coupling is to split the fourfold degenerate states corresponding to $|l| = 1$ and $|l| = 2$. For the case of $|l| = 1$ (Fig. 11(d)), the lowest energy state is linearly polarised azimuthally, while the highest energy state is polarised radially. Remarkably, if the molecule is excited out of resonance at high excitation density, polariton condensation takes place in the lowest energy of these states, polarised azimuthally as experimentally measured (Fig. 12(a)). It is interesting to analyse in detail the spin-phase structure of this state, which can be written: $(|-1, \sigma+\rangle - |+1, \sigma-\rangle)/\sqrt{2}$. It is formed by a current of spin up polaritons circulating clockwise and a current of spin down polaritons circulating counter-clockwise.

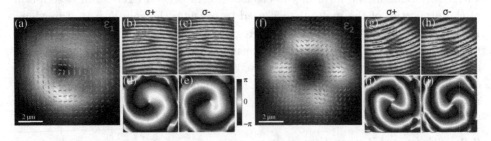

Fig. 12. (a) Emitted intensity when condensation takes place in one of the ε_1 substates of the benzene molecule. The green lines show the orientation of the local linear polarisation. (b), (c) Interferometric measurement of the $\sigma+$, $\sigma-$ circularly polarised emission, showing counterpropagating phase vortices of charge -2π, $+2\pi$ as evidenced in (d), (e), respectively. (f)–(j) Same as (a)–(e) when condensation takes place in an ε_2 substate. In this case, the vortices in circular polarisation have a charge $+4\pi$, -4π. From Ref. 60.

Experimentally, these currents can be measured by selecting the $\sigma\pm$ polarisation of emission. An interferometry measurement reveals clockwise and counter-clockwise vortices corresponding to $\sigma+$ and $\sigma-$, respectively (Fig. 12(b)–(e)).

If the excitation density is changed, condensation in higher orbital momentum states can be observed. This is the case shown in Fig. 12(f)–(j), where the excitation density favors the condensation in the state $|\psi\rangle = (|2,\sigma+\rangle + |-2,\sigma-\rangle)/\sqrt{2}$. Indeed, a direct measurement of the phase reveals that the currents are twice as large (the phase changes by $\pm 4\pi$ when going around the molecule), and of opposite sense, as compared to Fig. 12(a)–(e).

7. Polariton lattices

The coupling of few confined polariton structures can be extended to the fabrication of two dimensional polariton lattices. Several techniques have been used to fabricate such lattices. The surface acoustic waves technique mentioned in Section 3 has allowed the engineering of squared lattices made out of the interference of two acoustic waves propagating in orthogonal directions (Fig. 13(a)). Two dimensional minibands and minigaps have been reported as shown in Fig. 13(b), with gap energies controlled by the intensity of the surface waves. Another possibility is the evaporation of a layer of gold on top of the microcavity, which changes the energy of the photonic mode as described in Section 3. If a network of holes is designed on the gold layer, the polariton energy can be modulated in a 2D periodic lattice. This technique has been extensively used in the group of Y. Yamamoto to produce square[42] (Fig. 13(c)), triangular[134] and Kagome lattices.[135] Despite their shallowness, these lattices present the interest of allowing the study of condensation in excited states with a tailored momentum distribution. For instance, in the case of squared lattices,[42] condensation in so-called d-wave states was observed (Fig. 13(d)).

The coupling of etched micropillars used for the benzene analogue molecules described above can also be employed to generated two dimensional lattices. The advantage of this technique is that the strong lateral confinement provided by the micropillar structure gives rise to deep lattices in almost any geometry and with a well-controlled coupling. One example is the honeycomb lattice shown in Fig. 13(c).[162] This lattice is the same as that of graphene, and both share the same geometric properties. The honeycomb structure can be thought as two shifted triangular lattices (green and red points in Fig. 13(e)). Each sublattice gives rise to shifted periodic bands in momentum space which cross at singular points. Around these points the bands are conical and have a linear dispersion Fig. 13(f). This particular geometry in real and momentum space confers extraordinary electron transport properties to graphene. The two-sublattice character of the structures can be expressed in terms of a pseudospin. The pseudospin of electrons with momenta close to the Dirac points is conserved under smooth variations of the lattice potential. This results on the ballistic transport and the quenching of backscattering of

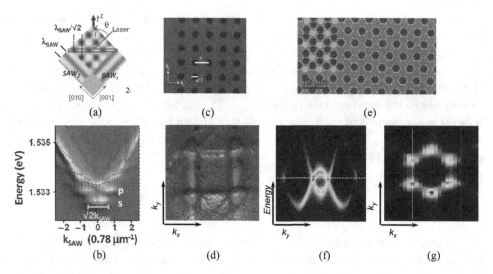

Fig. 13.   (a) Squared lattice formed by the superposition of two orthogonally propagating surface acoustic waves.[142] (b) The momentum resolved spectrum along the framed direction in (a) shows minibands and minigaps similar to those of Fig. 3(b) in a 1D system. (c) Squared lattice made out of a holed thin gold layer evaporated on top of the microcavity.[42] (d) Polariton condensation takes place in metastable d-wave states. (e) Honeycomb lattice made out of coupled micropillars etched out of a planar semiconductor microcavity at CNRS-Laboratoire de Photonique et Nanostructures. (f) Energy selected momentum space image showing emission from the Dirac points. (f) Energy-momentum cut along the dotted line in (g) shows the Dirac cones.

Dirac electrons. A remarkable manifestation of this effect is Klein tunnelling[136]: the perfect transmission of electrons through a potential barrier, with consequences on the localisation/anti-localisation of electrons when disorder is present in the structure.

All these phenomena have been indirectly studied in actual graphene samples. However, it is very difficult to control the vector momentum of the electrons propagating in the sample. The polariton analogue system shown in Fig. 13(f) provides an excellent platform to study these phenomena. One of the advantages of polaritons is that injection of particles with a well-defined momentum can be straightforwardly done by resonant pumping with a selected angle allowing the study of the polariton dynamics in the presence of engineered potential barriers.

The direct optical access to the wavefunction via the emitted light and the high degree of control in the engineering of structures place the polariton system as a remarkable platform to study 1D and 2D lattices. The combination of band gap engineering and polariton interactions provide new opportunities to study localisation in quasi-crystal structures, strongly correlated phases in lattices with flat bands or frustration effects when accounting for the polariton spin degree of freedom. Additionally, interesting Berry phase effects are expected in the honeycomb lattice when the structure is slightly altered. For example, when the two triangular sublattices are disymmetrised a Berry curvature appears at the momenta of the

Dirac cones.[137] This strongly affects the dynamics of wavepackets with momenta close to those points.[138] If instead an axial deformation of the lattice is engineered, a gauge field appears acting on particles as if they were charges in a magnetic field. This phenomenon has been demonstrated with photons in a quasi-hexagonal waveguide array.[139] The possibility to design synthetic gauge fields for photons in photonic and polaritonic lattices opens the way to the engineering of photonic topological insulators with edge states having directional circulation protected from disorder.[140,141]

8. Conclusion

As illustrated along this chapter, cavity polaritons present extraordinary physical properties inherited from their dual exciton-photon nature. On the one hand, they propagate like photons, and on the other hand they strongly interact with their environment, a key property which enables their manipulation.

Cavity polaritons can be injected in photonic circuits and manipulated via optical means. Many theoretical works propose now to make use of such a platform to develop new photonic devices, based on the giant polariton non-linearity.[143–145] Innovative polariton devices start to be implemented, such as polariton transistors,[83,146] polariton resonant tunneling diodes[147] or polariton interferometers.[54] These recent achievements illustrate the huge potential of cavity polaritons for non-linear optics. Up to now, these demonstrations have been implemented at low temperatures, but strong research efforts are also dedicated to push this system toward room temperature operation, using materials with more robust excitons. Also very important for applications is the possibility to electrically inject polaritons. In doped cavity samples, angle resolved electroluminescence have proved that cavity polaritons can be pumped electrically.[148–150] Up to now, polariton lasing under electrical injection has only been obtained using a strong magnetic field at low temperature.[151] Nevertheless further improvement of the optical quality of these laser diodes should allow operating the device at zero field.

From the fundamental point of view, cavity polaritons provide the analogous of a Bose Einstein condensate in a solid-state system, but with specificities due to their out of equilibrium character. For instance, their finite lifetime has been shown to stabilize dark solitons[152] so that these topological excitations have been first observed in a semiconductor cavity.[20] Also very interesting is the investigation of spatial coherence of out of equilibrium condensates. Recent theoretical investigation point out that the spatial coherence decay should strongly depend both on the dimensionality of the system and on its departure from equilibrium.[100,153]

We have shown that a rich variety of techniques have been developed to engineer the potential landscape in which polaritons are generated. Thus cavity polaritons appear as a versatile platform to implement non-linear Hamiltonians and study non-linear topological excitations. In addition, the spin currents measured for instance in the photonic molecules present similarities with the chiral edge states recently proposed in photonic topological insulators in birefringent metamaterials.[141] In the polariton system, the spin-orbit coupling is controlled via the engineering of the radial-azimuthal photonic confinement, opening new avenues to create photonic spin Hall states and spin topological insulators.[140]

Finally when brought to the ultimate regime of single polariton non-linearity, cavity polaritons are foreseen as a promising system to investigate the quantum phase of an out of equilibrium Bose Hubbard Hamiltonian.[154–158] Very interesting quantum effects such as quantum polariton blockade[159] or strongly correlated polariton states[155,160] are envisioned in this highly non-linear regime.

Acknowledgments

The authors would like to warmly acknowledge all their collaborators from Laboratoire de Photonique et de Nanostructures and from LASMEA, Institut Pascal.

References

1. E. M. Purcell, Spontaneous emission probabilities at radio frequencies, *Phys. Rev.* **69**, 681 (1946).
2. J. M. Gérard et al., Enhanced spontaneous emission by quantum boxes in a monolithic optical microcavity, *Phys. Rev. Lett.* **81**, 1110–1113 (1998).
3. P. Michler et al., A quantum dot single-photon turnstile device, *Science* **290**, 2282–2285 (2000).
4. E. Moreau et al., Single-mode solid-state single photon source based on isolated quantum dots in pillar microcavities, *Appl. Phys. Lett.* **79**, 2865–2867 (2001).
5. M. Pelton et al., Efficient source of single photons: A single quantum dot in a micropost microcavity, *Phys. Rev. Lett.* **89**, 233602 (2002).
6. J. J. Hopfield, Theory of the contribution of excitons to the complex dielectric constant of crystals, *Phys. Rev.* **112**, 1555–1567 (1958).
7. J. J. Hopfield and D. G. Thomas, Theoretical and experimental effects of spatial dispersion on the optical properties of crystals, *Phys. Rev.* **132**, 563–572 (1963).
8. C. Weisbuch and R. G. Ulbrich, Resonant polariton fluorescence in gallium arsenide, *Phys. Rev. Lett.* **39**, 654–656 (1977).
9. R. G. Ulbrich and C. Weisbuch, Resonant brillouin scattering of excitonic polaritons in gallium arsenide, *Phys. Rev. Lett.* **38**, 865–868 (1977).
10. C. Weisbuch, M. Nishioka, A. Ishikawa, and Y. Arakawa, Observation of the coupled exciton-photon mode splitting in a semiconductor quantum microcavity, *Phys. Rev. Lett.* **69**, 3314–3317 (1992).
11. M. Richard et al., Experimental evidence for nonequilibrium Bose condensation of exciton polaritons, *Phys. Rev. B* **72**, 201301(R) (2005).
12. J. Kasprzak et al., Bose-Einstein condensation of exciton polaritons, *Nature* **443**, 409–414 (2006).

13. R. Balili, V. Hartwell, D. Snoke, L. Pfeiffer, and K. West, Bose-Einstein condensation of microcavity polaritons in a trap, *Science* **316**, 1007–1010 (2007).

14. E. Wertz *et al.*, Spontaneous formation of a polariton condensate in a planar GaAs microcavity, *Appl. Phys. Lett.* **95**, 051108 (2009).

15. S. Christopoulos *et al.*, Room-temperature polariton lasing in semiconductor microcavities, *Phys. Rev. Lett.* **98**, 126405 (2007).

16. S. Kena-Cohen and S. R. Forrest, Room-temperature polariton lasing in an organic single-crystal microcavity, *Nature Phot.* **4**, 371–375 (2010).

17. I. Carusotto and C. Ciuti, Probing microcavity polariton superfluidity through resonant rayleigh scattering, *Phys. Rev. Lett.* **93**, 166401 (2004).

18. A. Amo *et al.*, Superfluidity of polaritons in semiconductor microcavities, *Nature Phys.* **5**, 805–810 (2009).

19. K. G. Lagoudakis *et al.*, Quantized vortices in an exciton-polariton condensate, *Nature Phys.* **4**, 706–710 (2008).

20. A. Amo *et al.*, Polariton superfluids reveal quantum hydrodynamic solitons, *Science* **332**, 1167–1170 (2011).

21. J. Keeling, F. M. Marchetti, M. H. Szymanska, and P. B. Littlewood, Collective coherence in planar semiconductor microcavities, *Semicond. Sci. Technol.* **22**, R1–R26 (2007).

22. R. M. Stevenson *et al.*, Continuous wave observation of massive polariton redistribution by stimulated scattering in semiconductor microcavities, *Phys. Rev. Lett.* **85**, 3680–3683 (2000).

23. P. G. Savvidis *et al.*, Angle-resonant stimulated polariton amplifier, *Phys. Rev. Lett.* **84**, 1547 (2000).

24. J. P. Karr, A. Baas, R. Houdré, and E. Giacobino, Squeezing in semiconductor microcavities in the strong-coupling regime, *Phys. Rev. A* **69**, 031802 (2004).

25. E. L. Bolda, R. Y. Chiao, and W. H. Zurek, Dissipative Optical flow in a nonlinear Fabry-Pérot cavity, *Phys. Rev. Lett.* **86**, 416–419 (2001).

26. W. Wan, S. Jia, and J. W. Fleischer, Dispersive superfluid-like shock waves in nonlinear optics, *Nature Phys.* **3**, 46–51 (2007).

27. R. Houdré *et al.*, Measurement of cavity-polariton dispersion curve from angle-resolved photoluminescence experiments, *Phys. Rev. Lett.* **73**, 2043 (1994).

28. G. Bastard, *Wave Mechanics Applied to Semiconductor Heterostructures* (Les Editions de Physique, Les Ulis, 1988).

29. A. Trichet *et al.*, Long range correlations in a 97% excitonic one-dimensional polariton condensate, *Phys. Rev. B* **88**, 121407(R) (2013).

30. T. Freixanet, B. Sermage, A. Tiberj, and R. Planel, In-plane propagation of excitonic cavity polaritons, *Phys. Rev. B* **61**, 7233–7236 (2000).

31. F. Tassone, C. Piermarocchi, V. Savona, A. Quattropani, and P. Schwendimann, Bottleneck effects in the relaxation and photoluminescence of microcavity polaritons, *Phys. Rev. B* **56**, 7554–7563 (1997).

32. G. Malpuech, A. Kavokin, A. Di Carlo, and J. J. Baumberg, Polariton lasing by exciton-electron scattering in semiconductor microcavities, *Phys. Rev. B* **65**, 153310 (2002).

33. F. Tassone and Y. Yamamoto, Exciton-exciton scattering dynamics in a semiconductor microcavity and stimulated scattering into polaritons, *Phys. Rev. B* **59**, 10830–42 (1999).

34. C. Ciuti, V. Savona, C. Piermarocchi, A. Quattropani, and P. Schwendimann, Role of the exchange of carriers in elastic exciton-exciton scattering in quantum wells, *Phys. Rev. B* **58**, 7926–7933 (1998).

35. C. Ciuti, V. Savona, C. Piermarocchi, A. Quattropani, and P. Schwendimann, Threshold behavior in the collision broadening of microcavity polaritons, *Phys. Rev. B* **58**, R10123 (1998).

36. R. Idrissi Kaitouni *et al.*, Engineering the spatial confinement of exciton polaritons in semiconductors, *Phys. Rev. B* **74**, 155311 (2006).

37. R. B. Balili, D. W. Snoke, L. Pfeiffer, and K. West, Actively tuned and spatially trapped polaritons, *Appl. Phys. Lett.* **88**, 031110 (2006).

38. D. Sanvitto *et al.*, Exciton-polariton condensation in a natural two-dimensional trap, *Phys. Rev. B* **80**, 045301 (2009).

39. A. Baas *et al.*, Synchronized and desynchronized phases of exciton-polariton condensates in the presence of disorder, *Phys. Rev. Lett.* **100**, 170401 (2008).

40. D. N. Krizhanovskii, *et al.*, Coexisting nonequilibrium condensates with long-range spatial coherence in semiconductor microcavities, *Phys. Rev. B* **80**, 045317 (2009).

41. C. W. Lai *et al.*, Coherent zero-state and π-state in an exciton–polariton condensate array, *Nature* **450**, 529 (2007).

42. N. Y. Kim *et al.*, Dynamical d-wave condensation of exciton-polaritons in a two-dimensional square-lattice potential, *Nature Phys.* **7**, 681–686 (2011).

43. B. Zhang *et al.*, Zero dimensional polariton laser in a sub-wavelength grating based vertical microcavity, *Arxiv:1304.2061* (2013).

44. Z. Chen, *PLMCN14 Conference* (2013).

45. O. El Daïf *et al.*, Polariton quantum boxes in semiconductor microcavities, *Appl. Phys. Lett.* **88**, 061105 (2006).

46. E. A. Cerda-Méndez *et al.*, Polariton condensation in dynamic acoustic lattices, *Phys. Rev. Lett.* **105**, 116402 (2010).

47. E. A. Cerda-Méndez *et al.*, Dynamic exciton-polariton macroscopic coherent phases in a tunable dot lattice, *Phys. Rev. B* **86**, 100301 (2012).

48. J. Bloch *et al.*, Strong-coupling regime in pillar semiconductor microcavities, *Superlatt. and Microst.* **22**, 371 (1998).

49. T. Gutbrod *et al.*, Weak and strong coupling of photons and excitons in photonic dots, *Phys. Rev. B* **57**, 9950–9956 (1998).

50. A. Kuther *et al.*, Confined optical modes in photonic wires, *Phys. Rev. B* **58**, 15744–15748 (1998).

51. G. Dasbach, M. Schwab, M. Bayer, D. N. Krizhanovskii, and A. Forchel, Tailoring the polariton dispersion by optical confinement: Access to a manifold of elastic polariton pair scattering channels, *Phys. Rev. B* **66**, 201201 (2002).

52. D. Tanese *et al.*, Polariton condensation in solitonic gap states in a one-dimensional periodic potential, *Nat. Commun.* **4**, 1749 (2013).

53. M. Bayer *et al.*, Optical demonstration of a crystal band structure formation, *Phys. Rev. Lett.* **83**, 5374–5377 (1999).

54. C. Sturm *et al.*, Giant phase modulation in a Mach-Zehnder exciton-polariton interferometer, *arXiv:1303.1649* (2013).

55. J. Bloch *et al.*, Strong-coupling regime in pillar semiconductor microcavities, *Superlattice. Microst.* **371**, 22 (1998).

56. G. Panzarini and L. C. Andreani, Quantum theory of exciton polaritons in cylindrical semiconductor microcavities, *Phys. Rev. B* **60**, 16799–16806 (1999).

57. G. Dasbach, M. Schwab, M. Bayer, and A. Forchel, Parametric polariton scattering in microresonators with three-dimensional optical confinement, *Phys. Rev. B* **64**, 201309 (2001).

58. M. Bayer *et al.*, Optical modes in photonic molecules, *Phys. Rev. Lett.* **81**, 2582–2585 (1998).

59. S. Michaelis de Vasconcellos *et al.*, Spatial, spectral, and polarization properties of coupled micropillar cavities, *Appl. Phys. Lett.* **99**, 101103 (2011).

60. V. G. Sala *et al.*, Topological spin currents in a benzene-like polariton molecule, *(submitted)* (2013).

61. M. Galbiati *et al.*, Polariton condensation in photonic molecules, *Phys. Rev. Lett.* **108**, 126403 (2012).

62. J. M. Blatt, K. W. Böer and W. Brandt, Bose-Einstein Condensation of Excitons, *Phys. Rev.* **126**, 1691-1692 (1962).

63. A. A. High *et al.*, Spontaneous coherence in a cold exciton gas, *Nature* **483**, 584–588 (2012).

64. M. Combescot, O. Betbeder-Matibet, and R. Combescot, Bose-Einstein condensation in semiconductors: The key role of dark excitons, *Phys. Rev. Lett.* **99**, 176403 (2007).

65. Z. Vörös, D. W. Snoke, L. Pfeiffer, and K. West, Direct measurement of exciton-exciton interaction energy, *Phys. Rev. Lett.* **103**, 016403 (2009).

66. M. Alloing *et al.*, Evidence for a Bose-Einstein condensate of excitons, *arXiv:1304. 4101* (2013).

67. G. Malpuech *et al.*, Room-temperature polariton lasers based on GaN microcavities, *Appl. Phys. Lett.* **81**, 412–414 (2002).

68. A. Imamoglu and R. Ram, Quantum dynamics of exciton lasers, *Phys. Lett. A* **214**, 193 (1996).

69. A. Imamoglu, R. J. Ram, S. Pau, and Y. Yamamoto, Nonequilibrium condensates and lasers without inversion: Exciton-polariton lasers, *Phys. Rev. A* **53**, 4250 (1996).

70. F. P. Laussy, G. Malpuech, A. Kavokin, and P. Bigenwald, Spontaneous coherence buildup in a polariton laser, *Phys. Rev. Lett.* **93**, 016402 (2004).

71. M. Saba *et al.*, High-temperature ultrafast polariton parametric amplification in semiconductor microcavities, *Nature* **414**, 731–735 (2001).

72. A. Amo *et al.*, Collective fluid dynamics of a polariton condensate in a semiconductor microcavity, *Nature* **457**, 291–295 (2009).

73. J. J. Baumberg *et al.*, Parametric oscillation in a vertical microcavity: A polariton condensate or micro-optical parametric oscillation, *Phys. Rev. B* **62**, R16247–R16250 (2000).

74. C. Diederichs *et al.*, Parametric oscillation in vertical triple microcavities, *Nature* **440**, 904 (2006).

75. C. Ciuti, Branch-entangled polariton pairs in planar microcavities and photonic wires, *Phys. Rev. B* **69**, 245304 (2004).

76. J. P. Karr, A. Baas, and E. Giacobino, Twin polaritons in semiconductor microcavities, *Phys. Rev. A* **69**, 063807 (2004).

77. S. Portolan, O. Di Stefano, S. Savasta, and V. Savona, Emergence of entanglement out of a noisy environment: The case of microcavity polaritons, *Europhys. Lett.* **88** (2009).

78. A. Baas, J. P. Karr, H. Eleuch, and E. Giacobino, Optical bistability in semiconductor microcavities, *Phys. Rev. A* **69**, 023809 (2004).

79. A. Baas, J.-P. Karr, M. Romanelli, A. Bramati, and E. Giacobino, Optical bistability in semiconductor microcavities in the nondegenerate parametric oscillation regime: Analogy with the optical parametric oscillator, *Phys. Rev. B* **70**, 161307 (2004).

80. N. A. Gippius *et al.*, Polarization multistability of cavity polaritons, *Phys. Rev. Lett.* **98**, 236401 (2007).

81. T. K. Paraïso, M. Wouters, Y. Leger, F. Mourier-Genoud, and B. Deveaud-Pledran, Multistability of a coherent spin ensemble in a semiconductor microcavity, *Nature Mater.* **9**, 655–660 (2010).

82. A. Amo, *et al.*, Exciton-polariton spin switches, *Nature Phot.* **4**, 361–366 (2010).
83. D. Ballarini *et al.*, All-optical polariton transistor, *Nat. Commun.* **4**, 1778 (2013).
84. C. Ciuti and I. Carusotto, Quantum fluid effects and parametric instabilities in microcavities, *Phys. Stat. Sol. (b)* **242**, 2224–2245 (2005).
85. G. Nardin *et al.*, Hydrodynamic nucleation of quantized vortex pairs in a polariton quantum fluid, *Nature Phys.* **7**, 635–641 (2011).
86. D. Sanvitto *et al.*, All-optical control of the quantum flow of a polariton superfluid, *Nature Phot.* **5**, 610–614 (2011).
87. R. Hivet *et al.*, Half-solitons in a polariton quantum fluid behave like magnetic monopoles, *Nature Phys.* **8**, 724–728 (2012).
88. I. Carusotto and C. Ciuti, Quantum fluids of light, *Rev. Mod. Phys.* **85**, 299 (2013).
89. D. Sanvitto *et al.*, Persistent currents and quantized vortices in a polariton superfluid, *Nature Phys.* **6**, 527–533 (2010).
90. D. N. Krizhanovskii *et al.*, Effect of interactions on vortices in a nonequilibrium polariton condensate, *Phys. Rev. Lett.* **104**, 126402 (2010).
91. K. B. Davis *et al.*, Bose-Einstein condensation in a gas of sodium atoms, *Phys. Rev. Lett.* **75**, 3969–3973 (1995).
92. M. H. Anderson, J. R. Ensher, M. R. Matthews, C. E. Wieman, and E. A. Cornell, Observation of bose-einstein condensation in a dilute atomic vapor, *Science* **269**, 198–201 (1995).
93. P. Senellart, J. Bloch, B. Sermage, and J. Y. Marzin, Microcavity polariton depopulation as evidence for stimulated scattering, *Phys. Rev. B* **62**, R16263 (2000).
94. A. I. Tartakovskii *et al.*, Relaxation bottleneck and its suppression in semiconductor microcavities, *Phys. Rev. B* **62**, R2283–R2286 (2000).
95. P. G. Lagoudakis *et al.*, Electron-polariton scattering in semiconductor microcavities, *Phys. Rev. Lett.* **90**, 206401 (2003).
96. M. Perrin, P. Senellart, A. Lemaitre, and J. Bloch, Polariton relaxation in semiconductor microcavities: Efficiency of electron-polariton scattering, *Phys. Rev. B* **72**, 075340–5 (2005).
97. A. Qarry *et al.*, Nonlinear emission due to electron-polariton scattering in a semiconductor microcavity, *Phys. Rev. B* **67**, 115320 (2003).
98. M. M. Glazov *et al.*, Polariton-polariton scattering in microcavities: A microscopic theory, *Phys. Rev. B* **80**, 155306 (2009).
99. G. Malpuech, Y. G. Rubo, F. P. Laussy, P. Bigenwald, and A. V. Kavokin, Polariton laser: Thermodynamics and quantum kinetic theory, *Semicond. Sci. Technol.* **18**, S395 (2003).
100. A. Chiocchetta and I. Carusotto, Non-equilibrium quasi-condensates in reduced dimensions, *EPL* **102**, 67007 (2013).
101. G. Christmann, R. Butte, E. Feltin, J.-F. Carlin, and N. Grandjean, Room temperature polariton lasing in a GaN/AlGaN multiple quantum well microcavity, *Appl. Phys. Lett.* **93**, 051102 (2008).
102. T. Guillet *et al.*, Polariton lasing in a hybrid bulk ZnO microcavity, *Appl. Phys. Lett.* **99**, 161104 (2011).
103. F. Li *et al.*, From excitonic to photonic polariton condensate in a ZnO-based microcavity, *Phys. Rev. Lett.* **110**, 196406 (2013).
104. A. Trichet *et al.*, One-dimensional ZnO exciton polaritons with negligible thermal broadening at room temperature, *Phys. Rev. B* **83**, 041302 (2011).
105. H. Franke, C. Sturm, R. Schmidt-Grund, G. Wagner, and M. Grundmann, Ballistic propagation of excitonic polariton condensates in a ZnO-based microcavity, *New J. Phys.* **14**, 013037 (2012).

106. M. Wouters and I. Carusotto, Excitations in a nonequilibrium bose-einstein condensate of exciton polaritons, *Phys. Rev. Lett.* **99**, 140402 (2007).

107. M. Wouters, I. Carusotto, and C. Ciuti, Spatial and spectral shape of inhomogeneous nonequilibrium exciton-polariton condensates, *Phys. Rev. B* **77**, 115340 (2008).

108. E. Wertz *et al.*, Spontaneous formation and optical manipulation of extended polariton condensates, *Nature Phys.* **6**, 860–864 (2010).

109. L. Ferrier *et al.*, Interactions in confined polariton condensates, *Phys. Rev. Lett.* **106**, 126401 (2011).

110. G. Tosi *et al.*, Sculpting oscillators with light within a nonlinear quantum fluid, *Nature Phys.* **8**, 190–194 (2012).

111. P. Cristofolini *et al.*, Optical superfluid phase transitions and trapping of polariton condensates, *Phys. Rev. Lett.* **110**, 186403 (2013).

112. A. Amo *et al.*, Light engineering of the polariton landscape in semiconductor microcavities, *Phys. Rev. B* **82**, 081301 (2010).

113. G. Tosi *et al.*, Geometrically locked vortex lattices in semiconductor quantum fluids, *Nat. Commun.* **3**, 1243 (2012).

114. W. Chen and D. L. Mills, Gap solitons and the nonlinear optical response of superlattices, *Phys. Rev. Lett.* **58**, 160–163 (1987).

115. H. S. Eisenberg, Y. Silberberg, R. Morandotti, A. R. Boyd, and J. S. Aitchison, Discrete spatial optical solitons in waveguide arrays, *Phys. Rev. Lett.* **81**, 3383–3386 (1998).

116. J. W. Fleischer, T. Carmon, M. Segev, N. K. Efremidis, and D. N. Christodoulides, Observation of discrete solitons in optically induced real time waveguide arrays, *Phys. Rev. Lett.* **90**, 023902 (2003).

117. J. W. Fleischer, M. Segev, N. K. Efremidis, and D. N. Christodoulides, Observation of two-dimensional discrete solitons in optically induced nonlinear photonic lattices, *Nature* **422**, 147–150 (2003).

118. B. J. Eggleton, R. E. Slusher, C. M. de Sterke, P. A. Krug, and J. E. Sipe, Bragg grating solitons, *Phys. Rev. Lett.* **76**, 1627–1630 (1996).

119. B. D. Josephson, Possible new effects in superconductive tunnelling, *Phys. Lett.* **1**, 251–253 (1962).

120. S. Raghavan, A. Smerzi, S. Fantoni, and S. R. Shenoy, Coherent oscillations between two weakly coupled Bose-Einstein condensates: Josephson effects, π oscillations, and macroscopic quantum self-trapping, *Phys. Rev. A* **59**, 620–633 (1999).

121. M. Albiez *et al.*, Direct observation of tunneling and nonlinear self-trapping in a single bosonic josephson junction, *Phys. Rev. Lett.* **95**, 010402 (2005).

122. T. Zibold, E. Nicklas, C. Gross, and M. K. Oberthaler, Classical bifurcation at the transition from Rabi to Josephson dynamics, *Phys. Rev. Lett.* **105**, 204101 (2010).

123. M. Abbarchi *et al.*, Macroscopic quantum self-trapping and Josephson oscillations of exciton polaritons, *Nature Phys.* **9**, 275–279 (2013).

124. K. G. Lagoudakis, B. Pietka, M. Wouters, R. André, and B. Deveaud-Plédran, Coherent oscillations in an exciton-polariton Josephson junction, *Phys. Rev. Lett.* **105**, 120403 (2010).

125. B. Juliá-Díaz *et al.*, Dynamic generation of spin-squeezed states in bosonic Josephson junctions, *Phys. Rev. A* **86**, 023615 (2012).

126. R. A. Webb, S. Washburn, C. P. Umbach, and R. B. Laibowitz, Observation of h/e Aharonov-Bohm oscillations in normal-metal rings, *Phys. Rev. Lett.* **54**, 2696–2699 (1985).

127. J. E. Hirsch, Spin-split states in aromatic molecules, *Mod. Phys. Lett. B* **04**, 739–749 (1990).

128. J. E. Hirsch, Spin-split states in aromatic molecules and superconductors, *Phys. Lett. A* **374**, 3777–3783 (2010).

129. D. Loss, P. Goldbart, and A. V. Balatsky, Berry's phase and persistent charge and spin currents in textured mesoscopic rings, *Phys. Rev. Lett.* **65**, 1655–1658 (1990).

130. A. Tomita and R. Y. Chiao, Observation of Berry's topological phase by use of an optical fiber, *Phys. Rev. Lett.* **57**, 937–940 (1986).

131. M. V. Berry, Quantal phase factors accompanying adiabatic changes, *Proc. R. Soc. Lond. A* **392**, 45–57 (1984).

132. J. N. Ross, The rotation of the polarization in low birefringence monomode optical fibres due to geometric effects, *Opt. Quantum Electron.* **16**, 455–461 (1984).

133. Y. K. Kato, R. C. Myers, A. C. Gossard, and D. D. Awschalom, Observation of the spin hall effect in semiconductors, *Science* **306**, 1910–1913 (2004).

134. N. Y. Kim *et al.*, Exciton-polariton condensates near the Dirac point in a triangular lattice, *New J. Phys.* **15**, 035032 (2013).

135. N. Masumoto *et al.*, Exciton-polariton condensates with flat bands in a two-dimensional kagome lattice, *New J. Phys.* **14**, 065002 (2012).

136. P. Allain and J. N. Fuchs, Klein tunneling in graphene: Optics with massless electrons, *EPJ B* **83**, 301–317 (2011).

137. G. Sundaram and Q. Niu, Wave-packet dynamics in slowly perturbed crystals: Gradient corrections and Berry-phase effects, *Phys. Rev. B* **59**, 14915–14925 (1999).

138. M. Cominotti and I. Carusotto, Berry curvature effects in the Bloch oscillations of a quantum particle under a strong (synthetic) magnetic field, *EPL* **103**, 10001 (2013).

139. M. C. Rechtsman *et al.*, Strain-induced pseudomagnetic field and photonic Landau levels in dielectric structures, *Nature Phot.* **7**, 153–158 (2013).

140. C. L. Kane and E. J. Mele, Z_2 Topological Order and the quantum spin hall effect, *Phys. Rev. Lett.* **95**, 146802 (2005).

141. A. B. Khanikaev *et al.*, Photonic topological insulators, *Nature Mater.* **12**, 233–239 (2013).

142. E. A. Cerda-Méndez *et al.*, Dynamic exciton-polariton macroscopic coherent phases in a tunable dot lattice, *Phys. Rev. B* **86**, 100301 (2012).

143. T. C. H. Liew, A. V. Kavokin, and I. A. Shelykh, Optical circuits based on polariton neurons in semiconductor microcavities, *Phys. Rev. Lett.* **101**, 016402 (2008).

144. T. C. H. Liew *et al.*, Exciton-polariton integrated circuits, *Phys. Rev. B* **82**, 033302 (2010).

145. T. Espinosa-Ortega and T. C. H. Liew, Complete architecture of integrated photonic circuits based on and and not logic gates of exciton polaritons in semiconductor microcavities, *Phys. Rev. B* **87**, 195305 (2013).

146. C. Anton *et al.*, Dynamics of a polariton condensate transistor switch, *Appl. Phys. Lett.* **101**, 261116 (2012).

147. H. S. Nguyen *et al.*, Realization of a double-barrier resonant tunneling diode for cavity polaritons, *Phys. Rev. Lett.* **110**, 236601 (2013).

148. D. Bajoni *et al.*, Polariton light-emitting diode in a GaAs-based microcavity, *Phys. Rev. B* **77**, 113303 (2008).

149. A. A. Khalifa, A. P. D. Love, D. N. Krizhanovskii, M. S. Skolnick, and J. S. Roberts, Electroluminescence emission from polariton states in GaAs-based semiconductor microcavities, *Appl. Phys. Lett.* **92**, 061107 (2008).

150. S. I. Tsintzos, N. T. Pelekanos, G. Konstantinidis, Z. Hatzopoulos, and P. G. Savvidis, A GaAs polariton light-emitting diode operating near room temperature, *Nature* **453**, 372–375 (2008).

151. P. Bhattacharya, B. Xiao, A. Das, S. Bhowmick, and J. Heo, Solid state electrically injected exciton-polariton laser, *Phys. Rev. Lett.* **110**, 206403 (2013).

152. A. M. Kamchatnov and S. V. Korneev, Condition for convective instability of dark solitons, *Phys. Lett. A* **375**, 2577–2580 (2011).

153. G. Roumpos *et al.*, Power-law decay of the spatial correlation function in exciton-polariton condensates, *PNAS* **109**, 6467 (2012).

154. M. P. A. Fisher, P. B. Weichman, G. Grinstein, and D. S. Fisher, Boson localization and the superfluid-insulator transition, *Phys. Rev. B* **40**, 546–570 (1989).

155. M. J. Hartmann, F. G. S. L. Brandao, and M. B. Plenio, Strongly interacting polaritons in coupled arrays of cavities, *Nature Phys.* **2**, 849–855 (2006).

156. A. D. Greentree, C. Tahan, J. H. Cole, and L. C. L. Hollenberg, Quantum phase transitions of light, *Nature Phys.* **2**, 856–861 (2006).

157. D. G. Angelakis, M. F. Santos, and S. Bose, Photon-blockade-induced Mott transitions and XY spin models in coupled cavity arrays, *Phys. Rev. A* **76**, 031805 (2007).

158. A. Le Boité, G. Orso, and C. Ciuti, Steady-state phases and tunneling-induced instabilities in the driven dissipative Bose-Hubbard model, *Phys. Rev. Lett.* **110**, 233601 (2013).

159. A. Verger, C. Ciuti, and I. Carusotto, Polariton quantum blockade in a photonic dot, *Phys. Rev. B* **73**, 193306 (2006).

160. I. Carusotto *et al.*, Fermionized photons in an array of driven dissipative nonlinear cavities, *Phys. Rev. Lett.* **103**, 033601 (2009).

161. B. Eiermann *et al.*, Bright Bose-Einstein gap solitons of atoms with repulsive interaction, *Phys. Rev. Lett.* **92**, 230401 (2004).

162. T. Jacqmin *et al.*, Direct observation of Dirac cones and a flatband in a honeycomb lattice for polaritons, arXiv:1310.8105 (2013).

Chapter 7

Quantum Plasmonics

Darrick Chang

ICFO - Institut de Ciencies Fotoniques
Mediterranean Technology Park
08860 Castelldefels (Barcelona), Spain
darrick.chang@icfo.es

Robust techniques to achieve and manipulate strong coupling between individual quantum emitters and single photons are a critical resource for quantum information processing and single-photon nonlinear optics. Despite remarkable theoretical and experimental progress, however, such techniques remain elusive. Here, we describe a novel approach of "quantum plasmonics" that has been introduced in recent years, in which quantum emitters interact strongly with the large fields of tightly confined, nanoscale surface plasmons. We present an overview of the underlying theoretical concepts of plasmonics and of strong coupling, describe examples of how plasmonic systems can be used for quantum information processing protocols, and provide an outlook for the field.

1. Introduction

Developing new techniques to control and manipulate the interaction between light and matter has been a central theme in science and engineering for many decades. These tools find wide-ranging applications in fields as diverse as spectroscopy and imaging, communications, metrology, and ultrafast science. Recently, much effort has been devoted to pushing these ideas down to the level of the constituent particles of light and matter — single atoms and single photons. This interest stems from the wide belief that quantum mechanical phenomena will play an increasingly important role in emerging fields and applications, such as quantum computing and quantum information science. Over the past decade, there has been tremendous progress in manipulating single-atom, single-photon interactions using cavity quantum electrodynamics (QED),[1,2] in which single atoms are coupled to a photonic mode of a high-finesse optical microcavity. In the "strong coupling" regime of cavity QED when the coupling strength between atom and cavity mode far exceeds dissipation into the environment, such a system comprises an elegant platform for observing fundamental processes such as reversible transfer of excitation between a single atom and cavity mode (Rabi oscillations).[3] When combined with quantum optical techniques for coherent manipulation, strong coupling also forms the basis for applications such as efficient single-photon generation[4-6] and realization of single-photon nonlinearities (*e.g.*, single-photon blockade.[7]). In addition, strongly coupled atom-cavity systems constitute the building blocks for important theoretical proposals including quantum state transfer and entanglement distribution,[8] controlled phase gates between photons,[9] and the simulation of exotic many-body condensed matter phenomena using photons.[10-12] However, despite the spectacular theoretical and experimental progress in cavity QED, it remains technically challenging and difficult to scale.

It is thus tempting to ask whether there exist alternative physical systems where individual atoms and photons can be strongly and coherently coupled. A recent, separate line of ideas and experiments has provided new insights into how efficient coupling can be achieved without a cavity.[13-17] Here, as illustrated schematically in Fig. 1(a), the interaction between a single atom and photon is made efficient by tightly focusing the optical spatial mode to near the diffraction limit, with a beam area $A \sim \lambda_0^2$, where λ_0 is the resonant wavelength of the atom. The increase in efficiency can be naively understood by noting that the resonant scattering cross section for an atom (within the paraxial optics approximation) is $\sigma_{sc} = \frac{3\lambda_0^2}{2\pi}$, and thus the scattering probability for a single photon is $P_{sc} \sim \sigma_{sc}/A$ and can come close to unity. A more subtle calculation beyond paraxial theory shows that the interaction probability can in principle reach unity, although in practice one will be limited by the degree of focusing possible with realistic optics.[13]

While tight focusing of free-space fields is technologically difficult, it is now routinely achieved using guided modes of nanophotonic devices (see Fig. 1(b)). This is in fact the foundation of the field of "plasmonics," which takes advantage of the special properties of optical excitations guided at a metal-dielectric interface.[18] In Section 2, we first provide a brief introduction to the properties of surface plasmons, and show how extremely tight field confinement or focusing can be achieved using

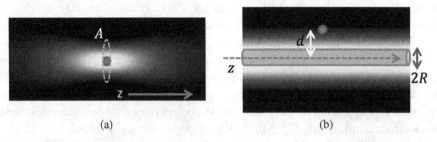

(a)   (b)

Fig. 1.   (a) In free space, a single photon propagating in a Gaussian beam (along z, with color-coded intensity profile) and focused to an area A interacts with a quantum emitter with probability $P \propto \lambda_0^2/A$, where λ_0 is the resonant wavelength. The diffraction limit $A \gtrsim \lambda_0^2$ makes it difficult to achieve high-efficiency interaction. (b) In contrast, surface plasmons guided along a metallic nanowire can be confined to transverse areas $A \sim R^2$, where R is the wire radius. For sub-wavelength wires, this ensures a highly efficient interaction between a single surface plasmon and a quantum emitter placed within the evanescent field of the wire.

metallic nanowires. In Section 3, we then discuss how the tight confinement enables highly efficient spontaneous emission of proximal quantum emitters into the surface plasmon modes. This efficient coupling between individual plasmons (*i.e.*, single photons) and quantum emitters can form the basis for quantum information and single-photon nonlinear devices. As an example, a protocol to implement a single-photon transistor is discussed in Section 4.

2. Introduction to surface plasmons

Metals that contain nearly free electrons have long been known to possess unique electromagnetic properties as compared to their normal, positive-dielectric counterparts. The frequency dependence of the electric permittivity of these metals is often well-approximated by a Drude model,

$$\epsilon(\omega) = 1 - \frac{\omega_p^2}{\omega^2 + i\omega\gamma_p}. \tag{1}$$

Here ω_p is the plasma frequency of the conductor, and γ_p is a parameter that characterizes material losses. Temporarily ignoring the losses, it can be seen that at frequencies below the plasma frequency, $\omega < \omega_p$, the permittivity of these metals is negative. The negative value of ϵ leads to a number of important consequences; for example, perhaps the most basic is that electromagnetic energy cannot propagate within the bulk of such materials.

Although the propagation of electromagnetic energy within the bulk is prohibited, the broken translational invariance of a metal surface does allow for the guided propagation of energy.[18] The resulting collective excitations of the electromagnetic field and charge density waves are known as surface plasmon polaritons, or more simply, surface plasmons (SPs). The simplest example is that of a flat, infinite interface that separates bulk regions of positive permittivity ϵ_1 and negative permittivity ϵ_2, as shown in Fig. 2(a). Guessing plane wave solutions on each side of the interface and enforcing appropriate boundary conditions for the fields, it is straightforward to show that the system supports guided modes of the form

$$\mathbf{E}_i \sim E_0 \left(\hat{z} + i\frac{k_\parallel}{\kappa_{i\perp}}\hat{x} \right) e^{ik_\parallel z - \kappa_{i\perp}|x| - i\omega t}. \tag{2}$$

Here $i = 1, 2$ denotes the regions of different permittivity. The wavevector $k_\parallel$ along the direction of propagation z satisfies

$$k_\parallel^2 = \left(\frac{\omega}{c} \right)^2 \frac{\epsilon_1 \epsilon_2}{\epsilon_1 + \epsilon_2}, \tag{3}$$

while the perpendicular components of the wavevector $k_{i\perp} = i\kappa_{i\perp}$ are imaginary and satisfy

$$\kappa_{i\perp}^2 = k_\parallel^2 - \epsilon_i \left(\frac{\omega}{c} \right)^2. \tag{4}$$

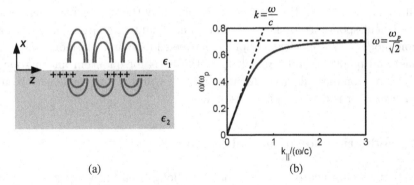

Fig. 2.   (a) Schematic of guided surface plasmon mode propagating on a flat metal-dielectric interface. The curves indicate electric field lines for these modes, while the + and − symbols illustrate the charge density wave component of these excitations. (b) Surface plasmon dispersion relation for a flat interface, taking $\epsilon_1 = 1$ and $\epsilon_2 = 1 - (\omega_p/\omega)^2$. The dispersion curve lies to the right of the light line $k = \omega/c$, indicating that these modes are guided. The upper frequency cutoff for these modes is given by $\omega = \omega_p/\sqrt{2}$.

The imaginary perpendicular wavevectors indicate that the field decays exponentially away from the interface, with the fields localized around a characteristic distance $d_\perp \sim \kappa_\perp^{-1}$. If $\epsilon_2(\omega)$ has the Drude-model frequency dependence of Eq. (1), the dispersion relation $k_\parallel(\omega)$ for the SPs takes the form shown in Fig. 2(b). It should be noted that the dispersion curve lies to the right of the light line, $k = \omega/c$, which indicates that these modes are guided. Also, at frequencies near the plasmon resonance $\omega = \omega_p/\sqrt{2}$, the wavevector $k_\parallel$ becomes very large, indicating that the SP wavelength $\lambda_{\mathrm{pl}} = 2\pi/k_\parallel$ can become much smaller than the vacuum wavelength at the same frequency. By use of Eq. (4) it also follows that at these frequencies, the SPs can be transversely confined near the interface to distances much smaller than the vacuum wavelength. Here, the presence of free charge plays a crucial role in the ability of SPs to circumvent the diffraction limit.

While we have considered a flat metal-dielectric interface in the above example, it turns out that the geometry of the interface can play a very important role in determining the localization of the SPs and the positions of plasmon resonances. The role of structure shape in determining the propagation, channeling, and focusing of electromagnetic energy has led to a number of dramatic observations involving SPs ranging from detection of single molecule fluorescence[19] to extraordinary light transmission through sub-wavelength apertures.[20] The potential applications of SPs to fields such as biosensing,[19] subwavelength imaging,[21] and photonic nano-circuitry[22] are being actively investigated.

A simple example of the effect of geometry on SP propagation, which is also relevant to coupling to quantum emitters, consists of a metallic nanowire, as illustrated in Fig. 1(b). As in the case of the planar interface, one can write down an ansatz for the field in each dielectric region, which enables separation of variables of the electromagnetic wave equation. In cylindrical coordinates, this takes the form

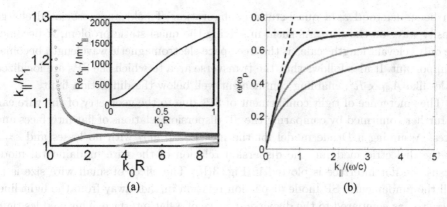

(a)   (b)

Fig. 3.   (a) SP wavevectors $k_\parallel/k_1$ for a metallic nanowire, as a function of dimensionless wire radius $k_0 R$. Modes with winding number $|m| \geq 1$ are cut off as $R \to 0$, while the wavevector of the $m = 0$ fundamental mode scales like $k_\parallel \sim 1/R$. Inset: the propagative losses for the fundamental mode, characterized by the ratio $\mathrm{Re}\,k_\parallel/\mathrm{Im}\,k_\parallel$. (b) Comparison of SP dispersion relations on a flat metal-dielectric interface versus the fundamental mode of a metallic nanowire, for the case of a Drude metal. Going from left to right, the curves correspond to a flat metal-dielectric interface, a wire of radius $k_0 R = 0.5$, and of radius $k_0 R = 0.2$. The dotted lines denote the light line $k_\parallel = \omega/c$ and upper frequency cutoff $\omega = \omega_p/\sqrt{2}$. Smaller wire sizes pull the fundamental mode dispersion relation further away from the light line, as compared to the flat interface.

$\mathbf{E}_i(\mathbf{r}) = \mathcal{E}_{i,m}\mathbf{E}_{i,m}(k_{i\perp}\rho)e^{im\phi}e^{ik_\parallel z}$, where $i = 1, 2$ denotes the regions outside and inside the cylinder, respectively, and m denotes the winding number of the mode. The vector fields $\mathbf{E}_{i,m}$ take the form of Bessel or Hankel functions. Enforcing field boundary conditions at the wire edge allows the SP dispersion relation and coefficients $\mathcal{E}_{i,m}$ to be uniquely determined.

In Fig. 3(a) we plot the allowed wavevectors $k_\parallel$ along the propagation direction (as obtained through the dispersion relation), as a function of dimensionless wire radius $k_0 R$ and for a few lowest-order modes in m. Here $k_0 = 2\pi/\lambda_0$ is the resonant wavevector in vacuum. For concreteness, the numerical results are for a silver nanowire and $\lambda_0 = 1\,\mu\mathrm{m}$, and with a surrounding dielectric $\epsilon_1 = 2$, although the physical processes described are not specific to silver or to some narrow frequency range. The electric permittivity of the silver nanowire at this frequency is assumed to correspond to its measured value in thin films, $\epsilon_2 \approx -50 + 0.6i$.[23] In plotting Fig. 3 we have temporarily ignored the dissipative imaginary part of ϵ_2, although we will address its effect later. Ignoring $\mathrm{Im}\,\epsilon_2$ results in purely real values of $k_\parallel$, indicating that these modes propagate without loss.

We highlight the important aspects of the SP modes illustrated in Fig. 3(a). First, all higher-order modes ($|m| \geq 1$) exhibit a cutoff as the wire radius $R \to 0$, much like a single-mode optical fiber, while the fundamental $m = 0$ mode exists for any wire size. Unlike a conventional fiber, however, the fundamental SP mode exhibits a unique $k_\parallel \propto 1/R$ scaling behavior.[24,25] This scaling indicates that the wavelength of these SPs can become strongly reduced relative to the free-space wavelength, just by using progressively smaller wires. Physically, the $m = 0$ mode can be understood as a quasi-static configuration of field and associated charge density wave on the wire. As such, in solving the quasi-static problem, R becomes the only relevant length scale, as the free-space electromagnetic wavelength becomes unimportant. It also follows that the transverse area to which the SPs are localized scales like $A_{\text{eff}} \propto R^2$, enabling confinement well below the diffraction limit.

The emergence of tight confinement of SPs due to the geometry of the wire can be further confirmed by comparing the SP dispersion relations of flat interfaces and wires. Assuming a Drude model for the metal (Eq. (1)) with no losses and $\epsilon = 1$ for the dielectric medium, the dispersion relation of the wire fundamental mode versus the flat interface is plotted in Fig. 3(b). The effect of small wire size is to pull the fundamental SP mode dispersion relation further away from the light line, $k_\parallel = \omega/c$, as compared to the dispersion curve of a flat interface. This enables tight confinement of SPs through geometric effects over a large bandwidth, as opposed to the tight confinement in flat interfaces that only occurs near the plasmon resonance frequency.

In practice ϵ_2 is not purely real but has a small imaginary part corresponding to losses in the metal (absorption) at optical frequencies. Its effect is to add a small imaginary component to $k_\parallel$ corresponding to dissipation as the SP propagates along the wire. In the inset of Fig. 3(a) we plot $\operatorname{Re} k_\parallel/\operatorname{Im} k_\parallel$ for the fundamental mode as a function of R. This quantity is proportional to the decay length in units of the SP wavelength $\lambda_{\text{pl}} \equiv 2\pi/\operatorname{Re} k_\parallel$. As R decreases, it can be seen that this ratio decreases monotonically but approaches a nonzero constant. For silver at $\lambda_0 = 1\,\mu\text{m}$ and room temperature and $\epsilon_1 = 2$ this constant is approximately 140. The fact that this ratio does not approach zero even as $R \to 0$ is important for potential applications, as it implies that the SPs can still travel several times the wavelength λ_{pl} for devices of any size.

3.   Efficient coupling between SPs and quantum emitters

The small mode volume associated with the fundamental SP mode of a nanowire offers a possible mechanism to achieve strong coupling with nearby optical emitters. This is in close analogy with the strong preferential spontaneous emission of an emitter into a high-finesse cavity, the well-known Purcell effect in cavity QED.[26] In this section we discuss the underlying mechanism for strong interactions and the fundamental limits for this process.

The theoretical techniques for calculating spontaneous emission near dielectric structures has been worked out elsewhere[27] and applied specifically to the case of metallic nanowires.[24, 25] This general calculation is quite cumbersome, however, and so we only give a heuristic approach and summarize the main results here.

Spontaneous emission rates can in fact be calculated classically, by replacing the atom with a classical oscillating dipole and considering the power loss of this system near a dielectric surface. In our specific case, we consider a oscillating dipole

$\mathbf{p}_0 e^{-i\omega t}$ positioned a distance d from the center of the wire (see Fig. 1(b)), and wish to calculate the total fields of the system. Some component of the dipole radiation will reflect off of the wire and interact again with the dipole. The work done by the reflected field on the dipole, when properly normalized, exactly yields the quantum spontaneous emission rates.

An excited quantum emitter placed near a metallic nanowire has three possible channels into which it can spontaneously emit a photon. The first channel consists of spontaneous emission into free space. The second channel consists of "non-radiative" emission, which can be understood classically as emerging from the Ohmic dissipation of eddy currents in the metal that are driven by the electric fields of the oscillating dipole. These first two channels are undesirable, as they cannot be efficiently collected into a well-defined spatial mode. The third channel consists of spontaneous emission into the SP modes. Emission into this well-defined mode offers opportunities for coupling the emitted photon with other photonic systems or quantum emitters, which would enable the scaling of quantum information devices.

We first focus on the undesirable decay channels. The radiative emission is not significantly changed near a nanowire compared to the vacuum emission rate Γ_0 of an isolated emitter,

$$\Gamma_{\text{rad}} \sim \Gamma_0. \tag{5}$$

On the other hand, the non-radiative rate diverges as the distance between the emitter and wire surface approaches zero,

$$\frac{\Gamma_{\text{non-rad}}}{\Gamma_0} \approx \frac{3}{8k_0^3(d-R)^3\epsilon_1^{1/2}} \text{Im}\,\epsilon_2 / \left(\text{Re}\,\epsilon_2\right)^2. \tag{6}$$

The $(d-R)^{-3}$ scaling reflects the amplitude of the electric near field of a dipole at the wire surface, while the dependence on $\text{Im}\,\epsilon_2$ (the lossy component of the electric permittivity) makes clear that the origin is Ohmic dissipation of fields generated in the nanowire.

A similar calculation also shows that the spontaneous emission rate into the SP modes behaves like

$$\Gamma_{\text{pl}} \propto \Gamma_0 \frac{K_1^2(\kappa_{1\perp}d)^2}{(k_0 R)^3}, \tag{7}$$

where the constant of proportionality depends only on the electric permittivity of the metal. Here $\kappa_{1\perp}^{-1}$ is the decay length of the SP evanescent field away from the wire surface, and K_1 is the first-order modified Bessel function, which decays roughly exponentially with increasing distance d. Physically, this function enforces that the coupling is only efficient when the emitter sits within the SP evanescent fields. A remarkable feature of the above expression is that the emission rate into SPs increases with decreasing wire radius like $1/R^3$ and thus becomes very strong for small wires. Here, we provide a heuristic derivation of this result based on Fermi's Golden Rule.

This rule states that, once the SP modes are quantized, the decay rate is given by

$$\Gamma_{\mathrm{pl}} = 2\pi g^2(\mathbf{r}, \omega)D(\omega),\tag{8}$$

where $g(\mathbf{r}, \omega)$ is the position-dependent, dipole-field interaction matrix element, and $D(\omega)$ is the SP density of states on the nanowire. Here, $g(\mathbf{r}, \omega) = \vec{\wp} \cdot \vec{\mathcal{E}}(\mathbf{r})/\hbar$ itself depends on the dipole matrix element $\vec{\wp} = \langle e|e\mathbf{r}|g\rangle$ of the emitter, where $|g\rangle$ ($|e\rangle$) is the ground (excited) state involved in the transition, and on the electric field per photon $\vec{\mathcal{E}}(\mathbf{r})$ for the SP modes. On dimensional grounds, the electric field per photon can be written in the form $\vec{\mathcal{E}}(\mathbf{r}) = \sqrt{\hbar\omega/\epsilon_0 V_{\mathrm{eff}}}(\mathbf{E}(\mathbf{r})/E_{\max})$, where V_{eff} is an effective mode volume characterizing the confinement of each mode, $\mathbf{E}(\mathbf{r})$ is the classical field profile for the SP modes, and $E_{\max} = \max|\mathbf{E}(\mathbf{r})|$. Having previously established that the fundamental SP mode has a transverse confinement proportional to R^2, the mode volume can be estimated as $V_{\mathrm{eff}} \propto R^2 L$. Here L is the quantization length, set by the wire length, which is assumed to be much longer than all other relevant length scales. In the end L will disappear from the physical quantities of interest (*e.g.*, the spontaneous emission rate). At the same time, the density of states for the SP modes is given by $D(\omega) \sim (L/2\pi)(dk_\parallel/d\omega)$, and is inversely proportional to the group velocity. The group velocity for SPs is strongly reduced due to the large SP wavevector $k_\parallel \propto 1/R$ on a nanowire, and consequently this gives rise to a $\sim 1/R$ enhancement in the density of states. Combining all of these results into Eq. (8) yields the desired scaling $\Gamma_{\mathrm{pl}} \propto \Gamma_0/(k_0 R)^3$.

Comparing the spontaneous emission rates given by Eqs. (5), (6), and (7), we now qualitatively discuss the behavior one should expect as the position of the emitter d from the wire is varied. In the limit that $d/R \gg 1$, the emitter feels no effect from the wire and the total spontaneous emission rate is close to the radiative rate Γ_0 in a uniform dielectric medium. As one brings the emitter closer to the wire surface and within a distance $d \sim 1/\kappa_{1\perp} \sim R$, the emitter starts to interact with the localized SP fields, with a corresponding rate of emission into SPs scaling with wire size like $1/R^3$. The emission rate into SPs continues to grow (but does not diverge) as the emitter is brought even closer to the wire edge, $d \to R$, due to the SP evanescent fields being strongest at the surface. However, the efficiency or probability of SP excitation eventually decreases due to the large non-radiative decay rate experienced by the dipole very near the wire, which diverges like $1/(d - R)^3$. We thus expect some optimal efficiency of spontaneous emission into the SP modes to occur when the emitter is positioned at a distance $\mathcal{O}(R)$ away from the wire edge, and for this optimal efficiency to improve as $R \to 0$.

The efficiency of coupling to the SP modes can be characterized by a "cooperativity factor" that is defined by the branching ratio $C = \Gamma_{\mathrm{pl}}/\Gamma'$, where $\Gamma' = \Gamma_{\mathrm{rad}} + \Gamma_{\mathrm{non\text{-}rad}}$ denotes the total emission rate into channels other than the fundamental SP mode. Like the well-known cooperativity factor from cavity QED, this parameter nearly universally characterizes the maximum achievable fidelity of a plasmonic system in quantum information processing tasks. Generally C depends

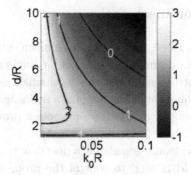

Fig. 4.   Cooperativity factor $C = \Gamma_{\mathrm{pl}}/\Gamma'$ for a silver nanowire, plotted on a $\log_{10}$ scale, as functions of dimensionless wire radius $k_0 R$ and dimensionless distance d/R of the emitter to the wire.

on both the wire size and on the position d of the emitter. In Fig. 4, we have numerically evaluated the spontaneous emission rates and plotted the cooperativity factor as a function of wire radius R and emitter position d. It can be clearly seen that an optimal position d exists for each wire size, with the maximum coopera- tivity increasing as R decreases. As $R \to 0$, the optimized cooperativity factor C exceeds $\sim 10^3$, indicating that the probability of emission into the SPs approaches almost unity. Examining this limit more carefully, it can be shown that the "error rate" in fact approaches a small quantity $\Gamma'/\Gamma_{\mathrm{total}} \propto \operatorname{Im} \epsilon_2/(\operatorname{Re} \epsilon_2)^2$, which explic- itly indicates that this process is ultimately limited by material losses. Again, we emphasize that these properties are specifically a result of the conducting properties of the nanowire. This can be contrasted with emission into the guided modes of a sub-wavelength optical fiber, which drops exponentially as $R \to 0$ due to the weak confinement of these guided modes.[28]

Of course, the strong coupling of a quantum emitter to SPs is ultimately only useful if the outgoing SP can be efficiently collected and transported toward other devices as a form of information (such as to a distant emitter for long-distance quantum information processing). Because of dissipative losses in metals, the SP modes are not directly suitable as carriers of information over long distances. In previous work,[24] however, it was shown a plasmonic device can under certain cir- cumstances be efficiently out-coupled to the modes of a co-propagating dielectric waveguide, thereby providing a link with conventional photonic and optical systems. There is intrinsically a trade off between wire size and out-coupling efficiency. In particular, while very small wires provide high cooperativity factors, their tight con- finement make it increasingly difficult to mode-match with dielectric nanophotonic systems. Taking this optimization into account, we find that ideally chosen wire sizes can enable up to $\sim 95\%$ collection efficiency of spontaneously emitted photons into a dielectric waveguide, including finite SP coupling efficiencies and waveguide coupling efficiencies. This high efficiency potentially makes plasmonic systems a promising platform to realizing quantum optical devices.

4.   Single-photon nonlinear optics

The previous section described a physical mechanism whereby efficient coupling between individual SPs (*i.e.*, single photons) and a single quantum emitter can be attained. With proper control techniques, this mechanism can be leveraged to realize a variety of quantum information or single-photon nolinear optical devices. Here, we provide an illustrative example, in which we propose a protocol for a single-photon transistor.

In analogy with the electronic transistor, a photonic transistor is a device where a small optical "gate" field is used to control the propagation of another optical "signal" field via a nonlinear optical interaction.[29, 30] Its fundamental limit is the single-photon transistor, where the propagation of the signal field is controlled by the presence or absence of a single photon in the gate field. Such a nonlinear device has many interesting applications from optical communication and computation[30] to quantum information processing.[9] However, its practical realization is challenging because the requisite single-photon nonlinearities are generally very weak.[29] However, the strong coupling between single SPs and quantum emitters constitutes a promising avenue toward realizing a single-photon transistor.

In the previous section, it was shown that a two-level quantum emitter near a metallic nanowire can emit a single SP with nearly unit efficiency. However, one can also consider the interaction efficiency of a complementary process, wherein a near-resonant single SP incident in the nanowire scatters off of a quantum emitter originally in its ground state. Generally, the SP can either be reflected, transmitted, or scattered out of the SP channels altogether (*e.g.*, re-emitted into free space). A simple limiting case to consider is when the emitter is perfectly coupled to the SPs, such that an initially excited emitter produces an outgoing SP in the left- or right-going directions with 50% probability. This equal coupling to two channels is known as "critical" coupling. For the complementary scattering problem, it is well-known that critical coupling produces *perfect* reflection of a resonant photon.[31] While this phenomenon is quite familiar for larger photonic systems, here we have a remarkable result that a single quantum emitter strongly coupled to a nanowire can act as a nearly perfect mirror for incident SPs. A more careful calculation yields that the reflection coefficient in the non-ideal case is[32]

$$r(\delta_k) = -\frac{1}{1 + \Gamma'/\Gamma_{\rm pl} - 2i\delta_k/\Gamma_{\rm pl}}, \tag{9}$$

where δ_k is the detuning of the incident SP from the atomic transition frequency. On resonance, $r \approx -(1 - 1/C)$, confirming that the emitter is an efficient single-photon reflector for large cooperativity factors.

While the quantum emitter can nearly perfectly reflect incoming single photons, its two-level nature ensures that it cannot scatter more than a single photon at a time. Thus, incident two-photon (or higher) wavepackets are likely to be transmitted by the emitter. A basic calculation to illustrate this mechanism consists of solving the system response to a resonant, incoming weak coherent state (the best approximation to a classical field). In particular, one we consider the second-order correlation function of the transmitted field

$$g^{(2)}(t) \equiv \langle \hat{E}^\dagger(\tau)\hat{E}^\dagger(\tau+t)\hat{E}(\tau+t)\hat{E}(\tau)\rangle / \langle \hat{E}^\dagger(t)\hat{E}(t)\rangle^2, \tag{10}$$

which physically describes the relatively likelihood of detecting a photon at time $t + \tau$ given the initial detection of a photon at time τ. For coherent states, one finds $g^{(2)}(t) = 1$, characteristic of Poissonian statistics. On the other hand, for the emitter coupled to a nanowire, the transmitted field obeys $g^{(2)}(0) = (C^2 - 1)^2$. For large C, this indicates a strong bunching of photons at the output. This bunching is due to the fact that single photon components of the incident field are efficiently reflected and filtered out by the emitter when the cooperativity is large. The strong coupling thus naturally gives rise to a rudimentary two-photon switching device.

A greater degree of coherent control over the field interaction can be gained by considering a multi-level emitter, such as the three-level configuration shown in Fig. 5(a). Here, a metastable state $|s\rangle$ is decoupled from the SPs due to, *e.g.*, a different orientation of its associated dipole moment, but is resonantly coupled to $|e\rangle$ via some classical, optical control field with Rabi frequency $\Omega(t)$. States $|g\rangle, |e\rangle$ remain coupled via the SP modes as discussed earlier. Such a system is capable of producing single SPs on demand. In particular, if the emitter is initialized in $|s\rangle$, the control field $\Omega(t)$ pumps the system to the excited state $|e\rangle$, which then emits an outgoing SP with high efficiency, as characterized by the system cooperativity factor. By time reversal symmetry, it also follows that an incoming SP can be reversibly absorbed with equally high efficiency, if the emitter is initialized in state $|g\rangle$ and an impedance-matched control field $\Omega(t)$ is applied.[32, 33]

We next consider the reflection properties of the emitter when the control field $\Omega(t)$ is turned off. If the emitter is in $|g\rangle$, the reflectance derived above for the two-level emitter remains valid. On the other hand, if the emitter is in $|s\rangle$, any incident fields will simply be transmitted with no effect since $|s\rangle$ is decoupled from the SPs. Therefore, with $\Omega(t)$ turned off, the three-level system effectively behaves as a conditional mirror whose properties depend sensitively on its internal state.

The techniques of state-dependent conditional reflection and single-photon storage can be combined to create a single-photon transistor, whose operation is illustrated in Fig. 5(b). The key principle is to utilize the presence or absence of a photon in an initial "gate" pulse to conditionally flip the internal state of the emitter during the storage process, and to then use this conditional flip to control the flow of subsequent "signal" photons arriving at the emitter. Specifically, we first initialize the emitter in $|g\rangle$ and apply the storage protocol for the gate pulse, which consists of either zero or one photon. The presence (absence) of a photon causes the emitter to flip to (remain in) state $|s\rangle$ ($|g\rangle$). Now, the interaction of each signal pulse arriving at the emitter depends on the internal state following storage. The

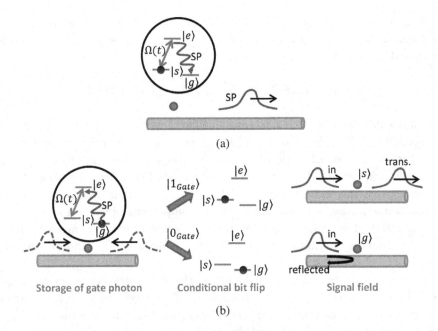

Fig. 5.   (a) A three-level quantum emitter, with ground state $|g\rangle$, metastable state $|s\rangle$, and excited state $|e\rangle$, can efficiently produce a single SP using the protocol shown. Here, the emitter is initialized in $|s\rangle$. An external field $\Omega(t)$ couples this state to $|e\rangle$. The excited state, which is strongly coupled to SPs in the nanowire, emits a single SP wavepacket with an efficiency characterized by the cooperativity C, while flipping the internal state to $|g\rangle$. By time reversal symmetry, it also follows that an emitter in state $|g\rangle$ can reversibly absorb an incoming SP wavepacket using a properly shaped control field $\Omega(t)$, while flipping the internal state to $|s\rangle$. (b) Single-photon transistor protocol. First, an incident gate pulse, consisting of either 0 or 1 SPs, is stored in an emitter initially in state $|g\rangle$ using a properly timed control field $\Omega(t)$. The presence of a gate photon causes a spin flip to $|s\rangle$, while the absence causes the emitter to remain in $|g\rangle$. Subsequently, signal SPs can be sent in. The emitter in state $|s\rangle$ is completely transparent to SPs, while $|g\rangle$ is highly reflecting. Thus, controlled propagation of the signal field depending on the gate pulse is achieved.

storage and conditional spin flip causes the emitter to be either highly reflecting or completely transparent depending on the gate, and the system therefore acts as an efficient switch or transistor for the subsequent signal field.

The ideal operation of the transistor is limited only by the characteristic time over which an undesired spin flip can occur. In particular, if the emitter remains in $|g\rangle$ after storage of the gate pulse, the emitter can eventually be optically pumped to $|s\rangle$ upon the arrival of a sufficiently large number of photons in the signal field. For strong coupling, the number of incident photons n that can be scattered before pumping occurs is given by the branching ratio of decay rates from $|e\rangle$ to these states, $n \sim \Gamma_{e \to g}/\Gamma_{e \to s}$, which can be large due to the large decay rate $\Gamma_{e \to g} \geq \Gamma_{\mathrm{pl}}$. Thus $n \gtrsim C$ and the emitter can reflect $\mathcal{O}(C)$ photons before an undesired spin flip occurs. The cooperativity thus corresponds to the effective "gain" of the single-photon transistor.

5. State-of-the-art and outlook

The physical mechanism for strong coupling between single SPs and quantum emitters, discussed in Section 3, can be quite robust and enables versatile potential applications in low-power nonlinear optics and quantum information processing. Here, we have described the applications of efficient single-photon generation and a single-photon transistor.

Efficient single-photon generation using SPs has now been experimentally observed in a variety of settings, in which quantum emitters such as quantum dot nanocrystals[34] or defect color centers in diamond[35] are coupled to chemically or lithographically fabricated metallic nanowires. Such experiments demonstrate not only the predicted enhancement of spontaneous emission near metallic nanowires, but the non-classical statistics associated with the single SPs being generated by the quantum emitter. In place of nanowires, other geometries such as plasmonic resonators[36] or antennas[37] are also being actively explored. While most of these experiments ultimately rely on out-coupling of SPs to free-space radiation as a detection mechanism, there have also been experiments where detection is done directly in the near-field,[38] which raises the interesting possibility of integrated quantum optical nano-circuitry.

A common challenge for quantum optics based on solid-state emitters is the complex environment to which these emitters are coupled. This often results in undesired effects such as phonon sidebands in spontaneous emission, spectral diffusion of the resonance line, and inhomogeneity of different emitters. This last issue is problematic in that it becomes difficult to couple different emitters together in a scalable fashion. Interestingly, in a separate set of experiments, trapping of cold neutral atoms near dielectric waveguides has also been achieved.[39, 40] While dielectric waveguides do not offer the possibility of large single atom cooperativities, there are many interesting conceptual parallels between these systems and plasmonic ones. Recently, for example, it has been theoretically proposed that highly reflecting atomic mirrors can be achieved by using many atoms and collective enhancement effects to compensate for the low single-atom coupling efficiencies.[41]

Once the robust mechanism for strong coupling is established, plasmonic systems can be utilized for many applications beyond those described here. For example, a number of interesting schemes have been proposed to entangle two emitters via surface plasmons[42, 43] or to observe plasmonic superradiance.[44, 45] The versatility and richness of quantum plasmonic systems ensures that this will be an active and interesting field for years to come.

References

1. R. Miller, T. E. Northup, K. M. Birnbaum, A. Boca, A. D. Boozer, and H. J. Kimble, Trapped atoms in cavity QED: coupling quantized light and matter, *J. of Phys. B: At. Mol. Opt. Phys.* **38**(9), S551–S565, (2005).

2. S. Haroche and J.-M. Raimond, *Exploring the Quantum: Atoms, Cavities, and Photons* (Oxford University Press, Oxford, 2006).

3. M. Brune, F. Schmidt-Kaler, A. Maali, J. Dreyer, E. Hagley, J. M. Raimond, and S. Haroche, Quantum rabi oscillation: A direct test of field quantization in a cavity, *Phys. Rev. Lett.* **76**(11), 1800–1803, (1996).

4. P. Michler, A. Kiraz, C. Becher, W. V. Schoenfeld, P. M. Petroff, L. Zhang, E. Hu, and A. Imamoglu, A quantum dot single-photon turnstile device, *Science* **290**(5500), 2282–2285, (2000).

5. M. Pelton, C. Santori, J. Vuckovi, B. Zhang, G. S. Solomon, J. Plant, and Y. Yamamoto, Efficient source of single photons: A single quantum dot in a micropost microcavity, *Phys. Rev. Lett.* **89**(23), 233602, (2002).

6. J. McKeever, A. Boca, A. D. Boozer, R. Miller, J. R. Buck, A. Kuzmich, and H. J. Kimble, Deterministic generation of single photons from one atom trapped in a cavity, *Science* **303**(5666), 1992–1994, (2004).

7. K. M. Birnbaum, A. Boca, R. Miller, A. D. Boozer, T. E. Northup, and H. J. Kimble, Photon blockade in an optical cavity with one trapped atom, *Nature* **436**(7047), 87–90, (2005).

8. J. I. Cirac, P. Zoller, H. J. Kimble, and H. Mabuchi, Quantum state transfer and entanglement distribution among distant nodes in a quantum network, *Phys. Rev. Lett.* **78**(16), 3221–3224, (1997).

9. L.-M. Duan and H. J. Kimble, Scalable photonic quantum computation through cavity-assisted interactions, *Phys. Rev. Lett.* **92**(12), 127902, (2004).

10. M. J. Hartmann, F. G. S. L. Brandao, and M. B. Plenio, Strongly interacting polaritons in coupled arrays of cavities, *Nature Phys.* **2**(12), 849–855, (2006).

11. A. D. Greentree, C. Tahan, J. H. Cole, and L. C. L. Hollenberg, Quantum phase transitions of light, *Nature Phys.* **2**(12), 856–861, (2006).

12. D. G. Angelakis, M. F. Santos, and S. Bose, Photon-blockade-induced mott transitions and XY spin models in coupled cavity arrays, *Phys. Rev. A* **76**(3), 031805, (2007).

13. S. J. van Enk and H. J. Kimble, Strongly focused light beams interacting with single atoms in free space, *Phys. Rev. A* **63**(2), 023809, (2001).

14. B. Darquie, M. P. A. Jones, J. Dingjan, J. Beugnon, S. Bergamini, Y. Sortais, G. Messin, A. Browaeys, and P. Grangier, Controlled single-photon emission from a single trapped two-level atom, *Science* **309**(5733), 454–456, (2005).

15. G. Wrigge, I. Gerhardt, J. Hwang, G. Zumofen, and V. Sandoghdar, Efficient coupling of photons to a single molecule and the observation of its resonance fluorescence, *Nature Phys.* **4**(1), 60–66, (2008).

16. M. K. Tey, Z. Chen, S. A. Aljunid, B. Chng, F. Huber, G. Maslennikov, and C. Kurtsiefer, Strong interaction between light and a single trapped atom without the need for a cavity, *Nature Phys.* **4**(12), 924–927, (2008).

17. G. Hetet, L. Slodicka, M. Hennrich, and R. Blatt, Single atom as a mirror of an optical cavity, *Phys. Rev. Lett.* **107**(13), 133002, (2011).

18. S. Maier, *Plasmonics: Fundamentals and Applications* (Springer, New York, 2007).

19. K. Kneipp, H. Kneipp, I. Itzkan, R. R. Dasari, and M. S. Feld, Surface-enhanced raman scattering and biophysics, *J. Phys.: Condens. Matter* **14**(18), R597–R624, (2002).

20. T. W. Ebbesen, H. J. Lezec, H. F. Ghaemi, T. Thio, and P. A. Wolff, Extraordinary optical transmission through sub-wavelength hole arrays, *Nature* **391**(6668), 667–669, (1998).

21. I. I. Smolyaninov, J. Elliott, A. V. Zayats, and C. C. Davis, Far-field optical microscopy with a nanometer-scale resolution based on the in-plane image magnification by surface plasmon polaritons, *Phys. Rev. Lett.* **94**(5), 057401, (2005).

22. M. L. Brongersma, J. W. Hartman, and H. A. Atwater, Electromagnetic energy transfer and switching in nanoparticle chain arrays below the diffraction limit, *Phys. Rev. B* **62**(24), R16356–R16359, (2000).

23. P. B. Johnson and R. W. Christy, Optical constants of the noble metals, *Phys. Rev. B* **6**(12), 4370–4379, (1972).

24. D. E. Chang, A. S. Sorensen, P. R. Hemmer, and M. D. Lukin, Quantum optics with surface plasmons, *Phys. Rev. Lett.* **97**(5), 053002, (2006).

25. D. E. Chang, A. S. Sorensen, P. R. Hemmer, and M. D. Lukin, Strong coupling of single emitters to surface plasmons, *Phys. Rev. B* **76**(3), 035420, (2007).

26. E. Purcell, Spontaneous emission probabilities at radio frequencies, *Phys. Rev.* **69**, 681, (1946).

27. G. S. Agarwal, Quantum electrodynamics in the presence of dielectrics and conductors. IV. general theory for spontaneous emission in finite geometries, *Phys. Rev. A* **12**(4), 1475–1497, (1975).

28. V. V. Klimov and M. Ducloy, Spontaneous emission rate of an excited atom placed near a nanofiber, *Phys. Rev. A* **69**(1), 013812, (2004).

29. R. W. Boyd, *Nonlinear optics* (Academic Press, San Diego, 2003).

30. H. M. Gibbs, *Optical bistability: controlling light with light* (Academic Press, Orlando, FL, 1985).

31. J. T. Shen and S. Fan, Coherent photon transport from spontaneous emission in one-dimensional waveguides, *Opt. Lett.* **30**(15), 2001–2003, (2005).

32. D. E. Chang, A. S. Sorensen, E. A. Demler, and M. D. Lukin, A single-photon transistor using nanoscale surface plasmons, *Nature Phys.* **3**(11), 807–812, (2007).

33. A. V. Gorshkov, A. Andre, M. Fleischhauer, A. S. Sorensen, and M. D. Lukin, Universal approach to optimal photon storage in atomic media, *Phys. Rev. Lett.* **98**(12), 123601, (2007).

34. A. V. Akimov, A. Mukherjee, C. L. Yu, D. E. Chang, A. S. Zibrov, P. R. Hemmer, H. Park, and M. D. Lukin, Generation of single optical plasmons in metallic nanowires coupled to quantum dots, *Nature* **450**(7168), 402–406, (2007).

35. R. Kolesov, B. Grotz, G. Balasubramanian, R. J. Stohr, A. A. L. Nicolet, P. R. Hemmer, F. Jelezko, and J. Wrachtrup, Wave-particle duality of single surface plasmon polaritons, *Nature Phys.* **5**(7), 470–474, (2009).

36. I. Bulu, T. Babinec, B. Hausmann, J. T. Choy, and M. Loncar, Plasmonic resonators for enhanced diamond NV- center single photon sources, *Opt. Express* **19**(6), 5268–5276, (2011).

37. A. W. Schell, G. Kewes, T. Hanke, A. Leitenstorfer, R. Bratschitsch, O. Benson, and T. Aichele, Single defect centers in diamond nanocrystals as quantum probes for plasmonic nanostructures, *Opt. Express* **19**(8), 7914–7920, (2011).

38. A. L. Falk, F. H. L. Koppens, C. L. Yu, K. Kang, N. de Leon Snapp, A. V. Akimov, M.-H. Jo, M. D. Lukin, and H. Park, Near-field electrical detection of optical plasmons and single-plasmon sources, *Nature Phys.* **5**(7), 475–479, (2009).

39. E. Vetsch, D. Reitz, G. Sague, R. Schmidt, S. T. Dawkins, and A. Rauschenbeutel, Optical interface created by laser-cooled atoms trapped in the evanescent field surrounding an optical nanofiber, *Phys. Rev. Lett.* **104**(20), 203603, (2010).

40. A. Goban, K. S. Choi, D. J. Alton, D. Ding, C. Lacroute, M. Pototschnig, T. Thiele, N. P. Stern, and H. J. Kimble, Demonstration of a state-insensitive, compensated nanofiber trap, *Phys. Rev. Lett.* **109**(3), 033603, (2012).

41. D. E. Chang, L. Jiang, A. V. Gorshkov, and H. J. Kimble, Cavity QED with atomic mirrors, *New J. Phys.* **14**(6), 063003, (2012).

42. D. Dzsotjan, A. S. Sorensen, and M. Fleischhauer, Quantum emitters coupled to surface plasmons of a nanowire: A greens function approach, *Phys. Rev. B* **82**(7), 075427, (2010).

43. A. Gonzalez-Tudela, D. Martin-Cano, E. Moreno, L. Martin-Moreno, C. Tejedor, and F. J. Garcia-Vidal, Entanglement of two qubits mediated by one-dimensional plasmonic waveguides, *Phys. Rev. Lett.* **106**(2), 020501, (2011).

44. D. Martin-Cano, L. Martin-Moreno, F. J. Garcia-Vidal, and E. Moreno, Resonance energy transfer and superradiance mediated by plasmonic nanowaveguides, *Nano Lett.* **10**(8), 3129–3134, (2010).

45. A. Manjavacas, S. Thongrattanasiri, D. E. Chang, and F. J. Garcia de Abajo, Temporal quantum control with graphene, *New J. Phys.* **14**(12), 123020, (2012).

Chapter 8

Quantum Polaritonics

S. Portolan*, O. Di Stefano† and S. Savasta†,‡

*Institute of Atomic and Subatomic Physics,
TU Wien, Stadionalle 2, 1020 Wien, Austria

and

*School of Physics and Astronomy,
University of Southampton, SO17 1BJ, United Kingdom

†Dipartimento di Fisica e Scienze della Terra,
Università di Messina, Viale F. Stagno d'Alcontres 31,
I-98166 Messina, Italy

‡ salvatore.savasta@unime.it

The excitonic polariton concept was introduced already in 1958 by J.J. Hopfield.[1] Although its description was based on a full quantum theory including light quantization, the investigation of the optical properties of excitons developed mainly independently of quantum optics. In this chapter we shall review exciton polariton quantum optical effects by means of some recent works and results appeared in the literatures on both bulk semiconductors and in cavity embedded quantum wells.

The first manifestation of excitonic quantum-optical coherent dynamics was observed experimentally 20 years later, in 1978, exploiting resonant Hyper-parametric scattering.[2] On the other hand, the possibility of generating entangled photon pairs by means of this resonant process was theoretically pointed out lately in 1999.[3] The experimental evidence for the generation of ultraviolet polarization-entangled photon pairs by means of biexciton resonant parametric emission in a single crystal of semiconductor CuCl has been reported only in 2004.[4] The demonstrations of parametric amplification and parametric emission in semiconductor microcavities,[5,6] together with the possibility of ultrafast optical manipulation and ease of integration of these micro-devices, have increased the interest on the possible realization of nonclassical cavity-polariton states. In 2005 an experiment that probes polariton quantum correlations by exploiting quantum complementarity has been proposed and realized.[7] These results had unequivocally proven that despite these solid state systems are far from being isolated systems, quantum optical effects at single photon level arising from the interaction of light with electronic excitations of semiconductors and semiconductor nanostructures were possible. The theoretical predictions that we shall revise are based on a microscopic quantum theory of the nonlinear optical response of interacting electron systems relying on the dynamics controlled truncation scheme [8] extended to include light quantization.[3,9,10] In addition, environment decoherence has been taken into account microscopically by Markov noise sources within a quantum Langevin framework.[11]

More recently the possibility to reach the light-matter strong coupling regime in ultracompact nanoscale systems,[12,13] by exploiting the ability of metallic nanoparticles to focus the electromagnetic field to spots much smaller than a wavelength, has been demonstrated.[14]

1.   Electronic excitation in semiconductor

A key role in the optical response of dielectric media at frequencies near the band gap is played by excitons. Indeed, when a direct band gap semiconductor is in the presence of an electromagnetic field of energy equal to the electronic band-gap, an electron can be promoted from the highest valence band to the lowest conduction band by the absorption of a photon, a hole is left in the valence band and a bound Coulomb-correlated electron-hole state is created. These bound two-particle states are known as excitons and modify substantially the optical properties of the semiconductor structure.

The standard model Hamiltonian near the semiconductor band-edge can thus be written as $\hat{H}_s = \hat{H}_0 + \hat{V}_{Coul}$,[8] which is composed of a free-particle part $\hat{H}_0$ and the Coulomb interaction $\hat{V}_{Coul}$. $\hat{H}_0$ is given by

$$
\hat{H}_0 = \sum_{m,\mathbf{k}} \epsilon^{(e)}_{m,\mathbf{k}} \hat{c}^{\dagger}_{m,\mathbf{k}} \hat{c}_{m,\mathbf{k}} + \sum_{m,\mathbf{k}} \epsilon^{(h)}_{m,\mathbf{k}} \hat{d}^{\dagger}_{m,\mathbf{k}} \hat{d}_{m,\mathbf{k}}, \tag{1}
$$

where $\hat{c}^{\dagger}$ and $\hat{d}^{\dagger}$ are the electron (e) and hole (h) creation operators and $\epsilon^{(e)}$ and $\epsilon^{(h)}$ the e and h energies. The Coulomb interaction is given by

$$
\hat{V}_{Coul} = \sum_{\mathbf{q} \neq 0} V_{\mathbf{q}} \sum_{m,\mathbf{k},m',\mathbf{k}'} \left[\frac{1}{2} \hat{c}^{\dagger}_{m,\mathbf{k}+\mathbf{q}} \hat{c}^{\dagger}_{m',\mathbf{k}'-\mathbf{q}} \hat{c}_{m',\mathbf{k}'} \hat{c}_{m,\mathbf{k}} \right.
$$
$$
\left. + \frac{1}{2} \hat{d}^{\dagger}_{m,\mathbf{k}+\mathbf{q}} \hat{d}^{\dagger}_{m',\mathbf{k}'-\mathbf{q}} \hat{d}_{m',\mathbf{k}'} \hat{d}_{m,\mathbf{k}} - \hat{c}^{\dagger}_{m,\mathbf{k}+\mathbf{q}} \hat{d}^{\dagger}_{m',\mathbf{k}'-\mathbf{q}} \hat{c}_{m',\mathbf{k}'} \hat{d}_{m,\mathbf{k}} \right], \tag{2}
$$

where $V_{\mathbf{q}}$ is the Fourier transform of the screened Coulomb interaction potential. The first two parts are the repulsive electron-electron and hole-hole interaction terms, while the third one describes the attractive eh interaction. The additional label m indicates the spin degrees of freedom. As we will see the spin degeneracy of the valence and conduction bands (subbands) play a relevant role in the nonlinear optical response of electronic excitations. Only eh pairs with total projection of angular momentum $\sigma = \pm 1$ are dipole active.

A many-body interacting state is usually very different from a product state, however a common way to express the former is by a superposition of uncorrelated product states. The physical picture that arises out of it expresses the *dressing* the interaction performs over a set of noninteracting particles. We shall consider the case of intrinsic semiconductor materials where the number of electron equals the number of holes and a good quantum number to label the states of the system is the total number of electron-hole pairs N. The state with $N = 0$ is the semiconductor ground state and corresponds to the full valence band. The $N = 1$ subspace is the exciton subspace. The optically active exciton states can be labelled with the additional quantum number $\alpha = (n, \sigma)$ where σ indicates the total spin projection and n spans all the exciton levels. Exciton eigenstates can be obtained by requiring that general one *eh* pair states be eigenstates of $\hat{H}_s$:

$$\hat{H}_s|N = 1, n\sigma, \mathbf{k}\rangle = \hbar\omega_{n\sigma}(\mathbf{k})|N = 1, n\ \sigma\mathbf{k}\rangle, \tag{3}$$

with $|N = 1, n\sigma, \mathbf{k}\rangle = \hat{B}^\dagger_{n,\sigma,\mathbf{k}}|N = 0\rangle$, $\hat{B}^\dagger_{n,\sigma,\mathbf{k}}$, being the exciton creation operator:

$$\hat{B}^\dagger_{n,\sigma,\mathbf{k}} = \sum_{\mathbf{k}'} \Phi^*_{n,\sigma,\mathbf{k}'} \hat{c}^\dagger_{\sigma,\mathbf{k}'+\mathbf{k}} \hat{d}^\dagger_{\sigma,-\mathbf{k}'}. \tag{4}$$

In order to simplify a bit the notation, the spin notation in Eq. (4) has been changed by using the same label for the exciton spin quantum number and for the spin projections of the electron and hole states forming the exciton. The exciton envelope wave function $\Phi_{n,\sigma\mathbf{k}}$ can be obtained solving the secular equation obtained from Eq. (3). It describes the correlated electron-hole relative motion in k-space. As it results from H_s, electrons and holes act as free particles with an effective mass determined by the curvature of the single-particle energies (bands or subbands) interacting via a (screened) Coulomb potential in analogy with an electron and a proton forming the hydrogen atom. The set of bound and unbound states with $N = 2$ *eh* pairs, determining the biexciton subspace, can be obtained solving the corresponding secular equation, or by using simpler interaction models. It is interesting to observe that the exciton subspace is spin degenerate. The exciton energy levels do not depend on the spin quantum number, hence the excitonic bands are doubly degenerate. The situation is more complex for the biexciton subspace. The structure of biexcitonic states and energies depends strongly on the spin of the involved carriers. In particular bound biexcitons that are the solid state analogue of the hydrogen molecules arises only when the two electrons and also the two holes have opposite spin. This is consequence of the Pauli exclusion principle. Bound biexcitons have energies that are usually some meV (or even more in large gap semiconductors) less then twice the energy of excitons. Biexcitonic states made by carriers with the same spin give rise only to a continuum starting at an energy that is twice that of the lowest energy exciton state.

The eigenstates of the Hamiltonian $\hat{H}_c$ of the cavity modes can be written as $|n, \lambda\rangle$ where n stands for the total number of photons in the state and $\lambda = (\mathbf{k}_1, \sigma_1; \ldots; \mathbf{k}_n, \sigma_n)$ specifies wave vector and polarization σ of each photon. Here we shall neglect the longitudinal-transverse splitting of polaritons[15] originating mainly from the corresponding splitting of cavity modes. It is more relevant at quite high in-plane wave vectors and often it results to be smaller than the polariton linewidths. The present description can be easily extended to include it. We shall treat the cavity field in the quasi-mode approximation, that is to say we shall quantize the field as the mirrors were perfect and subsequently we shall couple the cavity with a statistical reservoir of a continuum of external modes. This coupling is able to provide the cavity losses as well as the feeding of the coherent external impinging pump beam. The cavity mode Hamiltonian, thus, reads

$$\hat{H}_c = \sum_k \hbar \omega_k^c \hat{a}_k^\dagger \hat{a}_k, \tag{5}$$

with the operator $\hat{a}_k^\dagger$ which creates a photon state with energy $\hbar \omega_k^c = \hbar(\omega_{\text{exc}}^2 + v^2|\mathbf{k}|^2)^{1/2}$, v being the velocity of light inside the cavity and $k = (\sigma, \mathbf{k})$. The coupling between the electron system and the cavity modes is given in the usual rotating wave approximation [9,16]

$$\hat{H}_I = -\sum_{nk} V_{nk}^* \hat{a}_k^\dagger \hat{B}_{nk} + H.c., \tag{6}$$

$V_{n,k}$ is the photon-exciton coupling coefficient enhanced by the presence of the cavity[17] set as $V_{n,k} = \tilde{V}_\sigma \sqrt{A} \phi_{n,\sigma}^*(\mathbf{x} = 0)$, the latter being the real-space exciton envelope function calculated in the origin whereas A is the in-plane quantization surface, $\tilde{V}_\sigma$ is proportional to the interband dipole matrix element. Modeling the loss through the cavity mirrors within the quasi-mode picture means we are dealing with an ensemble of external modes, generally without a particular phase relation among themselves. An input light beam impinging on one of the two cavity mirrors is an external field as well and it must belong to the family of modes of the corresponding side (i.e. left or right). Being coherent, it will be the non zero expectation value of the ensemble. It can be shown [9–11] that for a coherent input beam, the driving of the cavity modes may be described by the model Hamiltonian[9,11]

$$\hat{H}_p = i\,t_c \sum_{\mathbf{k}} (E_{\mathbf{k}} \hat{a}_{\mathbf{k}}^\dagger - E_{\mathbf{k}}^* \hat{a}_{\mathbf{k}}), \tag{7}$$

where t_c determines the fraction of the field amplitude passing through the cavity mirror, $E_{\mathbf{k}}$ ($E_{\mathbf{k}}^*$) is a $\mathbb{C}$-number describing the positive (negative) frequency part of the coherent input light field amplitude.

2.   Linear and nonlinear dynamics

In any theory aiming at describing the dynamics of semiconductor electrons interacting with a light field an infinite hierarchy of dynamical variables is encountered, requiring consideration of an appropriate truncation procedure.

The idea is not to use a density matrix approach, but to derive directly expectation values of all the quantities at play. Thanks to the symmetry of the present system,[9,10] we can use the so-called dynamics controlled truncation scheme (DCTS). In the cases of coherent optical phenomena the DCTS scheme represents a classification of higher-order density matrices (or dynamical variables) according to their leading order scaling with respect to the applied laser field. It is a rigorous theorem inspired by a classification of nonlinear optical processes.[18] Apart from semiconductors, other systems such as Frenkel exciton systems (e.g. molecular aggregates, molecular crystals or biological antenna systems) are also commonly described in terms of the most widely used level of the DCTS theory is obtained by taking the cut for the truncation at the third order and using all available identities for reducing the number of dynamical variables.

Although the first theory describing quantum optical effects in exciton systems based on DCTS was developed in 1996,[9] only recently [10] it has been shown that under certain reasonable approximation the equations of motion for the exciton and photon operators can be cast in a form closely resembling those obtained within the semiclassical framework. In this way a unified picture of semiclassical and nonclassical nonlinear optical effects can be achieved.

The most common set-up for parametric emission is the one where a single coherent pump feed resonantly excites the structure at a given energy and wave vector, $\mathbf{k_p}$. The generalization to multi-pump set-up is straightforward. In order to be more *specific* we shall derive explicitly the case of input light beams activating only the $1S$ exciton sector with all the same circularly (e.g. σ^+) polarization, thus excluding the coherent excitation of bound two-pair coherences (biexciton) mainly responsible for polarization-mixing.[19] This situation can be realized, for instance, as soon as the biexciton resonance has been carefully tailored off-resonance with respect to the characteristic energies of the states involved in the parametric scattering.[a]

The Heisenberg equations for the exciton and photon operators has been derived in Ref. 10. They contains all the terms providing a complete third order nonlinear response (in the input light field). For simplicity, here we shall retain only the dominant contributions, namely those containing the semiclassical pump amplitude at k_p twice, thus focusing on the "direct" pump-induced nonlinear parametric scattering processes. The generalization at all wavevector is straightforward. Following Ref. 10, the lowest order ($\chi^{(3)}$) nonlinear optical response in SMCs is given by the following set of coupled equations:

$$\frac{d}{dt}\hat{a}_{\pm\mathbf{k}} = -i\omega_{\mathbf{k}}^c \hat{a}_{\pm\mathbf{k}} + i\frac{V}{\hbar}\hat{B}_{\pm\mathbf{k}} + t_c\frac{E_{\pm\mathbf{k}}}{\hbar}$$

$$\frac{d}{dt}\hat{B}_{\pm k} = -i\omega_{\mathbf{k}}\hat{B}_{\pm\mathbf{k}} + i\frac{V}{\hbar}\hat{a}_{\pm\mathbf{k}} - \frac{i}{\hbar}R_{\pm\mathbf{k}}^{NL}, \tag{8}$$

where $R_{\pm\mathbf{k}}^{NL} = (R_{\pm\mathbf{k}}^{sat} + R_{\pm\mathbf{k}}^{xx})$

$$R_{\pm\mathbf{k}}^{sat} = \frac{V}{n_{sat}}B_{\pm\mathbf{k}_p}a_{\pm\mathbf{k}_p}\hat{B}_{\pm\mathbf{k}_i}^\dagger$$

$$R_{\pm\mathbf{k}}^{xx} = \hat{B}_{\pm\mathbf{k}_i}^\dagger(t)\left(V_{xx}B_{\pm\mathbf{k}_p}(t)B_{\pm\mathbf{k}_p}(t) - i\int_{-\infty}^t dt' F^{\pm\pm}(t-t')B_{\pm\mathbf{k}_p}(t')B_{\pm\mathbf{k}_p}(t')\right).$$

$$\tag{9}$$

---

[a]From the point of view of experiments probing the quantum properties of the emitted radiation out of a SMC (our aim), in common structures (e.g. GaAs) the Rabi splitting amounts of a bunch of meV. As a consequence, in order to be safely off-resonance, the biexciton energy needs not to be in energy further than some dozens of meV.

The first nonlinear term describes the *saturation* dynamics related to the fermionic character of electrons and hole, while the other comes from Coulomb interaction. The contribution V_{xx} is the Hartee-Fock or mean-field term representing the first order treatment in the Coulomb interaction between excitons, the second term $(F^{\pm\pm}(t - t'))$ is a pure biexciton (four-particle correlation) contribution. This coherent memory may be thought of as a non-Markovian process involving the two-particle (excitons) states interacting with a bath of four-particle correlations.[19]

The strong exciton-photon coupling does not modify the memory kernel because four-particle correlations do not couple directly to cavity photons. As pointed out clearly in Ref. 20, cavity effects alter the phase dynamics of two-particle states during collisions, indeed, the phase of two-particle states in SMCs oscillates with a frequency which is modified with respect to that of excitons in bare QWs, thus producing a modification of the integral in Eq. (9). This way the exciton-photon coupling V affects the exciton-exciton collisions that govern the polariton amplification process.

Equations (8) are the main result of the present section. They can be considered the starting point for the microscopic description of quantum optical effects in SMCs. These equations extend the usual semiclassical description of Coulomb interaction effects, in terms of a mean-field term plus a genuine non-instantaneous four-particle correlation. Starting from here, in the strong coupling case, it might be useful to transform the description into a polariton basis (see Section 4).

3. Entangled photon pairs from the optical decay of biexcitons

Transient or frequency-resolved four-wave mixing (FWM) are among the most widely used techniques for probing the optical properties of electronic excitations in semiconductors.

The FWM process can be schematically described as follows: two incident pump photons with a given wave vector $\mathbf{k}_p$ propagating inside the crystal slab (or the taylored semiconductor structure) as excitonic polaritons, excite a virtual state with two electron-hole pairs with total wave vector $2\mathbf{k}_p$, while a probe beam with wave vector $\mathbf{k}_1$ stimulates the optical decay of the state with two e-h pairs into a final polariton pair at $\mathbf{k}_1$ and $\mathbf{k}_2 = 2\mathbf{k}_p - \mathbf{k}_1$. The states with two e-h pairs can be either bound biexcitons or two exciton scattering states. This stimulated process thus determines the generation of a new beam at $\mathbf{k}_2$ as well as the amplification of the probe beam at $\mathbf{k}_1$ (parametric gain) as observed in SMCs. If the pump beam is maintained while eliminating the probe beam, there is no coherent emission at $\mathbf{k}_2$ according to the semiclassical theory. Actually the quantum fluctuations of the light field can play the role of the probe beam *stimulating* the optical decay of states with two e-h pairs into a pair of final polariton states. In particular quantum fluctuations of modes at a generic wave vector $\mathbf{k}_1$ can determine light emission in the direction $\mathbf{k}_2 = 2\mathbf{k}_p - \mathbf{k}_1$ and vice-versa.

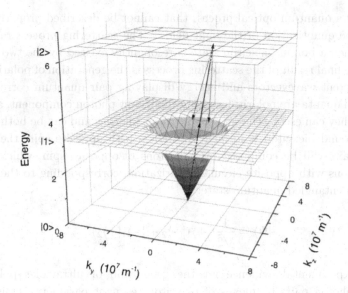

Fig. 1. Resonant two-photon Raman scattering process in CuCl involving the lower polariton branch, $|0\rangle, |1\rangle, |2\rangle$ are, respectively, the ground state, the 1s exciton stat and the bound biexciton state.

The process that we are going to describe is known as hyper-Raman scattering (HRS) or two-photon Raman scattering. HRS can be schematically described as follows (see Fig. 1): A first photon induces a transition from the ground state to the excitonic subspace, the second photon induces a further transition from the exciton to the biexciton subspace, then the excited biexciton may decay spontaneously into an exciton and a photon such that total energy and wave vectors are preserved. However, we have just mentioned that the allowed propagating modes inside the crystal slab are polaritons. The above picture should be modified accordingly: a pair of incident photons of given wavevector and energy propagate inside the crystal as polaritons of given energy $\hbar\omega(\mathbf{k}_i)$ and may give rise (according to the hierarchy of dynamic equations) to a biexciton excitation, such excitation can finally decay into a pair of final polaritons. Total energy and momentum are conserved in the whole process:

$$\omega(\mathbf{k}_1) + \omega(\mathbf{k}_2) = 2\omega(\mathbf{k}_i),$$

$$\mathbf{k}_1 + \mathbf{k}_2 = 2\mathbf{k}_i,$$

where final polariton states have been labelled with 1 and 2. The determination of the polariton dispersion has successfully been accomplished by HRS in several large-gap bulk semiconductors. HRS is a process related to the third-order nonlinear susceptibility and can be considered as a spontaneous resonant non-degenerate four-wave mixing (FWM). In FWM the optical decay of the coherently excited biexcitons is stimulated by sending an additional light beam, while in HRS the decay is determined by intrinsic quantum fluctuations. This subtle difference imply that HRS is a quantum optical process that cannot be described properly without including the quantization of the light field.[16] The scattering process rate largely increases when a bound biexciton state is resonantlly excited by the two polariton process. The final result of the scattering process is the generation of polaritons with slightly different wavevectors and energy displaying pair quantum correlations. If the two final quanta are polaritons with a significant photon component, at the end of the slab they can escape the semiconductor as photons and can be both detected. Since the bound biexciton is formed by carriers with opposite spin, the resulting polariton pairs will be composed by polaritons of opposite spin, which will give rise to photons with opposite circular polarization, corresponding to the following polarization entangled quantum state:

$$|\Psi\rangle = \frac{1}{\sqrt{2}}(|+\rangle_1|-\rangle_2 + |-\rangle_1|+\rangle_2). \tag{10}$$

In 2004, experimental evidence for the generation of ultraviolet polarization-entangled photon pairs by means of biexciton resonant parametric emission in a single crystal of semiconductor CuCl has been reported.[4]

4.   The picture of interacting polaritons

When the coupling rate V exceeds the decay rate of the exciton coherence and of the cavity field, the system enters the strong coupling regime. In this regime, the continuous exchange of energy before decay significantly alters the dynamics and hence the resulting resonances of the coupled system with respect to those of bare excitons and cavity photons. As a consequence, cavity-polaritons arise as the two-dimensional eigenstates of the coupled system. The coupling rate V determines the splitting ($\simeq 2V$) between the two polariton energy bands. This nonperturbative dynamics including the interactions (induced by $\hat{R}_{\mathbf{k}}^{NL}$) between different polariton modes can be accurately described by Eq. (8). Nevertheless there can be reasons to prefer a change of basis from excitons and photons to the eigenstates of the coupled system, namely polaritons. The resulting equations may provide a more intuitive description of nonlinear optical processes in terms of interacting polaritons. Moreover equations describing the nonlinear interactions between polaritons become more similar to those describing parametric interactions between photons widely adopted in quantum optics. Another, more fundamental reason, is that the standard second-order Born-Markov approximation scheme, usually adopted to describe the interaction with environment, is strongly basis-dependent, and using the eigenstates of the closed system provides more accurate results.

Equation (8) can be written in compact form as

$$\dot{\mathcal{B}}_{\mathbf{k}} = -i\Omega_{\mathbf{k}}^{xc}\mathcal{B}_{\mathbf{k}} + \mathcal{E}_{\mathbf{k}}^{in} - i\mathcal{R}_{\mathbf{k}}^{NL}; \tag{11}$$

where

$$\mathcal{B}_\mathbf{k} \equiv \begin{pmatrix} \hat{B}_\mathbf{k} \\ \hat{a}_\mathbf{k} \end{pmatrix}, \quad \Omega_\mathbf{k}^{\mathrm{xc}} \equiv \begin{pmatrix} \bar{\omega}_\mathbf{k}^x & -V \\ -V & \bar{\omega}_\mathbf{k}^c \end{pmatrix}, \quad \mathcal{E}_\mathbf{k}^{in} \equiv \begin{pmatrix} 0 \\ t_c E_\mathbf{k}^{in} \end{pmatrix}, \quad \text{and} \quad \mathcal{R}_\mathbf{k}^{NL} \equiv \begin{pmatrix} \hat{R}_\mathbf{k}^{NL} \\ 0 \end{pmatrix}.$$

In order to obtain the dynamics for the polariton system we perform unitary basis transformation on the exciton and photon operators

$$\mathcal{P}_\mathbf{k} = U_\mathbf{k} \mathcal{B}_\mathbf{k}; \tag{12}$$

being $\mathcal{P}_\mathbf{k} = \begin{pmatrix} \hat{P}_{1\mathbf{k}} \\ \hat{P}_{2\mathbf{k}} \end{pmatrix}$ and

$$U_\mathbf{k} = \begin{pmatrix} X_{1\mathbf{k}} & C_{1\mathbf{k}} \\ X_{2\mathbf{k}} & C_{2\mathbf{k}} \end{pmatrix}, \tag{13}$$

Diagonalizing $\Omega_\mathbf{k}^{\mathrm{xc}}$:

$$U_\mathbf{k} \Omega_\mathbf{k}^{\mathrm{xc}} = \tilde{\Omega}_\mathbf{k} U_\mathbf{k}, \tag{14}$$

where

$$\tilde{\Omega}_\mathbf{k} = \begin{pmatrix} \omega_{1\mathbf{k}} & 0 \\ 0 & \omega_{2\mathbf{k}} \end{pmatrix}.$$

$\omega_{1,2}$ are the eigenenergies (as a function of $\mathbf{k}$) of the lower (1) and upper (2) polariton states, while the X's and C's are the usual Hopfield coefficient. Introducing this transformation into Eq. (11), one obtains

$$\dot{\hat{P}}_{1\mathbf{k}} = -i\omega_{1\mathbf{k}} \hat{P}_{1\mathbf{k}} + \tilde{E}_{1,\mathbf{k}}^{in} - i\tilde{R}_{1\mathbf{k}}^{NL}, \tag{15}$$

$$\dot{\hat{P}}_{2\mathbf{k}} = -i\omega_{2\mathbf{k}} \hat{P}_{2\mathbf{k}} + \tilde{E}_{2,\mathbf{k}}^{in} - i\tilde{R}_{2\mathbf{k}}^{NL}; \tag{16}$$

where $\tilde{E}_{m\mathbf{k}}^{in} = t_c C_{m\mathbf{k}} E_\mathbf{k}^{in}$, and $\tilde{R}_{m\mathbf{k}}^{NL} = X_{i\mathbf{k}} \hat{R}_\mathbf{k}^{NL}$, $(m = 1, 2)$. Such a diagonalization is the necessary step when the eigenstates of the polariton system are to be used as starting states perturbed by the interaction with the environment degrees of freedom.[21] Equation (15) describes the coherent dynamics of a system of interacting cavity polaritons. The nonlinear term drives the mixing between polariton modes with different in-plane wave vectors and possibly belonging to different branches.

Analogous equations can be obtained starting from an effective Hamiltonian describing excitons as interacting bosons.[22] The resulting equations (usually developed in a polariton basis) do not include correlation effects beyond Hartree-Fock. Moreover the interaction terms due to phase space filling differs from those obtained within the present approach, not based on an effective Hamiltonian.

Only the many-body electronic Hamiltonian, the intracavity-photon Hamiltonian and the Hamiltonian describing their mutual interaction have been taken into account. Losses through mirrors, decoherence and noise due to environment interactions as well as applications of this theoretical framework, in the strong coupling regime, will be addressed in the next sections.

5.   Noise and environment: Quantum Langevin approach

Although spontaneous parametric processes involving polaritons in bulk semiconductors have been known for decades,[2] the possibility of generating entangled photons by these processes was theoretically pointed out only lately.[9] This result was based on a microscopic quantum theory of the nonlinear optical response of interacting electron systems relying on the DCTS[23] extended to include light quantization.[3,16,20] The above theoretical framework was also applied to the analysis of polariton parametric emission in semiconductor microcavities (SMCs).[16,20]

Previous descriptions of polariton parametric processes make deeply use of the picture of polaritons as interacting bosons. These theories have been used to investigate parametric amplifications, parametric luminescence, coherent control, entanglement and parametric scattering in momentum space.[22,24-27]

It is worth noting that in a realistic environment phase-coherent nonlinear optical processes involving real excitations compete with incoherent scattering as evidenced by experimental results. In experiments dealing with parametric emission, what really dominates emission at low pump intensities is the photoluminescence (PL) due to the incoherent dynamics of the single scattering events driven by the pump itself and the Rayleigh scattering of the pump due to the unavoidable presence of structural disorder. The latter is coherent and it can in principle be filtered out, on the contrary, PL cannot be easily separated from parametric emission.

In order to model the quantum dynamics of the polariton system in the presence of losses and decoherence we exploit the microscopic quantum Heisenberg-Langevin approach. We choose it because of its easiness in manipulating operator differential equations, and above all, for its invaluable flexibility and strength in performing even multitime correlation calculations, that are particularly important when dealing with quantum correlation properties of the emitted light.

In polariton language, we can start with Eq. (15) in the case of a single semiclassical pump feed resonantly exciting the lower polariton branch at a given wave vector $\mathbf{k}_p$. The generalization to a many-classical-pumps settings is straightforward. The nonlinear term couples pairs of wave vectors, let's say $\mathbf{k}$, the signal, and $\mathbf{k}_i = 2\mathbf{k}_p - \mathbf{k}$, the idler.

Following Lax's prescription we can promote Eqs. (15) to global bare-operator equations

$$\frac{d}{dt}\hat{P}_{\mathbf{k}} = -i\tilde{\omega}_{\mathbf{k}}\hat{P}_{\mathbf{k}} + g_{\mathbf{k}}\hat{P}_{\mathbf{k}_i}^{\dagger}\mathcal{P}_{\mathbf{k}_p}^2 + \hat{\mathcal{F}}_{\hat{P}_{\mathbf{k}}}$$

$$\frac{d}{dt}\hat{P}_{\mathbf{k}_i}^{\dagger} = i\tilde{\omega}_{\mathbf{k}_i}\hat{P}_{\mathbf{k}_i}^{\dagger} + g_{\mathbf{k}_i}^*\hat{P}_{\mathbf{k}}\mathcal{P}_{\mathbf{k}_p}^2 + \hat{\mathcal{F}}_{\hat{P}_{\mathbf{k}_i}^{\dagger}}, \qquad (17)$$

where the noise source verify the following fluctuation-dissipation relation.[11]

In the low and intermediate excitation regime, at pump densities where parametric emission starts to appear against the PL background, the main incoherent contribution to the dynamics is the PL the pump produces by itself e.g. by phonon scattering towards other states. The PL arising from incoherent scattering of polariton states populated by parametric processes can be neglected. In this case incoherent dynamics decouples from parametric processes. In particular one can solve the rate equations describing the incoherent dynamics without including parametric processes.

6. Quantum complementarity of cavity polaritons

When two physical observables are complementary, the precise knowledge of one of them makes the other unpredictable. The most known manifestation of this principle is the ability of quantum-mechanical entities to behave as particles or waves under different experimental conditions. For example, in the famous double-slit experiment, a single electron can apparently pass through both apertures simultaneously, forming an interference pattern. But if a "which-way" detector is employed to determine the particle's path, the particle-like behaviour takes over and an interference pattern is no longer observed. Quantum complementarity is rather an inherent property of a system, enforced by quantum correlations. The link between quantum correlations, quantum nonlocality and Bohr's complementarity principle was established in a series of "which-way" experiments using quantum-correlated photon pairs emitted via parametric down-conversion.[28–31]

Here we report on the theoretical and experimental investigation of quantum complementarity in semiconductor microcavities. Given a pump wave vector $\mathbf{k}_p$, the set of possible parametric processes that satisfy total energy and momentum conservation is represented by an "eight"-shaped curve in $\mathbf{k}$-space,[22] as displayed in Fig. 2. On this curve, a pair of signal- and idler-modes is defined by the intersection with a straight line passing through $\mathbf{k}_p$. As a convention, we designate as "idler" the modes with $k > k_p$. The experimental scheme that we devise is based on two energy-degenerate pump modes having momentum $\mathbf{k}_{p1}$ and $\mathbf{k}_{p2}$ respectively. In this configuration together with the two customary eight-shape curves involving a single pump, mixed-pump parametric processes are allowed depicting the "peanut"-shaped curve in $\mathbf{k}$-space in Fig. 2. Let us consider the two mutually coherent pump polariton fields as of equal energy and amplitude but with different in-plane wave vector $\mathcal{P}_{\mathbf{k}_{p2}} = \mathcal{P}_{\mathbf{k}_{p1}} e^{i\phi}$. Within this scheme, pairs of parametric processes sharing the signal mode are allowed.[7] Such a pair of processes involves two idler modes i1 and i2 and one common signal mode s.

The lowest polariton-number state would be cast in the form

$$|\Psi\rangle = \frac{1}{\sqrt{2}}|1\rangle_s(|1\rangle_{i1}|0\rangle_{i2} + \exp(-2i\phi)|0\rangle_{i1}|1\rangle_{i2}), \tag{18}$$

where the two entangled idler states are sharing the same signal. The polariton density at the signal mode $\langle\Psi|\hat{P}_{\mathbf{k}_s}^\dagger \hat{P}_{\mathbf{k}_s}|\Psi\rangle$ results to be clearly independent of ϕ. Interference is absent due to the orthogonality of the two idler states whose sum

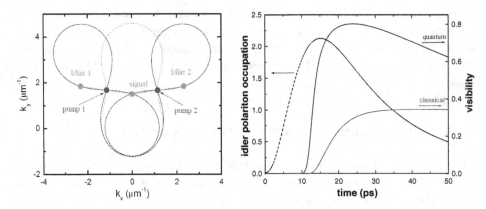

Fig. 2.   (left) **k**-space plot of the final states fulfilling energy and momentum conservation in a two-pump parametric process. The two "eight"-shaped full curves represent the single-pump processes determined by the conditions $2\mathbf{k}_{pj} = \mathbf{k}_s + \mathbf{k}_i$ and $2E_{\mathbf{k}_{pj}} = E_{\mathbf{k}_s} + E_{\mathbf{k}_i}$, with $j = 1, 2$. The dotted line describes the mixed-pump process defined by $\mathbf{k}_{p1} + \mathbf{k}_{p2} = \mathbf{k}_s + \mathbf{k}_i$ and $E_{\mathbf{k}_{p1}} + E_{\mathbf{k}_{p2}} = E_{\mathbf{k}_s} + E_{\mathbf{k}_i}$. Two parametric processes sharing $\mathbf{k}_s$, giving rise to mutual idler coherence are highlighted. (right) Calculated time-resolved idler polariton number per mode (dashed) originating from parametric emission; quantum and classical interference visibility when combining two idler beams sharing a common signal.

is enclosed in parentheses. This sum stores the "which-way" information, each term representing one possible idler path in the process. Hence the pair-correlation between signal and idler polaritons and the entanglement of the two idler paths are the reason for the absence of interference, as implied by expression (18). Interference is also absent from either idler-polariton density. A different result is found when looking at the mutual coherence of the two idler beams which is observable in the sum of the two idler polariton fields. The resulting particle density is $\langle\Psi|(\hat{p}^\dagger_{\mathbf{k}_{i1}} + \hat{P}^\dagger_{\mathbf{k}_{i2}})(\hat{P}_{\mathbf{k}_{i1}} + \hat{P}_{\mathbf{k}_{i2}})|\Psi\rangle = 2\langle\Psi|\hat{P}^\dagger_{\mathbf{k}_{i1}}\hat{P}_{\mathbf{k}_{i1}}|\Psi\rangle[1 + \cos(2\phi)]$. Interference now appears owing to the cross term $\langle\Psi|\hat{P}^\dagger_{\mathbf{k}_{i1}}\hat{P}_{\mathbf{k}_{i2}}|\Psi\rangle$.

In this case interference occurs because the two idler modes are pair-correlated with the same signal mode. Even by means of a signal-idler coincidence measurement, no "which-way" information could be retrieved. Hence the presence or absence of interference is a direct consequence of the complementarity principle enforced by the pair-correlation between signal and idler modes. A similar analysis on the more general many-polariton state leads to the same conclusions.[32,33]

The coherence properties of the polaritons are stored in the emitted photons, thus making it possible to carry out the devised "which-way" measurement within a standard optical spectroscopy setup. The investigated device[7] consists of a 25 nm GaAs/Al$_{0.3}$Ga$_{0.7}$As single quantum well placed in the center of a λ–cavity with AlAs/Al$_{0.15}$Ga$_{0.85}$As Bragg reflectors. Blocking one of the detection interferometer arms, the intensity $I_{\mathbf{k}} = |\mathcal{E}_{\mathbf{k}}|^2$ is measured (proportional to the polariton density $N_{\mathbf{k}} = \langle\Psi|\hat{P}^\dagger_{\mathbf{k}}\hat{P}_{\mathbf{k}}|\Psi\rangle$).

The resulting visibility, defined as

$$\mathcal{V} \equiv \frac{I_{\max} - I_{\min}}{I_{\max} + I_{\min}} \tag{19}$$

is $\mathcal{V} = (N_1 - N_2)/(N_1 + N_2)$. In the absence of incoherent processes $N_2 = 0$ and $\mathcal{V} = 1$. Incoherent effects thus lower the visibility. A classical model of parametric emission can be obtained from a system like Eqs. (17) replacing quantum operators with $\mathbb{C}$-numbers. In this case the quantum Langevin forces becomes classical stochastic noise terms. While the quantum parametric process is initiated by vacuum fluctuations, the classical version describes (in the absence of a coherent signal or idler input) essentially the parametric amplification of noise. Figure 2 displays the calculated time-resolved idler polariton number per mode (dashed) originating from parametric emission, and the calculated interference visibility. The pump is set so that the peak idler-polariton occupation number per mode is $N_{\mathbf{k}_{i1}} = N_{\mathbf{k}_{i2}} \simeq 2$. The figure also reports the corresponding classical calculation largely underestimating the measured visibility.[7] This calculation thus confirms the quantum nature of the measured interference. We observe that the present calculation includes a microscopic description of PL giving rise to a realistic description of its time dependence, while in Ref. 7 PL has been included as a constant phenomenological noise.

7. Emergence of entanglement out of a noisy environment: The case of microcavity polaritons

The concept of *entanglement* has played a crucial role in the development of quantum physics. It can be described as the correlation between distinct subsystems which cannot be reproduced by any classical theory (i.e. *quantum correlation*). In order to address entanglement in quantum systems,[4,34] the preferred experimental situation is the few-particle regime in which the emitted particles can be detected individually.[35,36] In a real system, environment always act as an uncontrollable and unavoidable continuous perturbation producing decoherence and noise. Even if polariton experiments are performed at temperature of few Kelvin,[7] polaritons created resonantly by the pump can scatter, by emission or absorption of acoustic phonons, into other states, acquiring random phase relations. These polaritons form an incoherent background (i.e. noise), responsible of *pump-induced photoluminescence* (PL), which competes with coherent photoemission generated by parametric scattering, as evidenced by experiments.[27] As a consequence, noise represents a fundamental limitation, as it tends to lower the degree of non-classical correlation or even completely wash it out.[7,37]

In this section, we present a microscopic study of the influence of time-dependent noise on the polarization entanglement of polaritons generated in parametric PL. Our treatment accounts for realistic features such as detectors noise background, detection windows, dark-counting etc., needed[11] in order to seek and limit all the unwanted detrimental contributions. We show how a tomographic reconstruction,[38,39] based on two-times correlation functions, can provide a quantitative assessment of the level of entanglement produced under realistic experimental conditions. In particular, we give a ready-to-use realistic experimental configuration able to measure the Entanglement of Formation (EOF),[40–42] out of a dominant time-dependent noise background, without any need for post-processing.[34]

7.1. Coherent and incoherent polariton dynamics

Third order nonlinear optical processes in quantum well excitons (with spin $\sigma = \pm 1$) can be described in terms of two distinct scattering channels: one involving only excitons (polaritons) with the same circular polarization (co-circular channel); and the other (counter-circular channel) due to the presence of both bound biexciton states and four-particle scattering states of zero angular momentum ($J = 0$).[19] Bound biexciton-based entanglement generation schemes,[3,4,43] producing entangled polaritons with opposite spin, need specific tunings for efficient generation, and are expected to carry additional decoherence and noise due to scattering of biexcitons. Moreover linearly polarized single pump excitation cannot avoid the additional presence of the co-circular scattering channel which can lower polarization entanglement. The experimental set-up that we will choose to calculate the emergence of polariton spin-entanglement is a two-pump scheme under pulsed excitation, involving the lower polariton branch only. The pumps (p_1 and p_2) are chosen with incidence angles below the *magic angle*[27] so that single-pump parametric scattering is negligible. In this setup, mixed-pump processes (signal at in-plane wave vector $\mathbf{k}$, idler at $\mathbf{k}_i = \mathbf{k}_1 + \mathbf{k}_2 - \mathbf{k}$) are allowed. We choose $\mathbf{k}_1 = (0., 0.)$, and $\mathbf{k}_2 = (0.9, 0.9)\mu m^{-1}$. As signal-idler pair, we choose to study the two energy-degenerate modes at $\mathbf{k} \simeq (k_{1x}, k_{2y})$ and $\mathbf{k}_i \simeq (k_{2x}, k_{1y})$, as shown in Fig. 3. Of course a number of different two-pump schemes can also conveniently be adopted. For istance, from an experimental viewpoint, a two energy-degenerate pumps setting can be more valuable. Then possible choices for signal-idler pairs within the circle of available parametrically-generated final states (see Fig. 3) would suffer only a slightly unbalance being anyway close to the origin of the polariton dispersion curve. For all the numerical simulations we will consider the sample investigated in Ref. 27. In particular, we shall employ two pump beams linearly cross polarized (then the angle θ will refer to the polarization of one of the two beams, see Fig. 3). This configuration is such that the counter-circular scattering channel (both bound biexciton and scattering states) is suppressed owing to destructive interference, while co-circular polarized signal-idler beams are generated. In the absence of the noisy environment, polariton pairs would be cast in the pure triplet entangled state $|\psi_\parallel\rangle = |+, +\rangle - \exp(i4\theta)|-, -\rangle$.

The advantages of this configuration are manyfold. First, processes detrimental for entanglement such as the excitation induced dephasing results to be largely suppressed.[9,44–46] Spurious coherent processes, e.g. Resonant Rayleigh Scattering,[47]

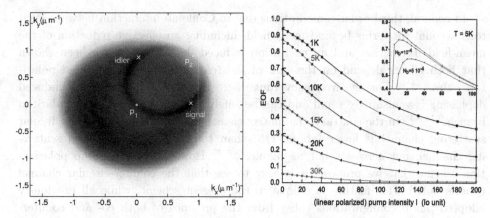

Fig. 3. (left) The simulated spectrally integrated polariton population in **k**-space. The parametric process builds up a circle passing through the two pumps (p_1 and p_2), signal and idler polariton states are represented by any two points on the circle connected by a line passing through its center. For illustration the pair of signal-idler polariton modes chosen for entanglement detection are depicted as crosses. The disc-shape contribution centered at the origin is the incoherent population background produced by phonon scattering. (right) Dependence of the EOF on pumping intensity. The laser intensity I is measured in units of $I_0 = 21$ photons μm^{-2}/pulse according to Ref. 27. In the inset the EOF for the case of $T = 5\,\text{K}$ against pump intensity is depicted for different temperature- and pump-independent noise background N_b representing other possible non-dominant (with respect to acoustic-phonons) noise channels, e.g. the photo-detection system (see text for comments and discussion).

are well separated in **k**-space from the signal and idler modes. In addition signal and idler close to the origin in **k**-space make negligible the longitudinal-transverse splitting of polaritons[15] (relevant at quite high in-plane wave vectors).

Following Refs. 38, 39 the tomographic reconstruction of the two-polariton density matrix is equivalent, in the $\sigma = \{+, -\}$ polarization basis, to the two-time coincidence

$$\rho_{\sigma\tilde{\sigma},\sigma'\tilde{\sigma}'} = \frac{1}{\mathcal{N}} \int_{T_d} dt_1 \int_{T_d} dt_2 \langle \hat{P}_{\mathbf{k}\sigma}^\dagger(t_1) \hat{P}_{\mathbf{k}_i\tilde{\sigma}}^\dagger(t_2) \hat{P}_{\mathbf{k}_i\tilde{\sigma}'}(t_2) \hat{P}_{\mathbf{k}\sigma'}(t_1) \rangle, \tag{20}$$

where $\hat{P}_{\mathbf{k}\sigma}^\dagger$ ($\hat{P}_{\mathbf{k}_i\tilde{\sigma}}^\dagger$) creates a signal polariton at **k** (an idler polariton at $\mathbf{k}_i = \mathbf{k}_1 + \mathbf{k}_2 - \mathbf{k}$), $\mathcal{N}$ is a normalization constant and T_d the detector window. We choose a very wide time window $T_d = 120\,\text{ps}$, allowing feasible experiments with standard photodetectors. In order to model the density matrix Eq. (20), we employ the dynamics controlled truncation scheme (DCTS), starting from the electron-hole Hamiltonian including two-body Coulomb interaction and radiation-matter coupling. In this approach nonlinear parametric processes within a third order optical response are microscopically calculated. The main environment channel is acoustic phonon interaction via deformation potential coupling.[11,21]

We use a DCTS-Langevin approach,[11] with noise sources given by exciton-LA-phonon scattering and radiative decay (treated in the Born-Markov approximation).

In general, third order contributions due to Coulomb interaction between excitons account for terms beyond mean-field, including an effective reduction of the mean-field interaction and an excitation induced dephasing. It has been shown that both effects depend on the sum of the frequencies of the scattered polariton pairs.[9,44–46] For the frequency range here exploited, the excitation induced dephasing is vanishingly small and can be safely neglected on the lower polariton branch.[9,44–46] On the contrary, the matrix elements of the ($J = 0$) counter-circular scattering channel is lower (about 1/3) than that for the co-polarized scattering channel, but certainly not negligible.[9,44–46] However, in the pump polarization scheme that we propose, it is easy to see that the counter-circular channel cancels out.[48] This feature is unique to the present scheme, while all previously adopted pump configurations suffer from the presence of both co- and counter-circular polarized scattering channels.

7.2. *Results*

Figure 3 shows a typical pattern of photoluminescence in **k**-space we can simulate with our microscopic model. We can neatly distinguish the disc-shape contribution centered at the origin due to the incoherent population produced by phonon scattering from the parametric ring dynamically emerging from the noisy background. The two pumps employed are marked as p_1 and p_2. Parametrically generated signal and idler polariton states are represented by any two points on the circle connected by a line passing through its center. For illustration the pair of signal-idler polariton modes chosen for entanglement detection are depicted as crosses. Early experiments in semiconductor microcavities[7,49] provided promising, though indirect, indications of polariton entanglement. In order to achieve a conclusive evidence of entanglement one has to produce its quantitative analysis and characterization, i.e. a *measure* of entanglement. Among the various measures proposed in the literature we shall use the *entanglement-of-formation* $E(\hat{\rho})$[40–42] for which an explicit formula as a function of the density matrix exists.[b] It has a direct operational meaning as the minimum amount of information needed to *form* the entangled state under investigation out of uncorrelated ones. The complete characterization of a quantum state requires the knowledge of its density matrix. Even though its off-diagonal elements are not directly related to physical observables, the density matrix of a quantum system composed by two two-level particles can be reconstructed using the recently developed *quantum state tomography*,[38,39] that has also been exploited in a bulk semiconductor.[4] It requires 16 two-photon coincidence measurements based on various polarization configurations.[38,39]

---

[b]Formally, the EOF is defined as the minimum average pure state entanglement over all possible pure state decompositions of the mixed density matrix. Easy speaking the minimum entanglement needed to construct the density matrix out of some pure states.

As Fig. 3 shows, there is a non-negligible region of the parameter space where, even in a realistic situation, high entanglement values are obtained. For increasing pump intensities, EOF decays towards zero. This is a known consequence of the relative increase of signal and idler populations[4,34] — dominating the diagonal elements of $\hat{\rho}$ — with respect to the two-body correlations responsible for the non-diagonal parts, which our microscopic calculation is able to reproduce. We expect entanglement to be unaffected by both intensity and phase fluctuations of the pump laser. The former is negligible,[48] while the latter only acts on the overall quantum phase of the signal-idler pair state. Different entanglement measures generally result in quantitatively different results for a given mixed state. However, they all provide upper bounds for the distillable entanglement,[50] i.e. the rate at which mixed states can be converted into the "gold standard" singlet state. Small EOF means that a heavily resource-demanding distillation process is needed for any practical purpose. Figure 3 shows how a relative small change in the lattice temperature has a sizeable impact on entanglement. As an example, for the pump intensity $I = 15I_0$, increasing the temperature from $T = 1\,K$ to $T = 20\,K$ means to corrupt the state from $E(\hat{\rho}) \simeq 0.88$ to $E(\hat{\rho}) \simeq 0.24$, whose distillation is nearly four times more demanding. For a fixed pump intensity, Fig. 3 shows that, above a finite temperature threshold, the EOF vanishes independently of the pump intensity, i.e. the influence of the environment is so strong that quantum correlation cannot be kept anymore. In physical terms, at about $30K$ the average phonon energy becomes comparable to the signal-pump and idler-pump energy differences, and the thermal production of signal-idler pairs is activated.

8. Quantum plasmonics: Individual quantum emitters coupled to metallic nanoparticles

Control over the interaction between single photons and individual optical emitters deserves great importance in quantum science and quantum engineering.[51] Recently, substantial advances towards the realization of solid state quantum optical devices have been made coupling single quantum dots (QDs) to high-finesse optical cavities.[52,53] An inherent limitation of cavity quantum electrodynamics (QED) is that the size of the cavity is at least half wavelength and practically much more than that owing to the presence of mirrors or of a surrounding photonic crystal. Unlike optical microcavities, metallic nanoparticles and metallic nanostructures are able to focus electromagnetic waves to spots much smaller than a wavelength. In this way it is possible to increase the local density of the electromagnetic modes as in microcavities but with ultra-compact structures. The ability of metallic nanoparticles and nanostructures to control the radiative decay rate of emitters placed in their near field has been widely demonstrated (see e.g.[54–56]). This ability stems from the existence of collective, wave-like motions of free electrons on a metal surface termed surface plasmons (SP).[57] Moreover the plasmon excitation covers a broad bandwidth and requires no special tuning to achieve resonance. An outstanding demonstration of the cavity-like behavior of metallic nanoparticles is the recent realization of a nanolaser based on surface plasmon amplification by stimulated emission of radiation (spaser).[58]

Optical nonlinearities enable photon-photon interaction and lie at the heart of several proposals for quantum information processing,[51,53,59] quantum nondemolition measurements of photons,[60] and single-photon switching[61] and transistors.[14] The nonlinear optical response of a semiconductor QD coupled with a metallic nanoparticle (MNP) has initially been theoretically investigated[62,63] by exploiting a semiclassical approach where the quantum emitter is treated quantum mechanically and the light field classically. Although this approach provides useful information on the absorption and elastic scattering of this system, the nonlinear optical properties of individual QDs display important quantum optical effects.[52,53,64,65] Moreover, in order to investigate the possible use of these hybrid artificial molecules as quantum devices for the control of individual light quanta, a full quantum mechanical description is required. The efficient coupling between an individual optical emitter and propagating SPs confined to a conducting nanowire has been demonstrated experimentally.[66] Non-classical photon correlations between the emission from the quantum dot and the ends of the nanowire demonstrate that the latter stems from the generation of single, quantized plasmons. The potential of this system as an ultracompact single-photon transistor has been theoretically demonstrated.[14]

Here we briefly present a nanoscale approach to cavity QED based on individual quantum emitters coupled to metallic nanoparticles.[12] In particular we describe the quantum optical properties of a QD-MNP hybrid artificial molecule (see Fig. 4). We also show that a single QD coupled to a nanoscale antenna constituted e.g. by a silver dimer can reach the strong coupling regime.[13] Such nanoscale polaritons known also as plexcitons have been already demonstrated in system constituted by metallic nanoparticles coupled with many optically active molecules (see e.g. Refs. 67–69). The strong coupling regime obtained at the nanoscale with an individual quantum emitter is highly desirable for the realization of quantum devices.[51] Laser cooling of the center-of-mass motion of such hybrid systems has been recently proposed.[70]

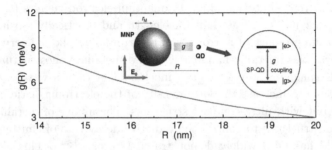

Fig. 4.   Interaction between a quantum dot and a silver nanosphere: the applied electromagnetic field induces a polarization that causes dipole-dipole coupling. States $|g\rangle$ and $|e\rangle$ are coupled via the localized surface plasmon dipole mode with a strength g. Dependence of the coupling g on the metallic nanoparticle-quantum dot distance R.

8.1. QE-MNP hybrid artificial molecule

We consider a spherical QD interacting with a spherical MNP of radius r_m, separated by a distance R (see Fig. 4). There is no direct tunneling between the MNP and the QD ($R - r_\mathrm{m} - r_\mathrm{QD} > 2$ nm, being r_QD the QD radius). The coupling mechanism is due to dipole-dipole interaction. The QE is modeled as a two level system with a dipole moment μ, that is a good approximation for high-quality QDs when studying optical processes at frequencies resonant with the lowest energy excitonic transition. The coupling mechanisms between the two units is the dipole-dipole interaction. Further details can be found in Ref. 12. Figure 4 displays the dependence of the coupling rate g on the QD-MNP distance R. Throughout this section we use a dipole moment $\mu = er_0$ with $r_0 = 0.7$ nm (corresponding to 33.62 Debye), being e the electron charge and $\epsilon_b = 3$. In the following we use for the QD transition a linewidth $\gamma_\mathrm{x} = 50\,\mu$eV.

Figure 5 (left panel) displays scattering spectra as function of the frequency of the incidence light obtained for different QD-MNP distances R as indicated in the panel. The spectra in Fig. 5 (left panel) have been calculated in the limit of very low excitation intensity, where the excitonic populations $\langle \sigma^\dagger \sigma \rangle \ll 1$. At $R = 14$ nm a

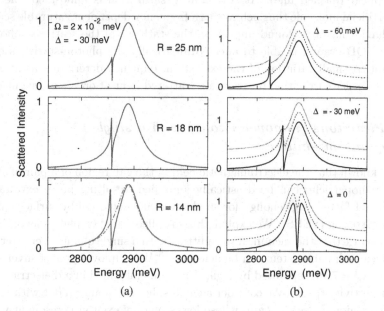

Fig. 5. (a) Scattered light intensity spectra (continuous line) calculated for different QD-MNP distances R at low density excitation power. For $R = 14$ nm, the scattered light without the presence of the QD (dot-dashed line) is also plotted. (b) Spectra calculated at $R = 14$ nm. Each panel shows calculations for a specific exciton-SP energy detuning (Δ) indicated in the figure. The black continuous line describes plots obtained for an input intensity field $\Omega = 0.02$ meV (corresponding to a photon flux $\Phi = 1.75\,\mu$m^{-2} ps^{-1}), the short-dashed line plots obtained at $\Omega = 0.4$ meV ($\Phi = 700\,\mu$m^{-2} ps^{-1}), and the short-dotted line plots at $\Omega = 1$ meV. Plots are peak-normalized. The plot at $\Omega = 0.4$ meV was vertically shifted by 0.2; the one at $\Omega = 1$ meV by 0.4.

Fano-like lineshape around the QD transition energy ω_x is evident. For a particular input frequency the scattered light is higly suppressed, while at slightly lower energy an enhancement of scattering due to constructive interference can be observed. For comparison the plot at $R = 14\,\text{nm}$ shows the scattering spectrum in the absence of the QD (dash-dotted line). Increasing the distance R, the Fano resonance narrows, due to the reduction of the MNP induced broadening of the QD linewidth. While at $R = 18\,\text{nm}$ the destructive interference remains almost complete, at larger distances ($R = 25\,\text{nm}$), the Fano interference effect lowers and the suppression as well as the increase of the scattered light are reduced. Figure 5 (right panel) displays scattering spectra obtained for QDs with different excitonic energy levels. In particular, each panel corresponds to different exciton-SP detunings $\Delta = \omega_x - \omega_{sp}$. Interestingly, the interference effect determining a strong suppression of scattering at specific wavelengths of the input field requires no special tuning unlike analogous effects in cavity QED. While SPs supported by the MNP can be described as harmonic oscillators, the single QD displays nonlinearities at single photon level. The left panels in Fig. 5 put forward the dependence of light scattering on the intensity of the input field. The continuous lines describe low-field spectra obtained for a Rabi energy $\Omega = 2\mu E_0 = 2 \times 10^{-2}\,\text{meV}$. Increasing the input field to $\Omega = 0.4\,\text{meV}$, saturation effects appear (dashed line). At $\Omega = 1\,\text{meV}$ saturation is almost complete. The hybrid artificial molecule thus behaves as a frequency dependent saturable scatterer. This behavior has a profound impact on the statistics of the scattered photons.[12] The MNP-QD system is able to affect dramatically the photon-statistics of scattered light. A small variation of the excitation frequency determines a variation of the second-order correlation function for scattered light of orders of magnitude.

8.2. Plexcitons (nanopolaritons) with a single quantum emitter

It is well known that the electromagnetic field in the gap region of a pair of strongly coupled nanoparticles can be drastically amplified, resulting in an extraordinary enhancement factor large enough for single-molecule detection by surface enhanced Raman scattering (SERS). We exploit this so-called hot spot phenomenon in order to demonstrate that the vacuum Rabi splitting with a single quantum emitter within a subwavelength nanosystem can be achieved.[13] We employ a pair of silver spheres of radius $r_{Ag} = 7\,\text{nm}$ separated by a gap $l = 8\,\text{nm}$ embedded in a dielectric medium with permittivity $\epsilon_r = 3$. We consider a small spherical quantum dot with radius of the active region $r_{rmQD} = 2\,\text{nm}$, whose lowest energy exciton is resonant with the dimer bonding mode. Figure 6a displays a sketch of the system and of the input field polarized along the dimer axis in order to provide the largest field enhancement at the dot position. Figure 6b shows the extinction cross section spectra calculated for different dipole moments $\mu = er_0$, being e the electron charge. For $r_0 = 0.1\,\text{nm}$ a narrow hole in the spectrum occurs which could be confused with the appearance of a small vacuum Rabi splitting. Only for higher dipole moments $r_0 = 0.3$ a significant

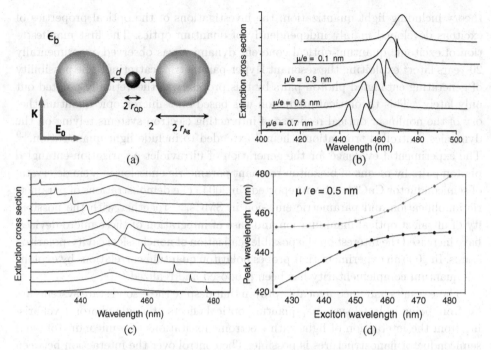

(a)

(b)

(c)

(d)

Fig. 6. Vacuum Rabi splitting with a single quantum dot in the center of a dimer nanoantenna. (a) Sketch of the system and of the excitation. (b) Calculated extinction cross sections as function of the wavelength of the input field obtained for different dipole moments of the quantum dot. (c) Extinction cross sections spectra obtained for different resonant energies E_0 of the quantum dot exciton ($\mu/e = 0.5$ nm). A clear anti-crossing is observed owing to strong coupling between the dot and the localized bonding surface plasmon dimer-mode. (d) Dependence of the two Rabi-peaks (extinction cross sections) wavelengths on the exciton transition wavelength $\lambda_0 = h c/E_0$ ($\mu/e = 0.5$ nm).

splitting can be observed. Figure 6c displays the extinction spectra for $r_0 = 0.5$ nm obtained changing the energy of the quantum dot exciton. At large detuning the peak arising from the quantum dot is significantly narrower from that originating from the SP bounding mode. Lowering the detuning increases the linewidth of the exciton-like peak while at the same time lowers that of the SP-like peak as a consequence of the strong coupling between the modes. Figure 6d shows the dependence of the two Rabi peaks (extinction cross sections) wavelengths on the exciton transition wavelength $\lambda_0 = h c/E_0$ ($\mu/e = 0.5$ nm). The anticrossing behaviour certifying true strong coupling is evident.

9.   Outlook

In this chapter we reviewed quantum optical effects with excitonic polaritons in both bulk semiconductors and in cavity embedded QWs. Although, the excitonic polariton concept, introduced in 1958 by J.J. Hopfield,[1] was based on a full quantum theory including light quantization, the investigations of the optical properties of excitons developed mainly independently of quantum optics. The first manifestation of excitonic quantum-optical coherent dynamics was observed experimentally 20 years later[2] exploiting the resonant Hyper-parametric scattering. The possibility of generating entangled photon pairs by this process was theoretically pointed out only lately.[3] This theoretical prediction was based on a microscopic quantum theory of the nonlinear optical response of interacting electron systems relying on the dynamics controlled truncation scheme[8] extended to include light quantization.[3,9] The experimental evidence for the generation of ultraviolet polarization-entangled photon pairs by means of biexciton resonant parametric emission in a single crystal of semiconductor CuCl has been reported[4] in 2004. The demonstrations of parametric amplification and parametric emission in SMCs,[5,6] together with the possibility of ultrafast optical manipulation and ease of integration of these micro-devices, have increased the interest on the possible realization of nonclassical cavity-polariton states. In 2005 an experiment that probes polariton quantum correlations by exploiting quantum complementarity has been proposed and realized.[7]

These results had unequivocably proven that despite these solid state systems are far from being isolated systems, quantum optical effects at single photon level arising from the interaction of light with electronic excitations of semiconductors and semiconductor nanostructures is possible. The control over the interaction between single photons and individual optical emitters is an outstanding problem in quantum science and engineering. In the last few years substantial advances have been made towards these goals by achieving the strong coupling regime for a single quantum dot embedded in a high-Q microcavity.[71–73] The quantum nature of this strong coupled system has been demonstrated.[7] Of great interest is also the interaction of quantum emitters with surface plasmon modes supported by metallic nano-particles.[14] These hybrid metal-semiconductor systems can display quantum optical effects promising for the realization of sub-wavelength devices.[12,13]

References

1. J. J. Hopfield, Theory of the contribution of excitons to the complex dielectric constant of crystals, *Phys. Rev.* **112**, 1555–1567 (Dec, 1958).
2. B. Hönerlage, A. Bivas, and V. D. Phach, Determination of the excitonic polariton dispersion in cucl by resonant two-photon raman scattering, *Phys. Rev. Lett.* **41**, 49–52 (Jul, 1978).
3. S. Savasta, G. Martino, and R. Girlanda, Entangled photon pairs from the optical decay of biexcitons, *Solid State Communications* **111**(9), 495–500, (1999). ISSN 0038-1098.
4. K. Edamatsu, G. Oohata, R. Shimizu, and T. Itoh, Generation of ultraviolet entangled photons in a semiconductor, *Nature* **431**, 167–170, (2004).
5. R. M. Stevenson, V. N. Astratov, M. S. Skolnick, D. M. Whittaker, M. Emam-Ismail, A. I. Tartakovskii, P. G. Savvidis, J. J. Baumberg, and J. S. Roberts, Continuous wave observation of massive polariton redistribution by stimulated scattering in semiconductor microcavities, *Phys. Rev. Lett.* **85**, 3680–3683 (Oct, 2000).

6. P. G. Savvidis, J. J. Baumberg, R. M. Stevenson, M. S. Skolnick, D. M. Whittaker, and J. S. Roberts, Angle-resonant stimulated polariton amplifier, *Phys. Rev. Lett.* **84**, 1547–1550 (Feb, 2000).

7. S. Savasta, O. D. Stefano, V. Savona, and W. Langbein, Quantum complementarity of microcavity polaritons, *Phys. Rev. Lett.* **94**, 246401 (Jun, 2005).

8. V. M. Axt and T. Kuhn, Femtosecond spectroscopy in semiconductors: A key to coherences, correlations and quantum kinetics, *Reports on Progress in Physics* **67**(4), 433, (2004).

9. S. Savasta and R. Girlanda, Quantum optical effects and nonlinear dynamics in interacting electron systems, *Phys. Rev. Lett.* **77**, 4736–4739 (Dec, 1996).

10. S. Portolan, O. Di Stefano, S. Savasta, F. Rossi, and R. Girlanda, Dynamics-controlled truncation scheme for quantum optics and nonlinear dynamics in semiconductor microcavities, *Phys. Rev. B* **77**, 195305 (May, 2008).

11. S. Portolan, O. Di Stefano, S. Savasta, F. Rossi, and R. Girlanda, Nonequilibrium langevin approach to quantum optics in semiconductor microcavities, *Phys. Rev. B* **77**, 035433 (Jan, 2008).

12. A. Ridolfo, O. Di Stefano, N. Fina, R. Saija, and S. Savasta, Quantum plasmonics with quantum dot-metal nanoparticle molecules: Influence of the fano effect on photon statistics, *Phys. Rev. Lett.* **105**, 263601 (Dec, 2010).

13. S. Savasta, R. Saija, A. Ridolfo, O. Di Stefano, P. Denti, and F. Borghese, Nanopolaritons: Vacuum rabi splitting with a single quantum dot in the center of a dimer nanoantenna, *ACS Nano* **4**(11), 6369–6376, (2010).

14. D. E. Chang, A. S. Sørensen, E. A. Demler, and M. D. Lukin, A single-photon transistor using nanoscale surface plasmons, *Nature Physics* **3**(11), 807–812, (2007).

15. K. V. Kavokin, I. A. Shelykh, A. V. Kavokin, G. Malpuech, and P. Bigenwald, Quantum theory of spin dynamics of exciton-polaritons in microcavities, *Phys. Rev. Lett.* **92**, 017401 (Jan, 2004).

16. S. Savasta and R. Girlanda, Hyper-raman scattering in semiconductors: A quantum optical process in the strong-coupling regime, *Phys. Rev. B* **59**, 15409–15421 (Jun, 1999).

17. V. Savona, L. Andreani, P. Schwendimann, and A. Quattropani, Quantum well excitons in semiconductor microcavities: Unified treatment of weak and strong coupling regimes, *Solid State Communications* **93**(9), 733–739, (1995). ISSN 0038-1098.

.18. P. N. Butcher and D. Cotter, *The Elements of Nonlinear Optics* (Cambridge University Press, 1991).

19. T. Östreich, K. Schönhammer, and L. J. Sham, Exciton-exciton correlation in the nonlinear optical regime, *Phys. Rev. Lett.* **74**, 4698–4701 (Jun, 1995).

20. S. Savasta, O. Di Stefano, and R. Girlanda, Many-body and correlation effects on parametric polariton amplification in semiconductor microcavities, *Phys. Rev. Lett.* **90**, 096403 (Mar, 2003).

21. F. Tassone, C. Piermarocchi, V. Savona, A. Quattropani, and P. Schwendimann, Bottleneck effects in the relaxation and photoluminescence of microcavity polaritons, *Phys. Rev. B* **56**, 7554–7563 (Sep, 1997).

22. C. Ciuti, P. Schwendimann, and A. Quattropani, Parametric luminescence of microcavity polaritons, *Phys. Rev. B* **63**, 041303 (Jan, 2001).

23. K. Victor, V. M. Axt, and A. Stahl, Hierarchy of density matrices in coherent semiconductor optics, *Phys. Rev. B* **51**, 14164–14175 (May, 1995).

24. C. Ciuti, P. Schwendimann, B. Deveaud, and A. Quattropani, Theory of the angleresonant polariton amplifier, *Phys. Rev. B* **62**, R4825–R4828 (Aug, 2000).

25. S. Kundermann, M. Saba, C. Ciuti, T. Guillet, U. Oesterle, J. L. Staehli, and B. Deveaud, Coherent control of polariton parametric scattering in semiconductor microcavities, *Phys. Rev. Lett.* **91**, 107402 (Sep, 2003).

26. C. Ciuti, Branch-entangled polariton pairs in planar microcavities and photonic wires, *Phys. Rev. B* **69**, 245304 (Jun, 2004).

27. W. Langbein, Spontaneous parametric scattering of microcavity polaritons in momentum space, *Phys. Rev. B* **70**, 205301 (Nov, 2004).

28. T. J. Herzog, P. G. Kwiat, H. Weinfurter, and A. Zeilinger, Complementarity and the quantum eraser, *Phys. Rev. Lett.* **75**, 3034–3037 (Oct, 1995).

29. X. Y. Zou, L. J. Wang, and L. Mandel, Induced coherence and indistinguishability in optical interference, *Phys. Rev. Lett.* **67**, 318–321 (Jul, 1991).

30. L. Mandel, Quantum effects in one-photon and two-photon interference, *Rev. Mod. Phys.* **71**, S274–S282 (Mar, 1999).

31. Y.-H. Kim, R. Yu, S. P. Kulik, Y. Shih, and M. O. Scully, Delayed "choice" quantum eraser, *Phys. Rev. Lett.* **84**, 1–5 (Jan, 2000).

32. S. Savasta and O. D. Stefano, Quantum optics with interacting polaritons, *physica status solidi (b)* **243**(10), 2322–2330, (2006). ISSN 1521-3951.

33. S. Savasta, O. Di Stefano, and S. Portolan, Quantum optics in semiconductor microcavities, *physica status solidi (c)* **5**(1), 334–339, (2008). ISSN 1610-1642.

34. G. Oohata, R. Shimizu, and K. Edamatsu, Photon polarization entanglement induced by biexciton: Experimental evidence for violation of bell's inequality, *Phys. Rev. Lett.* **98**, 140503 (Apr, 2007).

35. S. Portolan, O. Di Stefano, S. Savasta, F. Rossi, and R. Girlanda, Decoherence-free emergence of macroscopic local realism for entangled photons in a cavity, *Phys. Rev. A* **73**, 020101 (Feb, 2006).

36. R. J. Glauber, The quantum theory of optical coherence, *Phys. Rev.* **130**, 2529–2539 (Jun, 1963).

37. J.-W. Pan, C. Simon, C. Brukner, and A. Zeilinger, Entanglement purification for quantum communication, *Nature* **410**, 1067–1070, (2001).

38. A. G. White, D. F. V. James, P. H. Eberhard, and P. G. Kwiat, Nonmaximally entangled states: Production, characterization, and utilization, *Phys. Rev. Lett.* **83**, 3103–3107 (Oct, 1999).

39. D. F. V. James, P. G. Kwiat, W. J. Munro, and A. G. White, Measurement of qubits, *Phys. Rev. A* **64**, 052312 (Oct, 2001).

40. C. H. Bennett, D. P. DiVincenzo, J. A. Smolin, and W. K. Wootters, Mixed-state entanglement and quantum error correction, *Phys. Rev. A* **54**, 3824–3851 (Nov, 1996).

41. S. Hill and W. K. Wootters, Entanglement of a pair of quantum bits, *Phys. Rev. Lett.* **78**, 5022–5025 (Jun, 1997).

42. W. K. Wootters, Entanglement of formation of an arbitrary state of two qubits, *Phys. Rev. Lett.* **80**, 2245–2248 (Mar, 1998).

43. H. Oka and H. Ishihara, Highly efficient generation of entangled photons by controlling cavity bipolariton states, *Phys. Rev. Lett.* **100**, 170505 (Apr, 2008).

44. S. Schumacher, N. H. Kwong, and R. Binder, Influence of exciton-exciton correlations on the polarization characteristics of polariton amplification in semiconductor microcavities, *Phys. Rev. B* **76**, 245324 (Dec, 2007).

45. S. Savasta, O. D. Stefano, and R. Girlanda, Spectroscopy of four-particle correlations in semiconductor microcavities, *Phys. Rev. B* **64**, 073306 (Jul, 2001).

46. S. Savasta, O. D. Stefano, and R. Girlanda, Many-body and correlation effects in semiconductor microcavities, *Semiconductor Science and Technology* **18**(10), S294, (2003).

47. W. Langbein and J. M. Hvam, Elastic scattering dynamics of cavity polaritons: Evidence for time-energy uncertainty and polariton localization, *Phys. Rev. Lett.* **88**, 047401 (Jan, 2002).

48. S. Portolan, O. D. Stefano, S. Savasta, and V. Savona, Emergence of entanglement out of a noisy environment: The case of microcavity polaritons, *EPL (Europhysics Letters)* **88**(2), 20003, (2009).

49. J. P. Karr, A. Baas, R. Houdré, and E. Giacobino, Squeezing in semiconductor micro-cavities in the strong-coupling regime, *Phys. Rev. A* **69**, 031802 (Mar, 2004).

50. L. Amico, R. Fazio, A. Osterloh, and V. Vedral, Entanglement in many-body systems, *Rev. Mod. Phys.* **80**, 517–576 (May, 2008).

51. C. Monroe, Quantum information processing with atoms and photons, *Nature* **416**(6877), 238–246 (Mar 14, 2002). ISSN 0028-0836. doi: 10.1038/416238a.

52. K. Hennessy, A. Badolato, M. Winger, D. Gerace, M. Atatuere, S. Gulde, S. Faelt, E. L. Hu, and A. Imamoglu, Quantum nature of a strongly coupled single quantum dot-cavity system, *Nature* **445**(7130), 896–899, (2007).

53. M. A. Nielsen and I. L. Chuang, *Quantum Computation and Quantum Information (Cambridge Series on Information and the Natural Sciences)* (Cambridge University Press, Jan. 2004), 1 edition. ISBN 0521635039.

54. T. Härtling, P. Reichenbach, and L. Eng, Near-field coupling of a single fluores-cent molecule and a spherical gold nanoparticle, *Optics Express* **15**(20), 12806–12817, (2007).

55. S. Kühn, U. Håkanson, L. Rogobete, and V. Sandoghdar, Enhancement of single-molecule fluorescence using a gold nanoparticle as an optical nanoantenna, *Physical Review Letters* **97**(1), 17402, (2006).

56. A. G. Curto, G. Volpe, T. H. Taminiau, M. P. Kreuzer, R. Quidant, and N. F. van Hulst, Unidirectional emission of a quantum dot coupled to a nanoantenna, *Science* **329**(5994), 930–933, (2010).

57. S. A. Maier, *Plasmonics: Fundamentals and Applications* (Springer, 2007).

58. M. Noginov, G. Zhu, A. Belgrave, R. Bakker, V. Shalaev, E. Narimanov, S. Stout, E. Herz, T. Suteewong, and U. Wiesner, Demonstration of a spaser-based nanolaser, *Nature* **460**(7259), 1110–1112, (2009).

59. Q. A. Turchette, C. Hood, W. Lange, H. Mabuchi, and H. J. Kimble, Measurement of conditional phase shifts for quantum logic, *Physical Review Letters* **75**(25), 4710–4713, (1995).

60. G. Nogues, A. Rauschenbeutel, S. Osnaghi, M. Brune, J. Raimond, and S. Haroche, Seeing a single photon without destroying it, *Nature* **400**(6741), 239–242, (1999).

61. K. M. Birnbaum, A. Boca, R. Miller, A. D. Boozer, T. E. Northup, and H. J. Kimble, Photon blockade in an optical cavity with one trapped atom, *Nature* **436**(7047), 87–90, (2005).

62. W. Zhang, A. O. Govorov, and G. W. Bryant, Semiconductor-metal nanoparticle molecules: Hybrid excitons and the nonlinear fano effect, *Physical Review Letters* **97**(14), 146804, (2006).

63. R. D. Artuso and G. W. Bryant, Optical response of strongly coupled quantum dot- metal nanoparticle systems: Double peaked fano structure and bistability, *Nano Letters* **8**(7), 2106–2111, (2008).

64. E. Flagg, A. Muller, J. Robertson, S. Founta, D. Deppe, M. Xiao, W. Ma, G. Salamo, and C.-K. Shih, Resonantly driven coherent oscillations in a solid-state quantum emit-ter, *Nature Physics* **5**(3), 203–207, (2009).

65. N. Akopian, N. Lindner, E. Poem, Y. Berlatzky, J. Avron, D. Gershoni, B. Gerardot, and P. Petroff, Entangled photon pairs from semiconductor quantum dots, *Physical Review Letters* **96**(13), 130501, (2006).

66. A. Akimov, A. Mukherjee, C. Yu, D. Chang, A. Zibrov, P. Hemmer, H. Park, and M. Lukin, Generation of single optical plasmons in metallic nanowires coupled to quantum dots, *Nature* **450**(7168), 402–406, (2007).

67. N. T. Fofang, T.-H. Park, O. Neumann, N. A. Mirin, P. Nordlander, and N. J. Halas, Plexcitonic nanoparticles: Plasmon- exciton coupling in nanoshell- j-aggregate complexes, *Nano Letters* **8**(10), 3481–3487, (2008).

68. W. Ni, T. Ambjo rnsson, S. P. Apell, H. Chen, and J. Wang, Observing plasmonic-molecular resonance coupling on single gold nanorods, *Nano Letters* **10**(1), 77–84, (2009).

69. N. T. Fofang, N. K. Grady, Z. Fan, A. O. Govorov, and N. J. Halas, Plexciton dynamics: Exciton- plasmon coupling in a j-aggregate- au nanoshell complex provides a mechanism for nonlinearity, *Nano Letters* **11**(4), 1556–1560, (2011).

70. A. Ridolfo, R. Saija, S. Savasta, P. H. Jones, M. A. Iati, and O. M. Marago, Fano-doppler laser cooling of hybrid nanostructures, *ACS Nano* **5**(9), 7354–7361, (2011).

71. T. Yoshie, A. Scherer, J. Hendrickson, G. Khitrova, H. M. Gibbs, G. Rupper, C. Ell, O. B. Shchekin, and D. G. Deppe, Vacuum rabi splitting with a single quantum dot in a photonic crystal nanocavity, *Nature* **432**, 200–203, (2004).

72. E. Peter, P. Senellart, D. Martrou, A. Lemaître, J. Hours, J. M. Gérard, and J. Bloch, Exciton-photon strong-coupling regime for a single quantum dot embedded in a microcavity, *Phys. Rev. Lett.* **95**, 067401 (Aug, 2005).

73. J. P. Reithmaier, G. Sek, A. Löffler, C. Hofmann, S. Kuhn, S. Reitzenstein, L. V. Keldysh, V. D. Kulakovskii, T. L. Reinecke, and A. Forchel, Strong coupling in a single quantum dot semiconductor microcavity system, *Nature* **432**, 197–200, (2004).

Chapter 9

Optical Signal Processing with Enhanced Nonlinearity in Photonic Crystals

A. De Rossi* and S. Combrié

Thales Research and Technology
1, av. Augustin Fresnel, 91767 Palaiseau, France
* alfredo.derossi@thalesgroup.com

Owing to the very strong confinement of the optical field, nonlinear effects in photonic crystal cavities and waveguides are enhanced drastically. All optical processing, relying on these effects, is now feasible in integrated photonic devices, meeting specifications in terms of power consumption, speed and size for applications in signal processing.

1. Introduction: All optical processing

For decades, optical fibers have been the core of complex signal processing systems entailing laser sources, photodetectors, modulators, optical splitters and switches and where electronics was assigned the role of processing any signal before and after transmission. Very importantly, electronics is also required for deciding where signals have to go, e.g. the routing task; therefore, any time the simplest even operation is required, optical signals are detected, processed by electronics, and generated again.

Optics, however, can do more than low-loss transport of signal. In the spectral range suitable for optical fiber, the frequency of the carrier wave is about 200 THz. Compared to it, a bandwidth of several THz, such as that allowed in Erbium Doped Fiber Amplifiers (EDFAs), is small. In comparison, the bandwidth of an electric device, e.g. an amplifier, will hardly reach 100 GHz and, still, will even more hardly have a flat frequency response and acceptable gain. In contrast, is pretty trivial to amplify by 40 dB an optical signal with a bandwidth of several hundreds of GHz with an EDFA. The issue is rather to generate and detect such signals, e.g. the conversion from the electrical to the optical domain and *vice-versa*.

Recently, it has become clear that even in relatively short data link, namely within a microelectronic chip, high frequency signals will experience unbearable propagation losses. Thus, it has been hinted that replacing the metal interconnects with photonic links will solve this issue[1] and others: synchronization, clock delivery and crosstalk.[1]

The lack of a photonic device able to control the transmission of light *all-optically*, i.e. an "optical transistor" makes the Electric to Optical Conversion (EOC) necessary, to the extent that even elementary operations such as deciding where (which channel) to send a data packet has to be made by electronics. Amazingly, such a device was proposed long ago, and switching and bistablity (e.g. optical memory) were demonstrated in a Fabry Pérot resonator where a nonlinear response was produced by operating it near the absorption resonance of the Na vapor filling the cavity.[2] Although the operating power was relatively moderate (few mW), the device, especially in regard of its footprint was no match with integrated electronics. While the benefit of integrating photonic devices was realized early enough,[3] a photonic technology including a "transistor" has not yet emerged.

It is indeed clear that scaling the size of an optical switch down to the micrometer, e.g. 3 orders of magnitude with respect to a macroscopic optical resonator is crucial, as the energy required for switching is inversely proportional to the volume where the nonlinear interaction takes place. Thus, a nanophotonic optical switch would be about 9 orders of magnitude more energy-efficient than a macroscopic resonator. Interestingly, this could be traded, in part, with a weaker but faster and more practical nonlinear effects than what Gibbs *et al.* exploited in their experiment. Moreover, size reduction also implies energy efficiency and a faster response. There is a striking similarity with microelectronic, where it is not unfair to say that the "magic" of Moore's law has consisted mainly in the ability to scale down the size of the MOSFET, which then translated into a reduction of power consumption and into an increase of speed.

The point we will try to make in these notes is that Photonic Crystal (Fig. 1) is the technology that will allow the necessary improvement in energy efficiency and speed needed for all-optical processing.

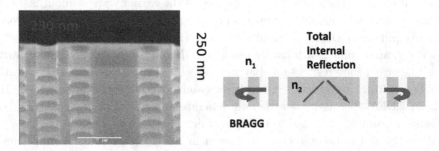

Fig. 1.   Left: SEM image of an air-suspended GaAs Photonic crystal slab fabricated at Thales.[4] Right: the confinement in an air-suspended PhC slab results both from the existence of a band gap (Bragg reflection) and the total internal reflection.

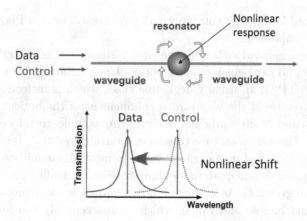

Fig. 2.   Principle of optical switching in a resonant cavity: the transmission of the signal ("data") depends on whether it is spectrally matched to the resonance of the cavity, which, in turns is controlled all-optically via the nonlinear effects.

2.   The optical transistor

The simplest all-optical switch (or "transistor") consists in a resonant filter, such as sketched in Fig. 2, where the resonant frequency is controlled optically. In the "OFF" state, the wavelength carrying "data" is set off-resonance, and the transmission is low. Upon the arrival of the optical control pulse, the cavity is spectrally shifted to become on-resonance with the "data", causing the transmission to increase. The control of the cavity resonance is achieved all-optically via an intensity-dependent refractive index. In semiconductors, when excited by photons with energy below the electronic gap, the dominant nonlinear effects are the Kerr and the multi-photon generation of free carriers, both resulting into a intensity-depended change in the refractive index.[5] The Kerr effect (or intensity-dependent refractive index) is related to the anharmonicity of the electronic polarizability and it is a general feature of optical materials. It is therefore a very fast phenomena and it is described by a contribution to the dielectric polarization of the third order in the electric field, namely $P_i^{NL} = \chi_{ijkl}^{(3)} E_j E_k E_l$, where $\chi^{(3)}$ is the third-order susceptibility tensor.[6]

As a matter of fact, the demonstration of switching based on the Kerr effect is challenging in semiconductors, as multi-photon absorption is dominant. Therefore, most of the all-optical switching experiments in semiconductors have exploited the carrier-induced change in the refractive index, or Free Carrier Dispersion (FCD), with the switching speed being ultimately related to the carrier lifetime.[7] In contrast to the Kerr effect, the FCD produces a blue shift of the cavity resonance and builds up as the photo-excited carriers accumulate, this also regulated by their lifetime. Eventually, when considering the trade-off between energy and speed for efficient all-optical signal processing, carrier-induced nonlinear effects appear to be a more promising candidate than the fast but weaker Kerr effect. In fact, very fast response times[8] and record low energy consumption[9] have been shown in Photonic Crystals made of III-V compounds.

Carriers are generated either by the direct absorption of radiation above the semiconductor band gap, which requires out of plane excitation, or by multi-phonon absorption (e.g. TPA), an usually negligible effect which is instead very strong in PhC cavities, because of the very strong confinement of the optical energy. Thus, TPA generates carriers efficiently and in a very tiny volume, thereby creating a high density plasma. The switch-off time (the relaxation of the system) is however limited by the carrier lifetime and it is a critical issue. In most high-quality semiconductors (e.g. Indium Phosphide and quaternary alloys based on it, Gallium Arsenide, but not low-temperature grown GaAs or some ion-bombarded semiconductor), the carrier lifetime is in the nanosecond range, which is unacceptably too long. A specific feature of sub-wavelength photonic structures, such as Photonic Crystals, is that the carriers are generated very close to the surface, such that the typical distance the generated carriers have to travel to reach the surface ($< 200\,nm$) is much smaller than the typical ambipolar diffusion length (typically larger than one micron). The dynamics, dominated by the surface properties, becomes faster.

That was demonstrated in Photonic Crystals (Fig. 3). With respect to earlier experiments,[10] three important changes were introduced:

— a different material: III-V semiconductor instead of Silicon;
— a design[11] providing a very small mode volume, namely. $0.25(\lambda/n)^3$;
— measurement by pump-probe technique with ps time resolution.

The main point was the measurement of the time recovery of the switch by a degenerate pump-probe technique, with time resolution given by the duration ($\approx 2\,ps$) of the nearly $sech^2$ pulses generated by a mode-locked laser. The measured

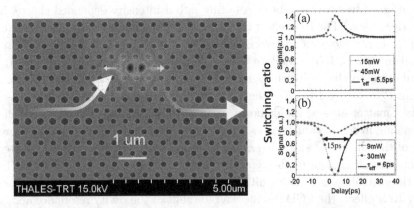

Fig. 3.   Left: all-optical switch based on a H0 cavity. SEM image of the GaAs sample (and two yellow arrows denoting the displacement of the holes generating the cavity), with superimposed the $|\vec{E}|^2$ field. Right: degenerate pump-probe measurement, normalized probe transmission (switching ratio) as a function of the pump-probe delay. (a) detuning is negative ($\Delta\lambda = -1.4\,nm$), (b) on resonance.

signal (Fig. 3) represents the transmitted probe power following the pump by a variable delay τ. For convenience this value is normalized to the probe signal without pump and thus represents the switching ratio $SR(\tau)$, which is in turn related to the differential transmission $\Delta T/T = SR - 1$, commonly used in the literature. First of all, it is verified that $SR(\tau)$ switches up when the pump and probe are blue detuned, consistently with the decrease of the refractive index induced by free carriers, and that it switches down when pump and probe are on resonance. The main result is a very short recovery time (about 6 ps), measured as the exponential fit of the trailing edge of $SR(\tau)$. This is a direct consequence of the sub-micron patterning of the structure. We note that the value of the differential transmission $\Delta T/T = 0.4$ is more than acceptable for a proof of principle. Indeed, it is still small for practical applications. Very recently, we reported a much larger switching contrast ($\Delta T/T > 14$). The optical peak power required for switching was about 50 mW (that is 100 fJ per pulse), which was also a very low value. By a smart optimization of the material (InGaAsP with band gap close to the optical pump), a similar experiment with basically the same cavity design and same topology, was performed at NTT and demonstrated a switching energy in the fJ range, which is very important given the awareness of energy efficiency in information technology.[9]

3. The optical memory

Optical bistability occurs when an optical system, fed with optical power P_{in}, can take any of a manifold of *stable* states (related to the energy U stored into it, playing here the role of the internal variable), as shown in Fig. 4(a). The simplest optical system entailing bistability consists in an optical resonator with a the frequency ω_0 depending on its internal energy, e.g. $\omega_0(U) = \omega_0 + \kappa U$, due for instance to the Kerr effect or FCD or also the thermal induced index change. When combined with the lorentzian spectral linewidth $U = P_{in}\Gamma^2/|(\omega - \omega_0(U) + i\Gamma|^2$, the typical S-shaped map in Fig. 4(b) is obtained. When the cavity is excited with a periodic (e.g. sinusoidal) input, the output reveals hysteresis, provided that the input power spans values above and below the two critical points (noted by arrows in Fig. 4(b)). While optical bistability was demonstrated experimentally by Gibbs in a macroscopic nonlinear Fabry-Pérot interferometer,[2] the group at NTT, about three decades later, demonstrated bistability in a microscopic optical device,[10] made of silicon. That important work marked the premises of a truly integrated optical memory.

An intriguing phenomena is the onset of oscillatory instability, predicted[12] when the cavity is excited at higher power levels and when the cavity and the carrier lifetimes are comparable. Under these conditions, the dynamics is described by a limit cycle in the phase space (Fig. 4). Thus, in spite of the wording, bistability is a dynamical phenomena.

An optical memory is a bistable switch, such as a nonlinear Fabry-Pérot. What makes it suitable for applications is meeting specifications in terms of size, power

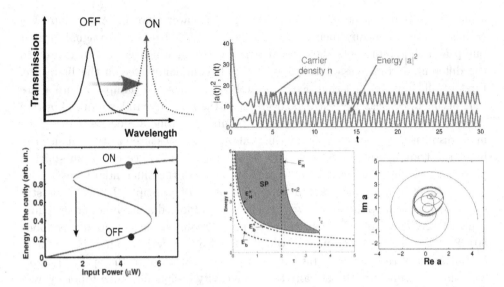

Fig. 4. Optical bistability in a resonant cavity. The two stable states (a) are associated to whether the excitation (arrow) is on resonance "ON" or off-resonance "OFF" and correspond to the same input power but different energy in the cavity (b). Here the nonlinear response is assumed to red-shift the resonance. Self-pulsing predicted for photonic crystal cavities based on a delayed resonance (e.g. from free carrier): time evolution (c), space of the parameters (d) and phase-space denoting a limit cycle(e). From Ref. 12.

consumption, speed and hold time. By suitably engineering the carrier lifetime and the thermal resistance, in order to minimize undesirable competition with thermal effects, III-V photonic crystal technology was proven to be a good candidate. A device as large as a few squared μm, switching in a few picosecond with fJ of energy, and holding its state for microseconds while consuming nanoWatts of power was demonstrated at NTT.[13]

On a more fundamental perspective, single photon transistor, that is the ultimate optical switch, has just been demonstrated, owing to the *strong coupling* regime between a PhC cavity mode and an exciton in a single quantum dot, which have been spectrally and spatially matched.[14] Recently, self pulsing has been demonstrated in graphene-loaded PhC cavities, lead by Prof. Chee Wei Wong.[15]

4.   Nonlinear waves on chip

In optical fibers, the weak Kerr nonlinearity of silica is enough to trigger a variety of phenomena such as Four-Wave-Mixing, Cross and Phase modulation, Soliton compression, Stimulated Raman Scattering,[16] because the very low propagation loss allows a long interaction length. Thus functions such as wavelength-conversion (e.g. generation of spectral replica centred at a prescribed wavelength), optical sampling, time domain demultiplexing can be performed extremely fast (on a ps or sub-ps time scale). Based on these functions, a single wavelength photonic link delivering 5.1 Tb/s has been demonstrated at DTU.[17]

Semiconductor nanophotonic devices, such as photonic wires and photonic crystals enable the confinement of light at the sub-wavelength level. This is related to the large index contrast associated to the semiconductor-air interface. Compared to the core of a standard single-mode fiber, the cross section A_{eff} of a photonic wire/crystal is more than 2 orders of magnitude smaller, which implies much larger density of the optical energy. Moreover, in semiconductors the third order susceptibility $\chi^{(3)}$, which is the origin of the nonlinear phenomena above, is about 3 orders of magnitude larger than in silica. Finally, as the group velocity in nanostructures can be substantially smaller than the phase velocity, this implies a further increase of the density of the electric field energy (which is proportional to the irradiance divided by the group velocity[18]). This last feature is particularly relevant to photonic crystals, as here low-loss propagation of slow-light ($v_g < c/30$) has been demonstrated by several groups,[19, 20] included us.[21]

Four-Wave-Mixing is an optical parametric effect related to the third order susceptibility $\chi^{(3)}$, where two pump photons at ω_p (in the degenerate case) decay into a signal-idler photon pair at ω_s and ω_i, such that $2\omega_p = \omega_s + \omega_i$, because of energy conservation. If a signal is combined with the pump at the input, then it can be amplified. Parametric amplification has some peculiar features which are of interest in quantum optics or in coherent optical communications. A decade ago, 50 dB parametric gain was already demonstrated[22] in optical fibers. In 2006, broadband parametric gain in a silicon on insulator (SOI) waveguide (L = 1.7 cm) was demonstrated at Cornell U. ,[23] somehow limited by nonlinear absorption (TPA). TPA-free self-phase modulation (SPM) was demonstrated in Gallium Indium Phoshpide (GaInP) Photonic Crystal waveguides (Fig. 5). The nonlinear parameter of the waveguide

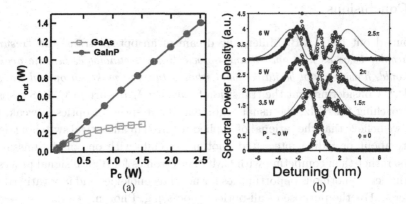

Fig. 5. Nonlinear properties of GaInP waveguides. (a) Transmission vs. input power demonstrating TPA-free nonlinear propagation (a measurement on a GaAs waveguide is added for comparison) and (b) output spectra revealing self-phase modulation, close to the expected outcome of a TPA-free nonlinear waveguide. 1 mm of PhC waveguide is equivalent to 1 Km of optical fiber, in term of all-optical processing capability. From Ref. 24.

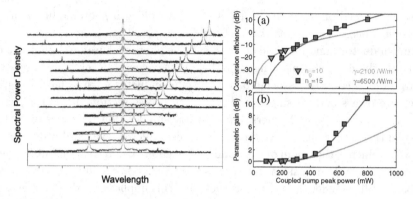

Fig. 6.   Left: Four-wave mixing spectra at the waveguide output as a function of the pump-probe detuning (measured at Technion). Right, demonstration of parametric gain on a 1 mm chip.[31]

$\gamma = 1 W^{-1} mm^{-1}$ was about six orders of magnitude larger than in a single mode standard optical fiber.[24] Four-wave-mixing was demonstrated in PhC waveguides by several groups independently in 2010:[25, 26] on a Silicon platform,[27] in chalcogenide and in GaInP.[28] Figure 6 (left) shows a series of output spectra as a function of probe detuning relative to the pump. The idler appears symmetrically relative to the pump. Owing to the development of "flat-band" (or dispersion-flattened) waveguides with large group index, and acceptable loss,[26, 29] all-optical processing based on wavelength conversion was demonstrated at CUDOS.[30] The first demonstration of FWM gain, e.g. with conversion efficiency > 100% was made in a GaInP PhC waveguide.[32] More recently, a gain of 11 dB (net gain 5 dB after subtracting the propagation loss) was achieved with only 800 mW of peak pump power.[31] That is shown in Fig. 6 (right).

5.   Conclusions

As pointed out by D.A.B. Miller, the dream of an optical transistor *"resurfaces on a regular basis as new optical and optoelectronic technologies become available. We should, however, not be naive about what it takes to make an optical transistor useful"*.[33] Considering that the transition frequency f_T of current CMOS transistor is approaching the THz, the usual argument about speed of optics is weak. Once again, we believe that the rational of applied research in all-optical switching devices is more about providing integrated photonic circuits with one of the missing key devices, rather than competing with CMOS electronics for digital signal processing. Still, this leave enormous opportunities for novel technologies and innovative devices to emerge. For the purpose of all-optical processing, Photonic Crystal waveguides inherit key functionalities of optical fibres, namely pulse compression, wavelength conversion and four-wave-mixing. The spatial scale is however reduced by orders of magnitudes, still keeping the required optical power level below 1 watt. This makes the integration on a single photonic chip feasible. In this respect, the GaInP PhC technology brings immunity to nonlinear absorption (more precisely within more comfortable limits), which recent experiments demonstrate to be a key advantage over silicon.

Acknowledgements

We thank G. Eisenstein, I. Cestier, M. Santagiustina, G. Bellanca, S. Trillo, S. Malaguti, C.W. Wong, E. Larkins, I. Sagnes, F. Raineri, R. Raj, Y. Moerk, S. Hughes, our collaborators E. Weidner, Q. N. V. Tran, C. Husko and P. Colman and collegues, G. Lehoucq, S. Xavier and J. Bourderionnet. We acknowledge financial support from the E.U. via the projects Copernicus (www.copernicusproject.eu) and Gospel (www.gospel-project.eu), and the French Research Agency (project L2CP).

References

1. D. Miller, Rationale and challenges for optical interconnects to electronic chips, *Proceedings of the IEEE* **88**(6), 728–749, (2000).
2. H. Gibbs, S. McCall, and T. Venkatesan, Differential gain and bistability using a sodium-filled fabry-perot interferometer, *Physical Review Letters* **36**(19), 1135–1138, (1976).
3. S. Miller, Integrated optics: an introduction., *Bell Syst. Tech. J.* **48**(7), 2059, (1969).
4. S. Combrié, *Etude et réalisation de structures en cristaux photoniques pour les applications de traitement du signal optique* PhD thesis, Université de Paris-Sud. Faculté des Sciences d'Orsay, (2006).
5. B. Bennett, R. Soref, and J. Del Alamo, Carrier-induced change in refractive index of inp, gaas and ingaasp, *Quantum Electronics, IEEE Journal of* **26**(1), 113–122, (1990).
6. R. Boyd, *Nonlinear Optics* (Academic press, 2002).
7. V. Van, T. Ibrahim, K. Ritter, P. Absil, F. Johnson, R. Grover, J. Goldhar, and P. Ho, All-optical nonlinear switching in gaas-algaas microring resonators, *Photonics Technology Letters, IEEE* **14**(1), 74–76, (2002).
8. C. Husko, S. Combrié, Q. V. Tran, C. W. Wong, and A. D. Rossi, Ultra-fast all-optical modulation in *gaas* photonic crystal cavities, *Applied Physics Letters* **94**, 021111, (2009).
9. K. Nozaki, T. Tanabe, A. Shinya, S. Matsuo, T. Sato, H. Taniyama, and M. Notomi, Sub-femtojoule all-optical switching using a photonic-crystal nanocavity, *Nature Photonics* **4**(7), 477–483, (2010).
10. T. Tanabe, M. Notomi, S. Mitsugi, A. Shinya, and E. Kuramochi, Fast bistable all-optical switch and memory on a silicon photonic crystal on-chip, *Optics Letters* **30**(19), 2575–2577, (2005).
11. Z. Zhang and M. Qiu, Small-volume waveguide-section high q microcavities in 2d photonic crystal slabs, *Optics Express* **12**(17), 3988–3995, (2004).
12. S. Malaguti, G. Bellanca, A. De Rossi, S. Combrie, and T. S, Self-pulsing driven by two-photon absorption in semiconductor nanocavities, *Physical Review A* **83**, –, (2011).
13. K. Nozaki, A. Shinya, S. Matsuo, Y. Suzaki, T. Segawa, T. Sato, Y. Kawaguchi, R. Takahashi, and M. Notomi, Ultralow-power all-optical ram based on nanocavities, *Nature Photonics* **6**(4), 248–252, (2012).
14. T. Volz, A. Reinhard, M. Winger, A. Badolato, K. Hennessy, E. Hu, and A. Imamoğlu, Ultrafast all-optical switching by single photons, *Nature Photonics* (2012).

15. T. Gu, N. Petrone, J. McMillan, A. van der Zande, M. Yu, G. Lo, D. Kwong, J. Hone, and C. Wong, Regenerative oscillation and four-wave mixing in graphene optoelectronics, *Nature Photonics* (2012).

16. G. Agrawal, *Nonlinear Fiber Optics* (Academic, 2007).

17. H. Mulvad, M. Galili, L. Oxenløwe, H. Hu, A. Clausen, J. Jensen, C. Peucheret, and P. Jeppesen, Demonstration of 5.1 tbit/s data capacity on a single-wavelength channel, *Optics Express* **18**(2), 1438–1443, (2010).

18. K. Sakoda, *Optical properties of photonic crystals* Vol. 80, (Springer, 2004).

19. S. Schultz, L. O'Faolain, D. Beggs, T. P. White, A. Melloni, and T. F. Krauss, Dispersion engineered slow light in photonic crystals: a comparison, *Journal of Optics* **12**(10), 104004, (Oct., 2010).

20. Y. Hamachi, S. Kubo, and T. Baba, Slow light with low dispersion and nonlinear enhancement in a lattice-shifted photonic crystal waveguide, *Opt. Lett.* **34**(7), 1072–1074, (Apr, 2009).

21. J. Sancho, J. Bourderionnet, J. Lloret, S. Combrié, I. Gasulla, S. Xavier, S. Sales, P. Colman, G. Lehoucq, D. Dolfi, and A. De Rossi, Integrable microwave filter based on a photonic crystal delay line, *Nature Communications* **3**, 1075, (2012).

22. J. Hansryd and P. Andrekson, Broad-band continuous-wave-pumped fiber optical parametric amplifier with 49-db gain and wavelength-conversion efficiency, *Photonics Technology Letters, IEEE* **13**(3), 194–196, (2001).

23. M. Foster, A. Turner, J. Sharping, B. Schmidt, M. Lipson, and A. Gaeta, Broad-band optical parametric gain on a silicon photonic chip, *Nature* **441**(7096), 960–963, (2006).

24. S. Combrie, N. Tran, A. De Rossi, C. Husko, and P. Colman, High quality *gainp* nonlinear photonic crystals with minimized nonlinear absorption, *Applied Physics Letters* **95**, 061105, (2009). ISSN 0003-6951.

25. J. McMillan, M. Yu, D. Kwong, and C. Wong, Observation of four-wave mixing in slow-light silicon photonic crystal waveguides, *Optics Express* **18**(15), 15484–15497, (2010).

26. C. Monat, M. Ebnali-Heidari, C. Grillet, B. Corcoran, B. Eggleton, T. White, L. OFaolain, J. Li, and T. Krauss, Four-wave mixing in slow light engineered silicon photonic crystal waveguides, *Optics Express* **18**(22), 22915–22927, (2010).

27. K. Suzuki, Y. Hamachi, and T. Baba, Fabrication and characterization of chalcogenide glass photonic crystal waveguides, *Optics Express* **17**(25), 22393–22400, (2009).

28. V. Eckhouse, I. Cestier, G. Eisenstein, S. Combrié, P. Colman, A. De Rossi, M. Santagiustina, C. Someda, and G. Vadalà, Highly efficient four wave mixing in gainp photonic crystal waveguides, *Optics Letters* **35**(9), 1440–1442, (2010).

29. L. OFaolain, S. Schulz, D. Beggs, T. White, M. Spasenović, L. Kuipers, F. Morichetti, A. Melloni, S. Mazoyer, J. Hugonin, *et al.*, Loss engineered slow light waveguides, *Optics Express* **18**(26), 27627–27638, (2010).

30. B. Corcoran, C. Monat, M. Pelusi, C. Grillet, T. White, L. OFaolain, T. Krauss, B. Eggleton, and D. Moss, Optical signal processing on a silicon chip at 640 gb/s using slow-light, *Optics Express* **18**(8), 7770–7781, (2010).

31. I. Cestier, S. Combrié, S. Xavier, G. Lehoucq, A. De Rossi, and G. Eisenstein, Chip-scale parametric amplifier with 11 db gain at 1550 nm based on a slow-light gainp photonic crystal waveguide, *Optics Letters* **37**(19), 3996–3998, (2012).

32. P. Colman, I. Cestier, A. Willinger, S. Combrié, G. Lehoucq, G. Eisenstein, and A. De Rossi, Observation of parametric gain due to four-wave mixing in dispersion engineered gainp photonic crystal waveguides, *Optics Letters* **36**(14), 2629–2631, (2011).

33. D. Miller, Are optical transistors the logical next step?, *Nature Photonics* **4**(1), 3–5, (2010).